PROCEEDINGS

OF

THE JOHNS HOPKINS WORKSHOP

ON

CURRENT PROBLEMS IN PARTICLE THEORY

22

PREVIOUS JOHNS HOPKINS WORKSHOP PROCEEDINGS [1]

1	BALTIMORE	1974	....
2	BALTIMORE	1978	....
3	FLORENCE	1979	....
4	BONN	1980	Lattice Gauge Theories. Integrable Systems.
5	BALTIMORE	1981	Unified Field Theories and Beyond.
6	FLORENCE	1982	Lattice Gauge Theories. Supersymmetry and Grand Unification.
7	BONN	1983	Lattice Gauge Theories. Supersymmetry and Grand Unification.
8	BALTIMORE	1984	Particles and Gravity.
9	FLORENCE	1985	New Trends in Particle Theory.
10	BONN	1986	Infinite Lie Algebras and Conformal Invariance in Condensed Matter and Particle Physics.
11	LANZHOU	1987	Frontiers in Particle Theory.
12	BALTIMORE	1988	TeV Physics.
13	FLORENCE	1989	Knots, Topology and Quantum Field Theory.
14	HUNGARY	1990	Nonperturbative Methods in Low Dimensional Quantum Field Theories.
15	BALTIMORE	1991	Particle Physics from Underground to Heaven.
16	GÖTEBORG	1992	Pathways to Fundamental Theories.
17	BUDAPEST	1993	Particles and the Universe.
18	FLORENCE	1994	Theory Meets Experiments.
19	BALTIMORE	1995	Particles, Strings and Cosmology.
20	HEIDELBERG	1996	Non-perturbative Particle Theory & Experimental Tests.

[1] The 21st workshop in LANZHOU (1997) is not published. However, two contributions to this workshop are published in this volume

PROCEEDINGS OF
THE JOHNS HOPKINS WORKSHOP ON
CURRENT PROBLEMS IN PARTICLE THEORY 22

NOVELTIES IN STRING THEORY

Göteborg, Sweden 20–22 August 1998

Edited by

L. BRINK
R. MARNELIUS

Department of Theoretical Physics
Chalmers University of Technology
University of Göteborg
Sweden

World Scientific
Singapore • New Jersey • London • Hong Kong

Published by

World Scientific Publishing Co. Pte. Ltd.
P O Box 128, Farrer Road, Singapore 912805
USA office: Suite 1B, 1060 Main Street, River Edge, NJ 07661
UK office: 57 Shelton Street, Covent Garden, London WC2H 9HE

Library of Congress Cataloging-in-Publication Data
Johns Hopkins Workshop on Current Problems in Particle Theory (22nd : 1998 : Göteborg, Sweden)
Novelties in string theory ; proceedings of the Johns Hopkins Workshop on Current Problems in Particle Theory 22, Göteborg, 1998 (August 20–22) / edited by L. Brink, R. Marnelius.
p. cm.
ISBN 9810240848 (alk. paper)
1. String models -- Congresses. 2. Particles (Nuclear physics) -- Congresses. I. Brink, Lars, 1945– . II. Marnelius, R. (Robert) III. Title. IV. Title: Proceedings of the Johns Hopkins Workshop on Current Problems in Particle Theory.
QC794.6.S85J64 1998
536.7'2 -- dc21 99-41242
CIP

British Library Cataloguing-in-Publication Data
A catalogue record for this book is available from the British Library.

Printed in Singapore.

THE JOHNS HOPKINS WORKSHOPS ON CURRENT PROBLEMS IN PARTICLE THEORY ARE ORGANIZED BY THE FOLLOWING INSTITUTIONS:

Chalmers University of Technology and University of Göteborg
Eötvös Loránd University and Central Research Institute for Physics
University of Florence
University of Heidelberg
The Johns Hopkins University
University of Lanzhou

ORGANIZING COMMITTEE

Lars BRINK (Göteborg)	**Roberto CASALBUONI** (Florence)	**Gabor DOMOKOS** (Johns Hopkins)
Yi-Shi DUAN (Lanzhou)	**Zalán HORVÁTH** (Eötvös)	**M. JAMIN** (Heidelberg)
Susan KOVESI–DOMOKOS (Johns Hopkins)	**Luca LUSANNA** (Florence)	**Robert MARNELIUS** (Göteborg)
Otto NACHTMANN (Heidelberg)	**András PATKÓS** (Eötvös)	**K. Q. YANG** (Lanzhou)

LOCAL ORGANIZERS

L. BRINK
M. CEDERWALL
A. HOFLING
R. MARNELIUS
B. NILSSON
P. SALOMONSON

FOREWORD

The twenty-second Johns Hopkins Workshop on Current Problems in Particle Theory was held in Göteborg, Sweden on August 20–22, 1998. This was the second time the group in Göteborg organized this workshop. As in the past the purpose was to provide a forum for theoretical physicists from around the world to discuss problems at the forefront of particle physics. The discussions were concentrated around invited talks with ample time for discussions also during lunches and evenings. The meeting was held at the very scenic Hotel Novotel in the harbour of Göteborg, where all the participants also stayed.

The title of the workshop was "Novelties in String Theory". In recent years we have witnessed remarkably many novelties in string theory. With the understanding of various forms of duality we have come to understand that the five different superstring models we have had since a number of years are but expansions around different vacua of an unknown underlying unifying theory which Ashoke Sen has named *U-theory.* In this picture of models dual to each other there is also the celebrated *M-theory* the low-dimensional theory of which is 11-dimensional supergravity. All these issues were of course discussed during the workshop. The latest form of duality is the one discovered by Maldacena between superstrings in anti-de Sitter backgrounds and conformal Yang-Mills theories in four dimensions of space-time. Much of the new development in the half-year leading up to the workshop came in this field which also was well represented in the workshop. All in all we feel that these proceedings give a balanced view of the situation in string theory in the year 1998 and that all the most recent novelties were represented in the talks.

The Workshop was financed mainly by the Nobel Committee for Physics. They paid the expenses for all lecturers. The local expenses were covered by grants from the Hasselblad Foundation, from Wilhelm och Martina Lundgrens Vetenskapsfond and the Royal Society of Arts and Sciences in Göteborg. The Johns Hopkins university has continued their support and has paid for the advertising. We are very grateful to all these contributors who made the workshop possible.

Last but not least we must thank Annika Hofling who single-handedly took care of almost all administration. She was the secretary, negotiator, travel agent and inspirator. She made life really easy for the organizers. We would also like to thank the members of the group in Göteborg for their assistance in organizing the workshop.

Finally we wish to thank Prof. K. K. Phua and Ms. E. H. Chionh of World Scientific Publishing Company for their skilful and enthusiastic handling of this volume.

The Organizers

CONTENTS

In addition to the above the following talk was presented orally at the Conference: D. GROSS "Can String Theory help to solve QCD?"

Talks presented at the 21st Workshop in Lanzhou 1997

PROCEEDINGS

OF

THE JOHNS HOPKINS WORKSHOP

ON

CURRENT PROBLEMS IN PARTICLE THEORY

22

Gauge Theory and Gravity — A New Synthesis

Hirosi Ooguri
366 Le Conte Hall, Department of Physics
University of California, Berkeley, CA 94720
and
Mail Stop 50A-5101, Physics Division
Lawrence Berkeley National Laboratory, Berkeley, CA 94720
E-mail: Hooguri@lbl.gov

1

At the Johns Hopkins Conference, I gave a brief review of the *AdS/CFT* correspondence and then discussed my work-in-progress with David Gross and Nadav Drukker on properties of the Wilson loop operator. Since the work on the Wilson loop is still in progress, I would like to contribute a written version of the first part of my talk to the proceedings.

For about a year or so, a new and exciting progress in string theory has been taking place. In particular, a surprising relation between string theory and gauge theory is now established. Using this relation, we have gained new insights into both string theory and gauge theory. Today I would like to give an intuitive explanation of this development.

String theory was originally proposed as a model of hadrons. It dates back to the discovery of the Veneziano amplitude in 1968. In those days, it was called Dual Resonance Model. Theorists at that time tried to identify mesons as open string states. This earlier proposal, however, had serious problems. When theorists tried to quantize this open string, they found that it can only be done consistently in certain dimensions, typically $(9+1)$ or $(25+1)$ dimensions, rather than in $(3+1)$ dimensions where we live. Another problem was that, when people analyzed the spectrum of the theory, they found that the consistency always requires an existence of a massless spin two particle. There is no such particle in the hadronic world and theorists did not know what to do with this.

While they were struggling with these problems, the asymptotic freedom of the gauge theory was discovered. It became clear that the gauge theory is a winner and the string theory was abandoned as a theory of strong interaction.

It seems that, in theoretical physics, an interesting idea would eventually become useful, although sometime not for its originally intended purpose.

String theory is an example of this. After the initial failure as a theory of hadrons, string theory was reinvented as a theory of quantum gravity.

Recently a close connection was discovered between these two directions. It was found that certain gauge theories are equivalent to string theories in some curved spacetime. This equivalence is called the *AdS/CFT* correspondence. This correspondence revived the earlier hope of using string theory to describe the strong interaction and other phenomena involving gauge theories.

2

The status of string theory as a unified theory was much enhanced in 1984 when three important discoveries were made. The anomaly cancellation mechanism of Green and Schwarz removed a crucial obstruction. The heterotic string construction combined with the Calabi-Yau compactification opened up a possibility of making realistic models of elementary particles out of string theory.

It has been recognized for a long time that quantizing gravity is not a straightforward task. There are many aspects of quantum gravity which we do not know how to deal with in the standard quantum field theory method. If the string theory truly unifies quantum mechanics and general relativity, string theory should be able to address the puzzles of quantum gravity. One of the important questions in quantum gravity is the information paradox of black hole. The problem is roughly as follows.

The black hole solution is characterized by the presence of an event horizon, which separates an observer at the asymptotic infinity from the interior of the black hole. The two regions are causally separated by the horizon. Now suppose we throw in some object in the black hole. Once the object passes the horizon, we won't be able to access to the information contained it. In classical general relativity, the information is not lost, but is just hidden behind the horizon. In quantum mechanics, however, the situation is different. Steven Hawking studied properties of quantum black hole in the semi-classical approximation, and showed that the black hole emits thermal radiation according to the Planckian distribution, namely with the maximum randomness. So it seems that the genetic information contained in the flower is completely dissipated as the thermal radiation. The loss of information is a serious problem in quantum mechanics.

String theory in fact has a large class of black hole solutions. They were called black p-branes for the following reason. p is an integer and it can be $0, 1, 2$, *etc.*

The 0-brane is the standard black hole, which is a point source. The spacetime becomes flat when one goes away from the center, but it is strongly curved near the center and there is a horizon. In $(9 + 1)$ dimensions, the geometry transverse to the black hole is 9 dimensions.

When $p = 1$, we have the black 1-brane. It is a string-like source. It is spread in one-spatial dimensions. Combined with the time direction, the source looks like a two-dimensional sheet. The directions transverse to the 1-brane is 8 dimensions. The 2-brane is a membrane. In fact it is where the name p-brane original came from. Its source is extended in 2-space and 1-time directions. In general, the p-brane is extended in p-spatial directions and 1-time direction.

So we have this zoo of black hole solutions. They are analogous to the solitons, monopoles and instantons in the standard quantum field theory. These black hole solutions have played an essential role in the recent development of string theory.

To address the question of the information paradox, one needs to know what happens if we throw in some matter into the black p-brane. There are two ways to study what happens to the string near the black p-brane, and they are equivalent.

One is to study the propagation of the string in the back ground geometry of black p-brane.

There is, however, another way to view this problem. The black p-brane is like a defect in the spacetime. It is very similar to the monopole solution in the gauge theory. Therefore we may try to identify collective degrees of freedom of the p-brane, which describe the motion and the fluctuation of the brane. One can then formulate the problem as interactions between the string and the collective coordinates of the p-brane. This approach was taken by Polchinski, and he called this description of the p-brane. In a certain situation, the collective coordinates of the p-brane are open strings ending on the brane (this description is called D-brane), whose low energy limit is a supersymmetric gauge theory in $(p + 1)$ dimensions.

So we have these two descriptions of the same object. In one way of viewing at it, we have strings in the background geometry of black p-brane. In another way of viewing at it, we have the collective coordinates of the p-brane, which is the $(p + 1)$-dimensional gauge theory, interacting with strings.

The equivalence of the collective coordinate approach and the background quantization is well-established in the standard quantum field theory. For example, the Callan-Rubakov effect of the monopole can be understood as interaction between the collective coordinates of the monopole, which includes the rotor degrees of freedom in the internal space, and matter fields.

The equivalence of these two descriptions is the origin of the connection between the quantum gravity and the gauge theory.

The relation between gauge theories and string theory has been one of the main focuses of string theory research in the past few years, and various important observations have been made. The underlying theme of these works is the equivalence of the two descriptions of the p-brane: One is in terms of the black p-brane as a geometric background in which strings can propagate, and another is a description in terms of the gauge theory.

3

These observations were crystallized in the work of Maldacena. His conjecture, called the *AdS*/*CFT* correspondence, can be summarized in the following diagram.

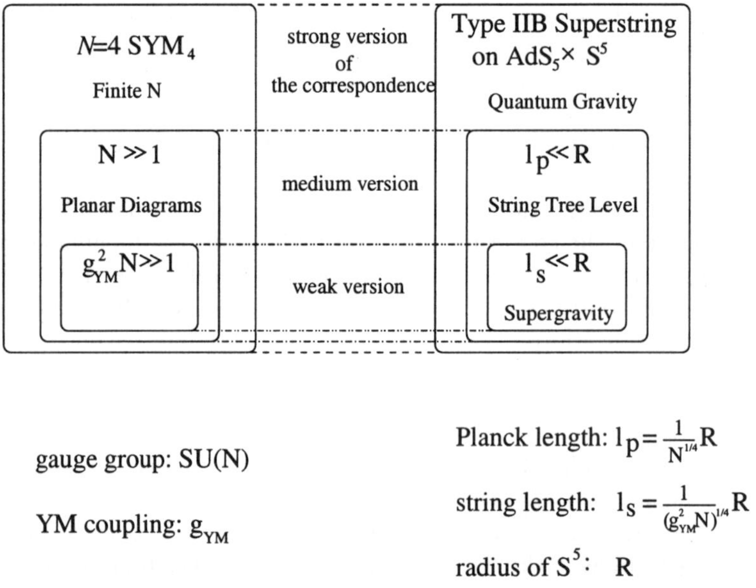

Figure 1:

What is striking about the AdS/CFT correspondence is that these two descriptions are claimed to be completely equivalent. Maldacena looked at the same object, which in this case is the black 3-brane, and took exactly the same limit. In the background quantization picture, the limit amplifies the region near the event horizon of the 3-brane. Its geometry is exactly that of $AdS_5 \times S^5$. What Maldacena showed was that it is exactly the same limit that extracts the gauge theory degrees of freedom from the collective coordinate description of the 3-brane. Thus the conjecture states that the string theory in 10 dimensions is completely equivalent to the gauge theory in $(3 + 1)$ dimensions.

Maldacena made this conjecture in November of 1997, but for a couple of months people did not know what to make of it since it seemed so strange. The string theory is in 10 and the corresponding gauge theory is in 4 dimensions. One has gravity and another don't. Gradually more precise formulation of the conjecture emerged and various checks have been made.

For example, Gary Horowitz and I thought that, if the string theory and the gauge theory are really equivalent as quantum theories, they must have the same Hilbert space. We studied how one can compare the Hilbert spaces of the two theories. Ferrara, Fronsdal and Zaffaroni went further, studied the Hilbert space structure in detail and came up with a complete table to the correspondence between chiral primary states.

If the two quantum theories are equivalent, they must have the same operator algebra. This issue was studied by Gubser, Klebanov and Polyakov and independently by Witten.

So I have explained to you how string theorists came to the conclusion that the string theory and the gauge theory are one and the same thing. This correspondence was found by studying quantum black hole in string theory and by looking at it in two different ways. One is to look at string in the curved black hole geometry. Another is to look at the collective coordinates of the black hole, which is the gauge theory. The equivalence of the two description leads to conclude that the string theory in the curved geometry is equivalent to the gauge theory. The logic seems reasonable, but the consequence is surprising.

First of all, string and the gauge theory live in different spacetime dimensions. I should point out that this is not the same as the Kaluza-Klein reduction. In the Kaluza-Klein mechanism, one curls up a part of the spacetime into a tiny circle. If we ignore the fluctuation of fields along the circle, the dimensionality of the space is reduced. Here we are not truncating a theory. The equivalence states that the string theory is fully equivalent to the gauge theory, without any reduction.

Second, string theory contains gravity and the gauge theory doesn't.

These two surprising features of the correspondence seem to be related to an earlier observation on quantum gravity by 't Hooft and Susskind. They suggested, based on various properties of black hole, that in quantum gravity the information of the theory can be stored in a lower dimensions. This idea is called Holography of Quantum Gravity. It seems that the AdS/CFT correspondence is an explicit realization of this idea. Clearly there are still lots of things that need to be understood in this direction.

What strikes me most about this development is that the string theory appears to be much more versatile and omnipresent than previously thought. The main purpose of the string theory is still a search for the unified theory. At the same time, we found that the string theory is useful to study various field theories, which may or may not be directly related to the unified theory. This seems to suggest that the string theory is a basic tool and perhaps a fundamental language to study physics in short distances.

Acknowledgments

I would like to thank the organizers of the conference for hospitality.

Interconnections between type II superstrings, M theory and $\mathcal{N} = 4$ Yang–Mills

Michael B. Green
Department of Applied Mathematics and Theoretical Physics, Silver Street, Cambridge CB3 9EW, UK
E-mail: M.B.Green@damtp.cam.ac.uk

Contents

1 Introduction

"The incalculable in pursuit of the ineffable"[a]

Over the past few years a number of very interesting nonperturbative aspects of string theory have emerged and are encapsulated in the term 'M theory'. The precise meaning of this term is open to a multitude of interpretations although a common theme is that M theory represents a framework for describing nonperturbative quantum gravity that reduces to perturbative string theory or classical eleven-dimensional supergravity in various limits. Although there have been attempts to give a microscopic definition of M theory it is not at all obvious that the correct concepts have yet been discovered. Nevertheless, certain fascinating features are emerging that will surely be of lasting significance. Foremost among these is the interplay between quantum gravity and Yang–Mills gauge theory, which has been suggested by the structure of string perturbation theory for many years but has recently come to the fore in the realization of the 'holographic' principle in the context of the AdS/CFT correspondence.

The point of view followed in these notes is that it is of interest to unravel the extent to which features of the theory emerge purely as a consequence of its very large symmetries and do not depend on the microscopic model. All present formulations of string theory and its M theory extensions depend in some manner on the background. Obviously this is not a satisfactory state of affairs for a quantum theory of gravity. In practise, this means that the properties of the theory depend on the number of moduli of the particular background

[a] Apologies to Oscar Wilde.

chosen. The number of moduli grows with the number of compact dimensions — there are no moduli in eleven dimensions while the ten-dimensional type IIA theory has a single modulus (the dilaton) and the ten-dimensional type IIB theory has two moduli (associated with the complex scalar field). In lower dimensions the number of moduli increases further and there is a corresponding increase in the size of the duality group. In order to avoid the possibility of terminological confusion the term 'M theory' will be restricted in these notes to apply to the low energy expansion of eleven-dimensional supergravity compactified on specified backgrounds[b].

Overview

In section 2 we will begin with a brief review of the classical relationships between M theory compactified on a two-torus, T^2, to nine dimensions and the two type II superstring theories compactified on a circle, S^1. These are configurations with maximal supersymmetry — there are 32 unbroken components in the supercharges. The nine-dimensional duality group is $SL(2,\mathbb{Z})$ and there are three scalar moduli fields that parameterize the three-dimensional coset space $SL(2,\mathbb{Z})\backslash SL(2,\mathbb{R})/U(1) \times R$. The scalars may be identified with the volume and the complex structure of T^2. The duality properties were originally motivated [1,2,3] by comparison of the classical (Einstein–Hilbert) eleven-dimensional action [4] with the classical low energy string type II string actions. Furthermore, the BPS states of the theories also coincide. For example, there are states in which the M2-brane (the membrane solution of eleven-dimensional supergravity) is wrapped around the two-cycles of T^2. Such states are interpreted in the string theories as fundamental string states with the string wound around the circle that forms the tenth dimension. This is a simple illustration of the more general network of dualities that relate all the different string perturbation theories and eleven-dimensional supergravity [5,1].

The duality that is manifested by the low energy effective actions of eleven-dimensional supergravity and type II string theories is supposed to extend to the complete quantum theories. We will consider the expansion of the quantum effective action in powers of derivatives on the fields. Thus, in string theory language the Wilsonian action will be expanded in inverse powers of the string tension[c],

$$\alpha'^{-4} S = \alpha'^{-4}(S^{(0)} + \alpha' S^{(1)} + \ldots + (\alpha')^n S^{(n)} + \ldots), \tag{1.1}$$

[b]The ultimate fundamental theory is perhaps worthy of representation by an unutterable third millenium TEX symbol instead of a mundane latin letter

[c]Recall that the string tension T_F has dimensions (length)$^{-2}$ and $\alpha' = (2\pi T_F)^{-1}$ has dimension (length)^2.

where, by dimensional analysis, each power of α' implicitly comes with two derivatives on bosonic fields whereas a fermionic field counts as a half power of a derivative, as usual. Obviously an expansion such as (1.1) is of limited value since it does not capture effects that are nonperturbative in α'. The leading term, $S^{(0)}$, contains the classical Einstein–Hilbert action and all the terms related by supersymmetry. We will use this notation even in the case of the type IIB theory where there is a well-known difficulty since there is no globally well-defined action for the self-dual five-form field strength, F_5. However, for the most part we will only use the notion of an action as a shorthand for packaging the equations of motion, which are well-defined. For such purposes a locally well-defined action, such as that in [6,7], is sufficient. The first nonzero terms beyond $S^{(0)}$ are those in $S^{(3)}$ which has a coefficient α'^{-1}. A striking example is the topological term in type IIA ten-dimensional string theory,

$$\frac{1}{(2\pi)^5\alpha'}\int d^{10}x\, B \wedge X_8(R), \tag{1.2}$$

where $X_8(R)$ is an eight-form made out of the curvature R and B is the two-form Neveu-Schwarz–Neveu-Schwarz ($NS \otimes NS$) antisymmetric tensor potential. This term, with its precisely determined overall coefficient, can be discovered directly from a one-loop string [8] calculation in which the stringy regulator plays a crucial rôle in regulating an otherwise quadratically divergent diagram. It can also be discovered by considering the cancellation of chiral gravitational anomalies in the type IIA five-brane. These anomalies arise from the coupling of the world-volume chiral fermions and self-dual antisymmetric three-form field strength to the pull-back of the target space gravitational field. In the presence of a five-brane 1.2 transforms anomalously under local diffeomorphisms and local Lorentz transformations in just the correct manner to cancel the world-volume anomaly. This is an example of the 'anomaly inflow mechanism' for cancelling local anomalies [9], which was the method used in [10] to derive the generalization of the term 1.2 to eleven-dimensional supergravity where it is the integral of an eleven-form, $C^{(3)} \wedge X_8$. Both of these arguments for the presence of the term 1.2 are outside the realm of perturbative quantum field theory – the first method invokes a stringy cutoff while the second method requires consistency of nonperturbative solitonic chiral five-branes.

Obviously 1.2 is not the only higher derivative term in the effective action. There will be terms with arbitrarily high numbers of derivatives as well as terms that are nonperturbative in α'. As we will see, the large amount of supersymmetry severely constrains the form of certain classes of 'protected' terms in the effective a ction. Whereas the classical type IIB supergravity (the supersymmetrized Einstein theory) is invariant under continuous SL(2,$\mathbb{R}$)

transformations the IIB superstring only possesses a discrete (local) $SL(2, \mathbb{Z})$. This means that the quantum effective action only possesses this discrete symmetry which must be a symmetry order by order in α'.

The first corrections to the Einstein–Hilbert term are of order α'^{-1} (α'^3 relative to the leading term). One such term is a specific contraction of four Riemann curvatures that only involves the Weyl tensor components (recall that the Weyl tensor is the part of the Riemann tensor that is trace-free). In the type IIB theory this term, which will be denoted by $\mathcal{R}^4$, must have the form (in string frame)[11],

$$\int d^{10}x\sqrt{-g}e^{-\phi/2} f^{(0,0)}(\tau,\bar{\tau})\mathcal{R}^4, \tag{1.3}$$

where τ is a complex scalar field,

$$\tau \equiv \tau_1 + i\tau_2 = C^{(0)} + ie^{-\phi}, \tag{1.4}$$

and $C^{(0)}$ is the Ramond–Ramond ($\mathcal{R} \otimes \mathcal{R}$) pseudoscalar field and ϕ is the type IIB dilaton field (which will sometimes be written as ϕ^B in order to distinguish it from the type IIA dilaton). The power of $e^{-\phi/2}$ in (1.3) cancels after transformation to the Einstein frame,

$$g \to g_E e^{\phi/2}, \tag{1.5}$$

where g_E is the Einstein-frame metric. In this frame the curvature is invariant under $SL(2,\mathbb{Z})$ so that the function $f^{(0,0)}(\tau,\bar{\tau})$ must be a scalar under $SL(2,\mathbb{Z})$ transformations which act as

$$\tau \to \frac{a\tau + b}{c\tau + d}, \tag{1.6}$$

with integer $a, b, c, d: \; ad - bc = 1$. The notation in (1.3) has been chosen with later generalizations in mind where we will encounter modular forms, $f^{(w,\hat{w})}(\tau,\bar{\tau})$, where the superscripts $(w, \hat{w})$ denote holomorphic and antiholomorphic weights. The string coupling constant is proportional to e^{ϕ} when ϕ is constant. It follows that the large $\tau_2 = e^{-\phi}$ expansion of $f^{(0,0)}$ must reproduce the terms that are known in the perturbative type IIB string expansion [11,12]. It will also include exponentially suppressed terms that correspond to the presence of arbitrary numbers of D-instantons that will be discussed in subsequent sections.

At the linearized level the physical fields of IIB supergravity package together into a constrained 'on-shell' superfield that is a function of a $SO(9,1)$ Grassmann spinor, θ[13]. In section 3 this will be used to express the $\mathcal{R}^4$ term, together with many other interaction terms of the same dimension, in terms of an integral over the sixteen components of θ, which is half the total number of components of the two type IIB supersymmetries. This makes manifest the fact that these terms of order ${\alpha'}^{-1}$ are related by supersymmetry at the linearised level. We will refer to terms such as these that are integrals over a fraction of the superspace as 'protected' terms. Generally these interactions violate the $U(1)$ R symmetry of the classical type IIB theory. For example, among such terms is one of the form [14]

$$\int d^{10}x\sqrt{-g}e^{-\phi/2}f^{(12,-12)}(\tau,\bar{\tau})\lambda^{16}, \tag{1.7}$$

where λ is the complex dilatino which is a sixteen-component chiral SO(9,1) spinor. The nonholomorphic modular form, $f^{(12,-12)}$, transforms with a phase under $SL(2,\mathbb{Z})$ in a manner that cancels the phase transformation of the dilatino. The characteristic feature of these terms is that the linearized approximation is a good approximation for the D-instanton contributions but not for the perturbative contributions of zero instanton number. The sixteen components of θ are identified as the components of supersymmetry that are broken in the D-instanton background [11,14].

The constraints imposed by the full nonlinear supersymmetry in the type IIB theory will be the subject of section 4 where the terms of order ${\alpha'}^{-1}$ in the derivative expansion of the effective action of type IIB supergravity will be deduced by requiring the closure of the superalgebra at this order [15]. This is a kind of Noether procedure that implements supersymmetry up to any given order in α'. In this manner terms we shall determine not only the terms in $S^{(3)}$ but certain of the modified supersymmetry transformations that relate $S^{(3)}$ to the classical action $S^{(0)}$. We will see that supersymmetry requires that the functions of the moduli $f^{(w,-w)}(\tau,\bar{\tau})$ are eigenfunctions of the Laplace operator on the Poincare upper-half plane with specific eigenvalues that depend on w. Assuming that the theory is invariant under $SL(2,\mathbb{Z})$ determines the solutions uniquely.

We will see in section 5 that the protected higher-derivative terms, such as $f^{(0,0)}\mathcal{R}^4$ can be deduced by consideration of one-loop Feynman diagrams in the eleven-dimensional theory compactified on T^2 [16]. Thus, the $\mathcal{R}^4$ term will be deduced by considering the one-loop scattering of four gravitons. This amplitude can be expressed as a sum over windings around the torus of the world-line of the particle circulating in the loop. The dependence on the moduli of the torus emerges from the sectors of nonzero winding while the ultraviolet (UV) divergence is contained in the zero-winding sector. Since the supergravity field theory is not renormalizable these UV divergences cannot be interpreted in the absence of a microscopic definition of M theory (just as string theory regulates the ultraviolet divergences in ten-dimensional supergravity). However, with any sensible regularization this term is proportional to the volume of the torus and it disappears in the limit in which the torus has zero volume. But this is the limit in which M theory is identified with the ten-dimensional type IIB theory. We will find that the expression for the $\mathcal{R}^4$ term that was deduced from supersymmetry in section 4 is reproduced by this one-loop calculation. More generally, for finite volume, there are volume-dependent terms that are ill-defined due to the UV divergence. However, if the condition of T-duality, that relates the type IIB and IIA string theories in nine dimensions is imposed, then the regularized coefficients are determined uniquely and there are no ambiguities. Having fixed the value of the regularized divergence the eleven-dimensional limit may be recovered with a specific finite coefficient for the $\mathcal{R}^4$ term[d].

In similar fashion the other terms of order α'^{-1} can be obtained from one-loop processes in M theory on T^2. The interaction of sixteen dilatini, λ^{16}, will also be considered in detail in section 5. This involves the evaluation of a sixteen-gravitino amplitude in which the gravitini are polarized in appropriate directions with respect to the two-torus so as to be interpreted as spin-1/2 dialtini in the IIB theory [14]. In this case there is no eleven-dimensional divergence at all and the loop amplitude is proportional to the modular form $f^{(12,-12)}$.

The structure of these $o(\alpha'^{-1})$ interactions will be analyzed further in section 6. The large τ_2 (small coupling) expansion of any of these terms is determined by writing the modular form, $f^{(w,-w)}(\tau,\bar{\tau})$, as a Fourier series in powers of $e^{2\pi i\tau_1}$, which is a series of D-instanton terms. The zero Fourier mode contains the perturbative contributions (powers of τ_2^{-1}) which consist of the tree-level and one-loop terms with no higher-order power behaved terms. The absence of perturbative terms beyond one loop points to a perturbative non-renormalization theorem which is a consequence of the fact that the protected interactions can be expressed as integrals over half the superspace.[e] The charge-K D-instanton contribution is correlated with a phase factor $e^{2\pi iK\tau_1}$. The significant τ_2 dependence of the charge-K D-instanton term is contained in the Bessel function $K_{-1}(2\pi|K|\tau_2)$ which has a large-τ_2 expansion that is suppressed by the exponential factor $e^{-2\pi|K|\tau_2}(1+o(\tau_2^{-1}))$. The coefficient of the leading charge-K D-instanton contribution will be found to be (with

[d] For related work in toroidally compactified type II theories see [17,18] and in type I theories see [19,20,21]. $SL(2,\mathbb{Z})$-invariant expressions for higher-dimensional terms in the type IIB effective action have also been proposed in [22,23,24,25,26].

[e] Some indirect arguments corroborate the validity of this 'theorem' [27,28,29].

suitable normalization)

$$\sum_{m|K} \frac{1}{m^2}, \tag{1.8}$$

where the notation indicates a sum over all the divisors of K. On the other hand, an extension of ideas in [30] suggest that this D-instanton measure should be identified with the partition function of the $SU(K)$ 'zero-dimensional matrix model', that is the compactification to zero dimensions of ten-dimensional supersymmetric Yang–Mills theory. In this manner we will be able to deduce the partition function for arbitrary K.

The K D-instanton measure is related to the problem of evaluating the Witten index for K D-particles which will be the subject of section 7 [31]. The chain of duality relations between string theory and eleven-dimensional supergravity requires that there be precisely one threshold bound state of K D-particles of charge K [1]. This means that the Witten index for the K D-particle system must be equal to one. The case of two D-particles ($K = 2$) was considered in detail in [32,33] where it was noted that the Witten index has two contributions – the 'bulk' term is equal to the zero-dimensional matrix model partition function while the 'boundary' term is obtained from the region of moduli in which the two D-particles can be described as identical free particles. We will see how these ideas generalize to arbitrary K. The bulk term in the index is now given by the zero-dimensional matrix model partition function for general K (1.8). Furthermore, if it is assumed that the boundary term in the index is again given by the free propagation of identical D-particles of any charge less than $K = mn$ then the total index is equal to one as expected.

In section 8 we turn to consider the effect of the higher derivative terms on the conjectured equivalence of type IIB string theory in $AdS_5 \times S^5$ (five-dimensional anti de Sitter space $\times S^5$) and four-dimensional $\mathcal{N} = 4$ supersymmetric $SU(N)$ Yang–Mills theory on the boundary of AdS_5 (with the S^5 appended). According to this correspondence [34] (see also [35,36]) the higher-derivative interactions of order α'^{-1} in the type IIB theory should correspond to contributions of correlation functions of composite operators in the Yang–Mills theory of order $N^{-3/2}$ relative to the leading terms. This leads to a prediction of the precise form of the Yang–Mills instanton contributions to classes of correlation functions that may be obtained from the protected supergravity processes. These exact expressions exhibit the Montonen–Olive duality $SL(2, \mathbb{Z})$ symmetry [37,38] that is inherited from the supergravity interaction terms discussed in the earlier sections.

Section 9 will focus on the correspondence between the D-instanton contributions to IIB supergravity in $AdS_5 \times S^5$ and the contribution of Yang–Mills instantons in the boundary supersymmetric Yang–Mills theory. The D-instanton is the stringy version of a classical solution to type IIB supergravity that is concentrated around a point in euclidean ten-dimensional space. In the Einstein frame this solution is one in which the metric is flat and $\nabla^2 e^\phi = 0$ and the BPS condition relates the euclidean continuation of $C^{(0)}$ to $e^{-\phi}$.

We will illustrate the AdS/CFT correspondence by comparing the one-instanton contribution to the sixteen-dilatino amplitude in three different ways. Firstly (method (a)), we will extract the leading one D-instanton term in $f^{(12,-12)}$ which is the coefficient of the sixteen-dilatino vertex. The amplitude with specified boundary data is obtained by contracting each leg of this vertex with a dilatino propagator from the vertex to a point on the boundary and integrating the position of the vertex in the bulk $AdS_5 \times S^5$ space. The AdS/CFT correspondence identifies the boundary value of the dilatino as the source of a particular fermionic supercurrent of the $\mathcal{N} = 4$ Yang–Mills theory. It is therefore of interest to compare the sixteen-dilatino amplitude with the correlation function of sixteen of these supercurrents. The second method (method (b)) is the direct determination of the leading one-instanton term in this correlator in $SU(2)$ $\mathcal{N} = 4$ Yang–Mills theory which we will see has the same form as the supergravity amplitude determined by method (a), up to an undetermined overall coefficient. The third method (method (c)), based on the semiclassical approximation to scattering in a D-instanton background, again gives the same expression. The agreement between the functional form of the correlators in $SU(2)$ Yang–Mills and the bulk amplitudes in IIB supergravity might appear somewhat surprising since the AdS/CFT correspondence is supposed to relate the leading α' piece of the $\mathcal{R}^4$ term to the large-N limit of $SU(N)$ Yang–Mills, rather than to the $N = 2$ case. However, recent work on the large-N limit of the one-instanton contributions of the Yang–Mills theory [39] shows that all the N-dependence enters in the overall coefficients of these correlation functions and their functional form is independent of N, thus explaining the agreement for $N = 2$. Furthermore, this overall coefficient has the predicted N-dependence for large N. This impressive agreement extends to the behaviour of the charge-K D-instanton contribution [40] to leading order in the coupling constant. This will also be described (with few details) in section 9.

2 M theory on T^2 and type II superstrings on S^1

The Einstein–Hilbert action of the eleven-dimensional theory has the form,

$$S_{EH} = \frac{1}{2\kappa_{11}^2} \int d^{11}x \sqrt{-G^{(11)}} R, \tag{2.1}$$

where the overall normalization has been expressed in terms of the coupling $\kappa_{11} \equiv (2\pi l_P)^{9/2}$ and l_P is the eleven-dimensional Planck length. The additional terms in the fully supersymmetric classical action of [4] involve the three-form potential, $C^{(3)}$, and the gravitino, $\psi_{\hat{\mu}}$ $(\hat{\mu}, \hat{\nu} = 0, 1, \ldots, 9, 11)$.

It was argued in [1] that eleven-dimensional M theory compactified on a circle of radius $R_{11} l_P$ [f] should be identified with ten-dimensional type IIA string theory with a coupling constant, $g = e^{\phi^A}$ (where ϕ^A is the IIA dilaton), that is related to R_{11}. This is seen by parameterizing the eleven-dimensional metric, $G_{\hat{\mu}\hat{\nu}}$,

$$ds^2 = G_{\mu\nu} dx^\mu dx^\nu + R_{11}^2 (dx^{11} - C_\mu^{(1)} dx^\mu)^2, \tag{2.2}$$

where $C_\mu^{(1)}$ is the $\mathcal{R} \otimes \mathcal{R}$ one-form potential of the type IIA theory. It is easy to check that with the identifications

$$R_{11} = e^{2\phi^A/3}, \tag{2.3}$$

and

$$G_{\mu\nu} = e^{-2\phi^A/3} g_{\mu\nu}^A \left(\frac{l_P}{l_S}\right)^2, \tag{2.4}$$

where $g_{\mu\nu}^A$ is the string-frame metric of type IIA string theory, the Einstein–Hilbert action (2.1) compactified on a circle becomes

$$S_{EH} = \frac{2\pi}{(2\pi l_S)^8} \int d^{10}x \sqrt{-g^A} e^{-2\phi^A} R. \tag{2.5}$$

In these equations l_S is the string scale which is related to the fundamental string tension, T_F, by

$$l_S^2 \equiv \alpha' = \frac{1}{2\pi T_F} \tag{2.6}$$

and the curvature in (2.5) is expressed in the string metric. The overall coefficient in (2.5) is often denoted by $(2\kappa_{10}^2)^{-1}$ where $\kappa_{10}^2 = (2\pi l_S)^8/4\pi$. With these conventions the tension in a Dp-brane is given by

$$T_p = \frac{1}{l_S} (2\pi l_S)^{-p} e^{-\phi^A} \tag{2.7}$$

---

[f] We will use a convention in which capital letters denote distances in eleven-dimensional Planck units while lower case letters denote distances in string units.

and so the tension of the D-string ($p = 1$) is $e^{-\phi^A} T_F$ [41]. The identification of the masses of the Kaluza–Klein modes of compactified M theory and the masses of the type IIA D-particles ($p = 0$ in (2.7)) leads to the relation between the scales,

$$\frac{1}{R_{11} l_P} = \frac{e^{-\phi^A}}{l_S}, \tag{2.8}$$

or

$$l_P = e^{\phi^A/3} l_S, \tag{2.9}$$

which expresses the eleven-dimensional Planck scale in terms of the string scale and coupling constant.

We will now generalize this description and obtain more insight by making use of the fact that in the nine-dimensional theory the $SL(2, \mathbb{Z})$ symmetry of the IIB string theory can be interpreted as a geometric symmetry of M-theory compactified on a torus [2,3]. Three scalar fields arise from the compactification of the eleven-dimensional theory (with coordinates labelled $x^0, \ldots, x^9, x^{11}$) on a two-torus, T^2, oriented in the (x^9, x^{11}) directions so that $G^2_{99} = R^2_9$. These scalars correspond to the volume of the torus, $\mathcal{V}$, and the complex structure, Ω. Using the ansatz (2.2) for the eleven-dimensional metric we have

$$\sqrt{-G^{(11)}} = \sqrt{G^{(2)}}\sqrt{-G^{(9)}} = \mathcal{V}\sqrt{-G^{(9)}}, \tag{2.10}$$

where $\mathcal{V} = R_9 R_{11}$ is the volume of T^2. The metric on the two-torus is

$$G^{(2)}_{IJ} = \frac{\mathcal{V}}{\Omega_2}\begin{pmatrix} |\Omega|^2 & \Omega_1 \\ \Omega_1 & 1 \end{pmatrix} \tag{2.11}$$

($I, J = 9, 11$) where the complex structure is given by

$$\Omega \equiv \Omega_1 + i\Omega_2 = \frac{G^{(2)}_{9\,11} + i\sqrt{G^{(2)}}}{G^{(2)}_{11\,11}} = C^{(1)} + i\frac{R_9}{R_{11}}. \tag{2.12}$$

In type IIA language the radius of the tenth dimension is identified as

$$r_A = R_9 \frac{l_P}{l_S} = R_9 R_{11}^{1/2} = \mathcal{V}^{\frac{3}{4}} \Omega_2^{\frac{1}{4}}. \tag{2.13}$$

Now T-duality on this circle can be used to determine the relation with the type IIB theory. This gives

$$r_B = \frac{1}{r_A} = R_9^{-1} R_{11}^{-1/2}, \qquad e^{-\phi^B} = r_A e^{-\phi^A} = \frac{R_9}{R_{11}} \tag{2.14}$$

and $C^{(0)} = C_9^{(1)} = \Omega_1$. These equivalences can be stated in the form,

$$\begin{aligned}\Omega &= \tau^A \equiv C_9^{(1)} + ir_A e^{-\phi^A} \\ &= \tau^B \equiv C^{(0)} + ie^{-\phi^B}.\end{aligned} \tag{2.15}$$

Large diffeomorphisms of the torus are $SL(2,\mathbb{Z})$ transformations under which Ω transforms by

$$\Omega \to \frac{a\Omega + b}{c\Omega + d}, \tag{2.16}$$

with $a, b, c, d \in \mathbb{Z}$ and $ad - bc = 1$.

The type IIB Einstein–Hilbert action is obtained by compactification of the type IIA theory on a circle of radius r_A followed by T-duality, leading to the type IIB expression (now in the type IIB string frame),

$$S_{EH} = \frac{2\pi}{(2\pi l_S)^8} \int d^{10}x \sqrt{-g^B} e^{-2\phi^B} R, \tag{2.17}$$

where $g^B_{\mu\nu}$ denotes the IIB string metric. Supersymmetry determines the remaining terms in the complete classical action in any of these parameterizations, subject to the caveat that the IIB theory does not have a globally well-defined action because of the special features of self-dual antisymmetric tensor fields [42,43]. It follows from simple dimensional analysis that all the terms that only involve bosonic fields have two derivatives. A pair of fermionic fields is dimensionally equivalent to a derivative on a bosonic field, so there are terms with two fermion fields and a single derivative as well as terms with four fermionic fields and no derivatives.

In the following we will be concerned with higher-derivative corrections to the lowest-order action that necessarily arise in the quantum theory. Certain of these terms are known from tree-level and one-loop string perturbation theory. For example, the exact expression for the four-graviton scattering amplitude in either of the type II superstring theories has a simple expansion in powers of $\alpha' = l_s^2$. The expression, which is reproduced in appendix A (A.1), has an overall kinematic factor of $\tau_2^2 \tilde{K}$ where [44],

$$\tilde{K} = t_8^{\mu_1 \ldots \mu_8} t_8^{\nu_1 \ldots \nu_8} \prod_{r=1}^{4} \zeta_{\mu_r \nu_r} k^{(r)}_{\mu_r} k^{(r)}_{\nu_2}, \tag{2.18}$$

and the eighth-rank tensor t_8 arises in open superstring amplitudes and is defined in [45]. The factor, $\tilde{K}$, is eighth-order in the external momenta and is the linearized approximation to $\mathcal{R}^4$. The leading term in the expansion of the amplitude in powers of α' in (A.1) summarizes the sum of the tree-level Feynman diagrams for on-shell four-graviton scattering in either of the ten-dimensional low energy type II supergravity theories. This consists of the sum of pole terms together with a four-graviton contact interaction with two derivatives that arise from the Einstein–Hilbert term. The next term, with coefficient $\zeta(3)$ is a contact interaction with eight powers of momenta (contained in $\tilde{K}$) that arises from the $\mathcal{R}^4$ term (which was deduced in this manner in [46] and from a four-loop sigma model term in [47]). The one-loop contribution to the four-graviton amplitude has coefficient $\tilde{K}$ with no dilaton dependence[g]. It does not possess massless particle poles and its leading low energy contribution is also to the $\mathcal{R}^4$ term [44]. The effect of the tree-level $\mathcal{R}^4$ term on Calabi–Yau compactifications has recently been reconsidered in [48,49] and its effect on black hole solutions in $AdS_5 \times S^5$ has been analyzed in [50] and [51].

After compactification to nine dimensions on a circle of radius r_A in the IIA units (or $r_B = 1/r_A$ in the IIB units) the one-loop term gets a simple $1/r_A^2$ (or $1/r_B^2$) correction due to the winding of the string around the compact tenth dimension. Putting these terms together with the appropriate coefficients leads to the expressions for the nine-dimensional $\mathcal{R}^4$ term in coordinates appropriate to the type IIA, IIB or M-theory parameterizations [12],

$$\begin{aligned} S_{R^4} &= \frac{1}{3\cdot(4\pi)^7 l_S}\int d^9x\sqrt{-g^{A(9)}}\,\mathcal{R}^4\, r_A\left[2\zeta(3)(\tau_2^A)^2 + \frac{2\pi^2}{3}(1+\frac{1}{r_A^2}) + \cdots\right] \\ &= \frac{1}{3\cdot(4\pi)^7 l_S}\int d^9x\sqrt{-g^{B(9)}}\,\mathcal{R}^4\, r_B\left[2\zeta(3)(\tau_2^B)^2 + \frac{2\pi^2}{3}(1+\frac{1}{r_B^2}) + \cdots\right] \\ &= \frac{1}{3\cdot(4\pi)^7 l_P}\int d^9x\sqrt{-G^{(9)}}\,\mathcal{R}^4\, R_9R_{11}\left[2\zeta(3)\frac{1}{R_{11}^3} + \frac{2\pi^2}{3} + \frac{2\pi^2}{3R_9^2R_{11}} + \cdots\right] \end{aligned} \tag{2.19}$$

where $g^{A(9)}$, $g^{B(9)}$ are the nine-dimensional metrics in the IIA and IIB theories. An important feature of the first two expressions is that they are related by T-duality. This is trivial for the leading terms but less so at one loop where the winding term of one of the type II theories (e.g., the $2\pi^2/3r_A^2$ term in the first expression) is interchanged with the nonwinding term of the other (e.g., the $2\pi^2/3$ term in the second expression).

The ellipsis in (2.19) indicate other terms that are yet to be determined. As we will see there are in fact no further perturbative terms (higher inverse pow-

[g] Here and in the following we will ignore the Gauss-Bonnet term which also arises in the one-loop amplitude and is proportional to the product of two eight-dimensional Levi–Civita symbols.

ers of τ_2) but there will be exponentially suppressed terms that are powers of $e^{2\pi i\tau^A} = e^{2\pi i\tau^B} = e^{2\pi i\Omega}$. Such contributions are associated with D-instantons. Generally, a *Dp*-brane is a stringy soliton extended in $p \geq 0$ spatial directions while the D-instanton corresponds to the value $p = -1$ and is associated with a point-like *space-time event* rather like a small Yang–Mills instanton. The classical D-instanton is a solution of euclidean type IIB supergravity in which e^{ϕ^B} and $C^{(0)}$ have nontrivial profiles while the other fields (including the Einstein-frame metric) are trivial [52]. The action for a D-instanton with charge K is $2\pi|K|/g$, where g is the asymptotic value of τ_2^{-1} so that the D-instanton contributions in (2.19) should be characterized by terms proportional to

$$e^{-2\pi(|K|/g \mp iKC^{(0)})}, \tag{2.20}$$

where the $-$ sign refers to a D-instanton and $+$ to an anti D-instanton. This classical configuration will be reviewed in section 9 where we will also see a direct correspondence between Yang–Mills instantons and D-instantons in the context of the AdS/CFT conjecture. String scattering amplitudes in a D-instanton background are described by world-sheets with boundaries on which the target-space coordinates satisfy Dirichlet boundary conditions [53,54,11]. Such pointlike closed string boundary conditions have a significant short distance effect on string scattering amplitudes – a fact that led to the suggestion [55] that they may be relevant in the formulation of a string theory of hadrons.

An explicit calculation of the leading effect of a single D-instanton was carried out in [11] by considering four-graviton scattering in the presence of Dirichlet world-sheet boundaries. This established the presence of the charge-one D-instanton term in the ten-dimensional type IIB effective action that arises in the $r_B \to \infty$ limit of $S^{IIB}_{R^4}$. However, in order to ensure the $SL(2,\mathbb{Z})$ duality invariance of the effective action the expressions in (2.19), there must be an infinite number of terms with D-instantons of arbitrarily high charge. In order to exhibit this modular invariance it is instructive to write the last line of (2.19) in terms of the complex structure as

$$\begin{aligned} S_{R^4} &= \frac{1}{3\cdot(4\pi)^7 l_P}\int d^9x\sqrt{-G^{(9)}}\,\mathcal{R}^4 \\ &\quad \left\{\mathcal{V}^{-1/2}\left[2\zeta(3)(\Omega_2)^{3/2} + \frac{2\pi^2}{3}(\Omega_2)^{-1/2} + \cdots\right] + \frac{2\pi^2}{3}\mathcal{V}\right\} \\ &= \frac{1}{3\cdot(4\pi)^7 l_P}\int d^9x\sqrt{-G^{(9)}}\,\mathcal{R}^4\left\{\frac{2\pi^2}{3}\mathcal{V} + \mathcal{V}^{-1/2} f^{(0,0)}(\Omega,\bar{\Omega})\right\}. \end{aligned} \tag{2.21}$$

We have here identified the expression in parentheses with the expansion of an as yet undetermined function, $f^{(0,0)}(\Omega,\bar{\Omega})$, that must transform as a scalar under $SL(2,\mathbb{Z})$ transformations of the complex structure, Ω. This translates into a function $f^{(0,0)}(\tau^B,\bar{\tau}^B)$ in the IIB theory and $f^{(0,0)}(\tau^A,\bar{\tau}^A)$ in the IIA theory.

The fact that this function must match the two known perturbative coefficients and that it must contain powers of $e^{2\pi i\tau^B}$ motivated the conjecture [11] that the complete coefficient of the $\mathcal{R}^4$ term in (1.3) is the modular function

$$f^{(0,0)}(\tau,\bar{\tau}) = \sum_{(m,n)\neq(0,0)} \frac{\tau_2^{3/2}}{|m+n\tau|^3} \equiv 2\zeta(3)\, E_{\frac{3}{2}}(\tau), \tag{2.22}$$

where the notation $E_{\frac{3}{2}}$ denotes a nonholomorphic Eisenstein series (as defined, for example, in [56]). The overall normalization in (2.22) has been chosen so that the leading term for large τ_2 is $2\zeta(3)\tau_2^{3/2}$. This coincides with the tree-level expression in parentheses in the second line of (2.19) (after allowing for the extra factor of $\tau_2^{1/2}$ due to the transformation between the string frame and Einstein frame). This expression for $f^{(0,0)}$ was further motivated by the study of the one-loop corrections to eleven-dimensional supergravity on T^2 [16] (see section 5) and will be derived from supersymmetry in section 4.

In the meantime it is worth noting that the perturbative contributions to the decompactification limits that give the ten-dimensional type IIA theories and eleven-dimensional M-theory are easily obtained from (2.19) and (2.21). In the ten-dimensional IIA limit, $r_A \to \infty$ in the first line in (2.19), the D-instanton terms vanish exponentially since $\tau^A \sim ir_A e^{-\phi^A} \to i\infty$, so $e^{2\pi i\tau^A} \to 0$ in this limit. This leaves only the perturbative terms. If we now consider the further decompactification to eleven dimensions $e^{\phi^A} = R_{11}^{3/2} \to \infty$ the tree-level IIA contribution vanishes and the one-loop term gives a finite contribution. This eleven-dimensional limit coincides with the limit $\mathcal{V} \to \infty$ in (2.21). In this limit the $\mathcal{V}^{-1/2}$ term vanishes and the finite eleven-dimensional term comes entirely from the term pr oportional to $\mathcal{V}$. It is important to note if there had been other terms of higher order in the type IIA coupling constant, $e^{\phi^A} = R_{11}^{3/2}$, the eleven-dimensional limit would have been singular order by order in IIA perturbation theory and the limit would not have been uniform. However, if $f^{(0,0)}$ is given by the expression (2.22) there are no further perturbative terms and the eleven-dimensional limit is simply

$$\begin{aligned} S_{\mathcal{R}^4} &= \frac{1}{18\cdot(4\pi)^7\cdot l_P^3}\int d^{11}x\sqrt{-G^{(11)}}\,\mathcal{R}^4 \\ &= \frac{1}{9\cdot 2^{12}\cdot\pi^4\cdot\kappa_{11}^{2/3}}\int d^{11}x\sqrt{-G^{(11)}}\,\mathcal{R}^4. \end{aligned} \tag{2.23}$$

As argued in [12] the coefficient of this term is in precise agreement with supersymmetry which relates the $\mathcal{R}^4$ term to the eleven-form $C^{(3)} \wedge X_8$.

3 Linearized supersymmetry and higher derivative terms

The existence of a large number of interactions in the IIB theory that are of the same dimension as the $\mathcal{R}^4$ interaction can be motivated very simply by using linearized supersymmetry. This can be implemented by packaging the physical fields or their field strengths into a constrained superfield $\Phi(x^\mu - i\bar{\theta}\gamma^\mu\theta, \theta)$ where θ^a $(a = 1, \ldots, 16)$ is a complex Grassmann coordinate that transforms as a Weyl spinor of $SO(9,1)$. This superfield is taken to satisfy the holomorphic condition [13],

$$D^*\Phi = 0, \tag{3.1}$$

and the constraints,

$$D^4\Phi = D^{*4}\Phi^*, \tag{3.2}$$

where

$$D_A = \frac{\partial}{\partial\theta^A} + 2i(\gamma^\mu\theta^*)_A\partial_\mu, \qquad D^*_A = -\frac{\partial}{\partial\theta^{*A}} \tag{3.3}$$

are the holomorphic and anti-holomorphic covariant derivatives that anticommute with the rigid supersymmetries

$$Q_A = \frac{\partial}{\partial\theta^A}, \qquad Q^*_A = -\frac{\partial}{\partial\theta^{*A}} + 2i(\bar{\theta}\gamma^\mu)_A\partial_\mu. \tag{3.4}$$

The constraints 3.1 and (3.2) ensure that the field Φ has an expansion in powers of θ (but not θ^*), that terminates after the θ^8 term and contains the 256 fields in an 'on-shell' supermultiplet [h],

$$\begin{aligned}\Phi &= \tau_0(1+\Delta)\\ &= \tau_0(1 + a + i\bar{\theta}^*\lambda + \hat{G}_{\mu\nu\rho}\bar{\theta}^*\gamma^{\mu\nu\rho}\theta + i\bar{\theta}^*\gamma^{\mu\nu\sigma}\theta\,\bar{\theta}^*\gamma_\nu\partial_\sigma\psi_\mu\\ &+ \mathcal{R}_{\mu\sigma\nu\tau}\bar{\theta}^*\gamma^{\mu\nu\rho}\theta\bar{\theta}^*\gamma^{\sigma\tau}{}_\rho\theta + \partial_\mu\hat{F}_{5\,\nu\rho\sigma\tau\omega}\,\bar{\theta}^*\gamma^{\mu\nu\rho}\theta\bar{\theta}^*\gamma^{\sigma\tau\omega}\theta + \cdots + \theta^8\partial^4\bar{\tau})\\ &\equiv \sum_{r=0}^{8}\Phi^{(r)}\theta^r,\end{aligned} \tag{3.5}$$

where $\tau_0\Delta$ is the linearized fluctuation around a flat background with a constant scalar, $\tau_0 = \tau - \tau_0 a = C_0^{(0)} + ig^{-1}$. The normalizations that will not

[h] We are using the usual convention that $\gamma^{\mu_1\cdots\mu_p}$ is the antisymmetrized product of p gamma matrices, normalized so that $\gamma^{\mu_1\cdots\mu_p} \equiv \gamma^{\mu_1}\ldots\gamma^{\mu_p}$ when $\mu_1 \neq \ldots \neq \mu_p$.

concern us here. The symbol $\hat{G}_{\mu\nu\rho}$ $(\mu,\nu,\rho=0,\ldots,9)$ denotes the 'supercovariant' combination of G and fermion bilinears defined in appendix C, where $G_{\mu\nu\rho}$ and $G^*_{\mu\nu\rho}$ are complex combinations of the field strengths of the $\mathcal{R}\otimes\mathcal{R}$ and $NS\otimes NS$ two-form potentials. The θ^4 terms are $\mathcal{R}$, the Weyl curvature, and $\hat{F}_{5\,\rho_1\cdots\rho_5}$, which is the field strength of the fourth-rank $\mathcal{R}\otimes\mathcal{R}$ potential (where the hat again denotes the supercovariant combination defined in appendix C). The fermionic field λ is the dilatino and ψ_μ is the gravitino. The gamma matrices with world indices are defined by $\gamma^\mu = e^\mu_m\gamma^m$, where $m=0,\cdots,9$ is the $SO(9,1)$ tangent-space index and e^μ_m is the inverse zehnbein. A barred Weyl spinor, such as $\bar{\theta}$, is defined by

$$\bar{\theta}_a \equiv \theta^*_b(\gamma^0)_{ba}. \tag{3.6}$$

The terms indicated by $\cdots$ in (3.5) fill in the remaining members of the ten-dimensional $N=2$ chiral supermultiplet, comprising (in symbolic notation) $\partial\psi^*_{\mu\nu}$, $\partial^2 G^*_{\mu\nu\sigma}$, $\partial^3\lambda^*$ and $\partial^4\tau^*$.

The $U(1)$ R-symmetry charge u_r of any of the component fields $\Phi^{(r)}$ in the expansion (3.5) can easily be determined since they are correlated with the powers of θ. Assigning a charge $-1/2$ to θ and an overall charge 2 to the superfield leads to the charge for the field with r powers of θ,

$$u_r = 2 - \frac{r}{2} \tag{3.7}$$

(as in[57,58]). For example, $u_{\partial\tau}=2$; $u_\lambda = 3/2$; $u_G=1$; $u_\psi = 1/2$; $u_R = u_{F_5} = 0$; ....

Although the linearized theory cannot capture the full structure of the terms in the effective action it can be used to relate various terms in the limit of weak coupling, Im $\tau_0 = g^{-1}\to\infty$ (where $g=e^{\phi_0}$ is the string coupling constant). The linearized approximations to the complete interactions are those that arise by integrating a function of Φ over the sixteen components of θ,

$$S^{(3)}_{linear} = \int d^{10}x d^{16}\theta\, F[\Phi] + \text{ c.c.}, \tag{3.8}$$

which is manifestly invariant under the rigid supersymmetry transformations, (3.4). The various component interactions contained in (3.8) are obtained from the θ^{16} term in the expansion,

$$F[\Phi] = F(\tau_0) + \Delta\frac{\partial}{\partial\tau_0}F(\tau_0) + \frac{1}{2}\Delta^2\left(\frac{\partial}{\partial\tau_0}\right)^2 F(\tau_0) + \cdots. \tag{3.9}$$

Using the expression for Δ in (3.5) and substituting into (3.8) leads to all the possible interactions at this order in α' [14,25],

$$S^{(3)} = \int d^{10}x \ \det e \ \Big(f^{(12,-12)}\lambda^{16} + f^{(11,-11)}\hat{G}\lambda^{14} + \ldots \\ + f^{(8,-8)}\hat{G}^8 + \ldots + f^{(0,0)}\mathcal{R}^4 + \ldots + f^{(-12,12)}\lambda^{*\,16} \Big) , \qquad (3.10)$$

where $\det e = \det e_\mu^m$ is the determinant of the zehnbein.

The $\mathcal{R}^4$ interaction comes from the Δ^4 term while the λ^{16} comes from the Δ^{16} term. We see, in particular, that the tensor structure of the contractions between the four Riemann tensors in the $\mathcal{R}^4$ term is summarized by the Grassmann integration [59],

$$R^4 = \int d^{16}\theta (R_{\theta^4})^4, \qquad (3.11)$$

where

$$R_{\theta^4} \equiv \bar\theta \gamma^{\mu\nu\sigma}\theta \bar\theta \gamma^{\rho\tau}{}_{\sigma}\theta \ R_{\mu\nu\rho\tau} . \qquad (3.12)$$

It follows from a standard Fierz transformation that all possible contractions of the Riemann tensor appearing in R_{θ^4} vanish, so that only the trace-free part — the Weyl tensor — survives.

The superscripts that label the coefficients $f^{(w,-w)}$ are related to the violation of the $U(1)$ charge. Thus, the linearized form of the general term in (3.10) contains a product of p fields,

$$\int d^{10}x \ \det e \ f^{(w,-w)} \prod_{k=1}^{p} \Phi^{(r_k)}, \qquad (3.13)$$

which violates the $U(1)$ charge by

$$2w = \sum_{k=1}^{p} u_{r_k} = 2p - 8 \qquad (3.14)$$

units, where we have used (3.7) and the fact that the total power of θ must be $\sum_k r_k = 16$. For example the $\mathcal{R}^4$ term ($w = 0$) conserves the $U(1)$ charge while the λ^{16} term ($w = 12$) violates the $U(1)$ charge by 24 and there are many other terms that violate the charge by any even number.

In the linearized approximation, $g \to 0$ ($\tau_2 \to \infty$), the coefficients $f^{(w,-w)}$ are constants that are related to each other by use of the Taylor expansion,

(3.9). For example, the $\mathcal{R}^4$ term has coefficient $\partial^4_{\tau_2} F$ while the λ^{16} term has coefficient $\partial^{16}_{\tau_2} F$ so that, at the linearized level,

$$f^{(12,-12)} \sim \left(\tau_2 \frac{\partial}{\partial \tau_2}\right)^{12} f^{(0,0)}, \tag{3.15}$$

where for the moment we are not concerned about the overall constant. In writing this we have used the fact that the linearized approximation is valid only if the inhomogeneous term in the modular covariant derivative, $\mathcal{D}$ (defined in appendix B), is negligible which requires that

$$2\tau_2 \partial_{\tau_0} f^{(w,-w)} >> w f^{(w,-w)} \tag{3.16}$$

since only in this case does the modular covariant derivative reduce to the ordinary derivative. This inequality is obviously not satisfied by terms in the expansion of $f^{(w,-w)}$ that are powers of τ_2, such as the perturbative tree and one-loop terms in (2.19) (with $\tau_2 \to \tau_0$). However, when acting on a factor such as $\tau_0^n e^{-2\pi|K|\tau_0}$ (where n is any constant) which is characteristic of a charge-K D-instanton, the inhomogeneous term may be neglected in the limit $\tau_0 \to \infty$ and the covariant derivative linearizes. Therefore, a linearized superspace expression such as (3.8) should contain the exact leading multi-instanton contributions to the $\mathcal{R}^4$ and related terms. These leading instanton terms arise by substituting the expression

$$F_K = c_K e^{2\pi i|K|\Phi} \tag{3.17}$$

into (3.8).

In the exact theory that will be considered in detail in the next section, the $SL(2,\mathbb{Z})$ symmetry of the IIB theory requires that the $f^{(w,-w)}(\tau,\bar\tau)$ are modular forms with holomorphic and anti-holomorphic weights as indicated in the superscripts. In that case the derivative $\tau_0 \partial_{\tau_0}$ in (3.15) will be replaced by the covariant derivative, D_w, that maps a modular form of weight $(w,\hat w)$ into a modular form of weight $(w+1, \hat w - 1)$[i],

$$D_w F^{(w,\hat w)} = i\left(\tau_2 \frac{\partial}{\partial \tau} - i\frac{w}{2}\right) F^{(w+1,\hat w-1)} = F^{(w+2,\hat w)}. \tag{3.18}$$

The covariant version of (3.15) is

$$\begin{aligned} f^{(12,-12)}(\tau,\bar\tau) &= D^{12} f^{(0,0)}(\tau,\bar\tau) \\ &\equiv D_{11} D_{10} \ldots D_0 f^{(0,0)}(\tau,\bar\tau). \end{aligned} \tag{3.19}$$

[i]Appendix B summarizes some relevant properties of modular covariant derivatives and laplacians.

We will take this relation to define the relative normalizations of $f^{(0,0)}$ and $f^{(12,-12)}$.

4 Nonlinear supersymmetry constraints on terms of order α'^{-1}

In this section, which is based on [15] where further details may be found, the constraints imposed by the full nonlinear supersymmetry of the type IIB theory will be used to derive the detailed expressions for all the terms in $S^{(3)}$ — the terms in the derivative expansion of the action of the type IIB theory that are of order α'^{-1}.

The procedure will be to impose supersymmetry order by order in α' by expressing the supersymmetry transformations on an arbitrary field Ψ as the series,

$$\delta_\epsilon \Psi = \left(\delta^{(0)} + \alpha' \delta^{(1)} + \ldots + \alpha'^n \delta^{(n)} + \ldots\right) \Psi, \tag{4.1}$$

while the effective action has the expansion (1.1). In principle, the action can be constructed by imposing the conditions,

$$\left(\sum_{m=0}^{r} \alpha'^m \delta^{(m)}\right) \sum_{n=o}^{r} \alpha'^n S^{(n)} = 0, \tag{4.2}$$

order by order in α'. In addition, it is important to impose closure of the supersymmetry algebra on all the fields. In theories of this type the supersymmetry algebra only closes on a field Φ up to the equations of motion and also includes specific local symmetry transf ormations, $\delta_{\mathrm{local}}\Phi$. The generic structure of the superalgebra is therefore of the form,

$$(\delta_1\delta_2 - \delta_2\delta_1)\Phi = \xi^\mu D_\mu \Phi + \Phi \text{ equations of motion} + \delta_{\mathrm{local}}\Phi, \tag{4.3}$$

where the first term is the usual translation term with parameter

$$\xi^\mu = -2\,\mathrm{Im}\,\bar{\epsilon}_2 \gamma^\mu \epsilon_1. \tag{4.4}$$

Imposing the supersymmetry conditions (4.2) together with closure of the superalgebra (4.3) simultaneously determines both the action and the supersymmetry transformations on the fields. The α'^0 ($n = m = 0$) term in (4.2) corresponds to the supersymmetry of the classical theory. These terms in the action and in the supersymmetry transformations were determined in [58], ignoring terms that are quartic in fermion fields[j]. Some of these zeroth order

[j] As stated in the introduction, it is really the equations of motion rather than the action that will be considered in the following.

supersymmetry transformations and action are reviewed in appendix C. One particular term quartic in the fermion fields is needed in the following (the term $L_1^{(0)}$ below) with a coefficient that is derived in [15].

There are no $n = 1$ or $n = 2$ terms at tree-level or one-loop, and there is strong evidence that supersymmetry precludes the presence of any such terms in (1.1). We will *assume* this to be the case so that the first corrections are the terms in $S^{(3)}$, which are of order $(\alpha')^3$ relative to $S^{(0)}$. These terms are eighth order in derivatives.

The procedure of imposing supersymmetry and closure can become very complicated unless one makes a judicious choice of starting point. We will start by selecting two special terms in $L^{(3)}$ that involve sixteen fermionic fields[k]. The choice is motivated by the fact that these two terms mix with each other and with no other terms under the classical supersymmetry transformations. The two terms are the sixteen-fermion terms contained in

$$\begin{aligned} L_1^{(3)} &= \det e \left(f^{(12,-12)}(\tau,\bar\tau)\lambda^{16} + f^{(11,-11)}(\tau,\bar\tau)\hat G \lambda^{14} \right) \\ &= \det e \left(f^{(12,-12)}\lambda^{16} - 3\cdot 144 f^{(11,-11)} \left(\lambda^{15}\gamma^\mu \psi_\mu^*\right) + \ldots \right), \end{aligned} \tag{4.5}$$

where the ellipsis represents other terms in $\hat G$ which do not affect the subsequent argument[l] (details of the spinor algebra that leads to the precise coefficients in this and following equations can be found in appendix C and [15]).

From the lowest order supersymmetry transformations given in appendix C we obtain terms that transform (4.5) into $\det e\, \lambda^{16}\, \psi_\mu^*\, \epsilon$,

$$\delta_1^{(0)} L_1^{(3)} = -i \det e\ \left(\bar\epsilon^* \gamma^\mu \psi_\mu^*\right) \lambda^{16} \left(8 f^{(12,-12)} + 6\cdot 144\, D_{11} f^{(11,-11)} \right), \tag{4.6}$$

where we have only kept terms proportional to $\lambda^{16}\,\psi_\mu^*\,\epsilon$. It is important to check whether there could also be a contribution of the same form as (4.6) arising from a $(\alpha')^3\delta^{(3)}$ variation of the fields in the lowest order action $S^{(0)}$. However, it is easy to see by inspection that no term with $\lambda^{16}\psi_\mu^*$ can arise from the variation of any term in $S^{(0)}$. This means that we must require $\delta_1^{(0)} L_1^{(3)} = 0$, which implies that

$$D_{11} f^{(11,-11)} = -\frac{4}{3\cdot 144} f^{(12,-12)}. \tag{4.7}$$

This condition is consistent with the modular weights assigned to the functions $f^{(w,-w)}$.

---

[k] Our notation is chosen so that $(\alpha')^{n-4}\int d^{10}x L^{(n)} = S^{(n)}$.

[l] The overall normalization of the action does not affect the arguments of this section.

We now consider the term in the variation of (4.5) that is proportional to $\det e\, \lambda^{16}\, \lambda^*\, \epsilon^*$,

$$\delta_2^{(0)} L_1^{(3)} = -2i \det e\, \lambda^{16} (\bar{\epsilon}\lambda^*) \left(\bar{D}_{-12} f^{(12,-12)} + 3 \cdot 144 \cdot \frac{15}{2} f^{(11,-11)} \right) + \ldots, \tag{4.8}$$

where we have made explicit only the terms containing $\lambda^{16}\lambda^*\epsilon^*$. In this case, there is another contribution of the same form as $\delta_2^{(0)} L_1^{(3)}$ that arises from the $(\alpha')^3 \delta^{(3)}$ variation of terms in the lowest order IIB Lagrangian $L^{(0)}$. Even though the complete set of interactions in the classical theory is not tabulated explicitly in the literature (it is implicit in the superspace formulation [13]), it is easy to convince oneself that the only possible term that can vary into $\delta_2^{(0)} L_1^{(3)}$ is a term of the form,

$$L_1^{(0)} = \frac{1}{256} \det e\, \bar{\lambda}\gamma^{\mu\nu\rho}\lambda^*\, \bar{\lambda}^* \gamma_{\mu\nu\rho}\lambda, \tag{4.9}$$

which is the unique tensor structure containing $\lambda^2\lambda^{*2}$. The coefficient of this term was determined by the lowest order supersymmetry transformations in [15] by considering the mixing of $L_1^{(0)}$ with other terms in the classical action.

We see that $L_1^{(0)}$ can vary into the same form as $\delta_2^{(0)} L_1^{(3)}$ if we assume a higher-order variation of λ^* of the form,

$$\delta^{(3)} \lambda_a^* = -\frac{1}{6}\, i g(\tau,\bar{\tau})\, (\lambda^{14})_{cd} (\gamma^{\mu\nu\rho}\gamma^0)_{dc}\, (\gamma_{\mu\nu\rho}\epsilon^*)_a, \tag{4.10}$$

where $g(\tau, \bar{\tau})$ is another function to be determined. Substituting in (4.9) gives a contribution,

$$\begin{aligned} \delta^{(3)} L_1^{(0)} &= -\frac{1}{768}\, i \det e\, g(\tau,\bar{\tau})\, \bar{\lambda}\gamma^{\mu\nu\rho}\gamma_{\rho_1\rho_2\rho_3}\epsilon^*\, (\lambda^{14})_{cd} (\gamma^{\rho_1\rho_2\rho_3}\gamma^0)_{dc}\, \bar{\lambda}^*\gamma_{\mu\nu\rho}\lambda \\ &= -30\, i \det e\, g(\tau,\bar{\tau}) \lambda^{16} (\bar{\epsilon}\lambda^*). \end{aligned} \tag{4.11}$$

Comparing with (4.8) we see that in order for the total contribution to δL_1 to vanish at order $(\alpha')^3$, there must be a linear relation between the function g and the functions $f^{(11,-11)}$ and $\bar{D}_{-12} f^{(12,-12)}$,

$$\bar{D}_{-12} f^{(12,-12)} + 3 \cdot 144 \cdot \frac{15}{2} f^{(11,-11)} + 15\, g = 0. \tag{4.12}$$

At this stage we have two equations ((4.7) and (4.12)) relating the three unknown functions, $f^{(12,-12)}$, $f^{(11,-11)}$ and g. A further constraint on these functions is obtained by requiring the closure of the superalgebra. We need to consider closure of the supersymmetry transformations on the field λ^*. First, in the classical theory (and keeping only the terms linear in λ derivatives) it was found in eq. (4.5) of [58] that

$$(\delta_1^{(0)}\delta_2^{(0)} - \delta_2^{(0)}\delta_1^{(0)})\lambda^* = \xi^\mu D_\mu\lambda^* - \frac{3}{8}i[\bar\epsilon_2\gamma^\rho\epsilon_1 - (1\leftrightarrow 2)]\gamma_\rho\gamma^\mu D_\mu\lambda^*$$
$$- \frac{1}{96}i[\bar\epsilon_2\gamma^{\rho_1\rho_2\rho_3}\epsilon_1 - (1\leftrightarrow 2)]\gamma_{\rho_1\rho_2\rho_3}\gamma^\mu D_\mu\lambda^* \quad (4.13)$$
$$\ldots.$$

This has the form of (4.3). The first term on the right-hand-side is of the form expected for the commutator of two supersymmetry transformations. The other terms that have been exhibted explicitely are proportional to the linear terms in the λ^* equation of motion. Many other terms that are not needed are indicated by .... These terms contribute to the commutator to complete the low-energy λ^* field equation as well as generating local transformations of λ^* [58].

The higher order terms in $L^{(3)}$ modify the equations of motion and this should also be apparent by considering the closure of the algebra. Therefore, we now consider terms that enter at order α'^3 from the commutator of a $\delta^{(0)}$ with a $\delta^{(3)}$. More precisely, we shall consider terms in the commutator involving only ϵ_2^* and ϵ_1,

$$\left(\delta^{(0)}_{\epsilon_1}\delta^{(3)}_{\epsilon_2^*} - \delta^{(3)}_{\epsilon_2^*}\delta^{(0)}_{\epsilon_1}\right)\lambda_a^* = -\frac{1}{3}\left(\tau_2\frac{\partial}{\partial\tau} - i\frac{45}{8}\right) i\, g\,(\bar\epsilon_1^*\lambda)(\lambda^{14})_{cd}(\gamma^{\mu\nu\rho}\gamma^0)_{dc}(\gamma_{\mu\nu\rho}\epsilon_2^*)_a$$
$$= 32\, D_{11}g\,\lambda_b^{15}\left[\frac{3}{8}\bar\epsilon_2\gamma^\mu\epsilon_1(\gamma_\mu)_{ba} + \frac{1}{96}\bar\epsilon_2\gamma^{\mu\nu\rho}\epsilon_1(\gamma_{\mu\nu\rho})_{ba}\right] + \delta_{\hat\epsilon}\lambda^*. \quad (4.14)$$

In the last line, we have separated a local supersymmetry transformation, $\delta_{\hat\epsilon}\lambda^*$, which is to be identified with a supersymmetry transformation of the form (4.10) with a particular field dependent coefficient, $\hat\epsilon = \frac{i}{4}\epsilon_2^*\,(\bar\epsilon_1^*\lambda)$.

Combining (4.13) and (4.14) (including the powers of α') we see that in order for the right-hand side of the commutator to vanish the λ^* field equation must be of the form,

$$i\gamma^\mu D_\mu\lambda^* - (\alpha')^3\, 32\, D_{11}g\,\lambda^{15} + \ldots = 0, \quad (4.15)$$

where the ellipsis indicates terms with different structure that we have not considered. This equation has to be identified with the appropriate sum of terms in the λ^* equation of motion that is obtained by varying the action with respect to λ. At the same order in α' this is given by,

$$i\gamma^\mu D_\mu\lambda^* - (\alpha')^3 f^{(12,-12)}\,\lambda^{15} + \ldots = 0, \quad (4.16)$$

where we have only made explicit the term that is proportional to λ^{15}. Comparing (4.15) and (4.16) gives the relation,

$$32 D_{11} g = f^{(12,-12)}. \tag{4.17}$$

Substituting (4.17) into (4.7) gives,

$$g = -\frac{3 \cdot 144}{128} f^{(11,-11)}. \tag{4.18}$$

There is no ambiguity in this relation between g and $f^{(11,-11)}$ because there is no solution to $D_{11} g = 0$. Substituting (4.18) into (4.12) gives,

$$\bar{D}_{-12} f^{(12,-12)} = 3 \cdot 144 \left(-\frac{15}{2} + \frac{45}{64} \right) f^{(11,-11)}. \tag{4.19}$$

The two simultaneous first-order differential equations, (4.19) and (4.7) are simply reduced to the independent second-order eigenvalues equations,

$$\nabla^2_{(-)\,12} f^{(12,-12)} = 4 D_{11} \bar{D}_{-12} f^{(12,-12)} = \left(-132 + \frac{3}{4} \right) f^{(12,-12)} \tag{4.20}$$

and

$$\nabla^2_{(+)\,11} f^{(11,-11)} \equiv 4 \bar{D}_{-12} D_{11} f^{(11,-11)} = \left(-132 + \frac{3}{4} \right) f^{(11,-11)}, \tag{4.21}$$

where the laplacians are defined in appendix B.

By applying $\bar{D}^{12}$ to (4.20) we can define the function,

$$\tilde{f}^{(0,0)}(\tau, \bar{\tau}) \equiv \bar{D}^{12} f^{(12,-12)}(\tau, \bar{\tau}). \tag{4.22}$$

Applying D^{12} to this equation and using (4.20) together with properties of covariant derivatives in appendix B leads to

$$f^{(12,-12)}(\tau, \bar{\tau}) = k D^{12} \tilde{f}^{(0,0)}, \tag{4.23}$$

where k is an irrelevant constant. Therefore, from (3.19), we can make the identification

$$\tilde{f}^{(0,0)} = k^{-1} f^{(0,0)}. \tag{4.24}$$

Putting these relations together we see that the Laplace eigenvalue equation for $f^{(12,-12)}$ (4.20) implies

$$\nabla^2_0 f^{(0,0)}(\tau, \bar{\tau}) \equiv 4 \tau_2^2 \frac{\partial}{\partial \tau} \frac{\partial}{\partial \tau^*} f^{(0,0)}(\tau, \bar{\tau}) = \frac{3}{4} f^{(0,0)}(\tau, \bar{\tau}). \tag{4.25}$$

It is an important fact that we will not prove here (see, for example, exercise 2 in section 3.5 of [56]) that (4.25) has the unique solution (2.22) provided it is assumed to transform as a scalar under $SL(2,\mathbb{Z})$ and has polynomial growth as $\tau_2 \to \infty$ (which is the perturbative regime).

Substituting (4.25) in (4.23) and (4.24) determines $f^{(12,-12)}$ to be

$$f^{(12,-12)}(\tau,\bar\tau) = \frac{1}{2^{12}}\frac{\Gamma\left(\frac{27}{2}\right)}{\Gamma\left(\frac{3}{2}\right)} \sum_{(m,n)\neq(0,0)} \tau_2^{3/2}\frac{(m+n\bar\tau)^{24}}{|m+n\tau|^{27}}. \qquad (4.26)$$

Similarly, (4.21) gives a unique expression for the modular form $f^{(11,-11)}$. The expressions for all the coefficients, $f^{(w,-w)}$, in (3.10) should also emerge from a more detailed application of the Noether procedure that considers all the possible mixing of terms in $S^{(3)}$ with arbitrary $U(1)$ charges. These are given by applying w covariant derivatives to $f^{(0,0)}$ which results in the expression

$$\begin{aligned} f^{(w,-w)}(\tau,\bar\tau) &= \frac{1}{2^{w}}\frac{\Gamma\left(w+\frac{3}{2}\right)}{\Gamma\left(\frac{3}{2}\right)} \sum_{(m,n)\neq(0,0)} \frac{\tau_2^{3/2}}{|m+n\tau|^3}\left(\frac{m+n\bar\tau}{m+n\tau}\right)^{w} \\ &= \frac{1}{2^{w}}\frac{\Gamma\left(w+\frac{3}{2}\right)}{\Gamma\left(\frac{3}{2}\right)} \sum_{(m,n)\neq(0,0)} \tau_2^{3/2}\frac{(m+n\bar\tau)^{2w}}{|m+n\tau|^{2w+3}}. \end{aligned} \qquad (4.27)$$

This transforms under $SL(2,\mathbb{Z})$ transformations by a phase

$$f^{(w,-w)}(\tau,\bar\tau) \to \left(\frac{c\tau+d}{c\bar\tau+d}\right)^{w} f^{(w,-w)}(\tau,\bar\tau), \qquad (4.28)$$

as expected for a $(w,-w)$ modular form.

The Noether procedure rapidly escalates in complication when applied at higher orders in the derivative expansion and it is probably not a practical procedure for terms with many more derivatives. It may just about be feasible to deduce terms of order α' that are of order α'^5 beyond the Einstein–Hilbert term. For example, this may be a method of determining the higher derivative terms suggested in [26] which were motivated by exactly known perturbative contributions [60,61].

The dimensions of the terms of yet higher order in α' suggested in [22,26] are so high that it is not obvious why they should be protected by supersymmetry. Nevertheless, they probably are protected based on the evidence for the exact form of these terms presented in these references.

There is also a suggestion [17,62] for how the structure of the protected $o(\alpha'^{-1})$ terms generalizes to compactifications to lower dimensions, where there are more moduli. The suggestion is that the functions $f^{(w,-w)}$ are generalized Eisenstein series for the appropriate duality groups.

5 The eleven-dimensional perspective

We shall now review the manner in which the $o(\alpha'^{-1})$ terms in the type II string actions can be deduced from eleven-dimensional supergravity. This was the subject of [16] where it was pointed out that the complete $\mathcal{R}^4$ term arises from the one-loop quantum contribution to four-graviton scattering in eleven-dimensional supergravity compactified on T^2. Similarly, the λ^{16} interaction can be deduced from the one-loop scattering amplitude of sixteen gravitini in eleven-dimensional supergravity [14]. The type IIB dilatino, λ, is identified with the gravitino of eleven-dimensional supergravity with its polarizations in appropriate directions. These calculations and generalizations were reviewed in [63] where some further details may be found, including a discussion of higher derivative contributions (which were also the subject of [22]).

5.1 The $\mathcal{R}^4$ term from one loop in eleven dimensions

We will begin with the $\mathcal{R}^4$ term which will be determined from the scattering amplitude for four gravitons at one loop in eleven-dimensional supergravity [16]. A fully covariant calculation would involve the sum of loop diagrams with the various component fields (the graviton, $g_{\hat{\mu}\hat{\nu}}$, the third-rank antisymmetric potential, $C^{(3)}_{\hat{\mu}\hat{\nu}\hat{\rho}}$, and the gravitino, $\Psi^a_{\hat{\mu}}$) and their ghosts circulating around the loop. An efficient method for summing all these contributions is to use a manifestly supersymmetric formalism based on the eleven-dimensional point particle [16] (details of the superparticle formalism outlined below have yet to appear [64]).

This method is modeled on the method used to construct string one-loop amplitudes in the light-cone gauge which is the parameterization in which

$$X^+(t) = p^+ t + x^+, \tag{5.1}$$

where t labels the world-line of the superparticle, $X^{\pm}(t) = (X^0(t) \pm X^8(t))/\sqrt{2}$ and p^+ is a constant (the directions x^9, x^{11} will continue to be the T^2 directions). The physical states are described in terms of the transverse $SO(9)$ representations (instead of the $SO(8)$ that is relevant for the transverse space in the case of the superstring). The quantum states are constructed in terms of fermionic sixteen-component spinors, $\mathcal{S}^A$ $(A = 1, \ldots, 16)$, satisfying the anticommutation relations,

$$\{\mathcal{S}^A, \mathcal{S}^B\} = \delta^{AB}. \tag{5.2}$$

The 32 components of the supercharges decompose the components that are realized linearly in the light-cone gauge,

$$Q^A = \sqrt{p}^+ \mathcal{S}^A, \tag{5.3}$$

and those that are nonlinearly realized,

$$\tilde{Q}^A = \frac{1}{\sqrt{p^+}}(\Gamma \cdot \dot{X}\, \mathcal{S})^A, \tag{5.4}$$

where Γ^I_{AB} are the 16×16 $SO(9)$ gamma matrices and $X^I(t)$ are the transverse coordinates of the superparticle.

The vertex operators that describe the linearized interactions can be derived by imposing the conditions that the g, ψ and $C^{(3)}$ vertices transform into each other under supersymmetry in the appropriate manner [64] which mimics the way in which light-cone superstring vertices can be derived. The resulting light-cone graviton vertex operator is given by

$$V_g^{(r)} = \zeta_{IJ}^{(r)}(\dot{X}^I \dot{X}^J - 2\dot{X}^I \mathcal{S}\gamma^{JL}\mathcal{S}k^L + 2\mathcal{S}\gamma^{IL}\mathcal{S}k^L\mathcal{S}\gamma^{JM}\mathcal{S}k^M)e^{ik\cdot X}, \tag{5.5}$$

where, as usual in the light-cone gauge operator formalism, a special frame has been chosen in which $k^+ = 0$ and $\zeta_{IJ}^{(r)}$ is the transverse polarization tensor of the eleven-dimensional graviton ($I = 1, \cdots, 7, 9, 11$). This expression reduces to the graviton vertex in the type IIA superparticle theory when reduced on a circle to ten dimensions[64]. In addition to the cubic graviton vertex higher-order contact interactions are necessary in order to reproduce general gauge-invariant amplitudes. Such contact terms compensate for short distance singularities in products of $V_g^{(r)}$'s that involve factors such as $\langle \dot{X}(\tau)\dot{X}(\tau')\rangle$ but are not needed for the amplitudes we are interested in.

The loop amplitude compactified on a two-torus is given by

$$A_{R^4} = \frac{1}{\mathcal{V}}\int \prod_{r=1}^{4} dt_r \int d^9p \sum_{m,n} \mathrm{tr}_{\mathcal{S}}(V_g^{(1)}(k^1)V_g^{(2)}(k_2)V_g^{(3)}(k_3)V_g^{(4)}(k_4)), \tag{5.6}$$

where $\mathrm{tr}_{\mathcal{S}}$ denotes the trace over the fermionic variables. The integers m, n are the Kaluza–Klein charges in the T^2 directions (x^9, x^{11}). For convenience we will only consider the situation in which all the external polarizations and momenta are oriented in the directions $x^1, \ldots, x^7$ which are transverse to the light-cone as well as the T^2 directions. The fermionic trace is saturated by the term with sixteen $\mathcal{S}$ factors, which picks out the last term in 5.5 (the term with four $\mathcal{S}$'s) so that no contractions of $\dot{X}$'s in the prefactor of the vertices. This picks out an overall kinematic factor which can be shown to be the same as the overall factor in the ten-dimensional one-loop superstring amplitude defined in (2.18) and can be expressed in a compact manner as an integral over a sixteen-component Grassmann spinor, as described in section 3.

After extracting the overall kinematic factor the remaining bosonic factor is simply the loop amplitude for the four-point amplitude in scalar ϕ^3 field theory. The full momentum dependence of this amplitude was discussed in[22,63]. Expanding in a power series in the momenta leads to higher derivatives acting on the fourth power of the curvature. However, here we are only interested in the leading momentum dependence. Therefore, after extracting the kinematic factor the external momenta can be set equal to zero in the remainder of the expression. The result can be written as

$$A_{R^4} = \frac{1}{\mathcal{V}}\tilde{K}\int_0^\infty dt t^3 \int d^9p \sum_{l_1,l_2} \exp\left(-t(p^2 + G^{IJ}l_I l_J)\right), \tag{5.7}$$

where G^{IJ} is the inverse metric on T^2 and $l_I = (m,n)$ $(I,J = 1,2)$. Substituting for the inverse metric,

$$G^{IJ}l_I l_J = \frac{1}{\mathcal{V}}\frac{|m+n\Omega|^2}{\Omega_2}, \tag{5.8}$$

and performing a double Poisson resummation converts the sum over Kaluza–Klein charges (m,n) to a sum over the windings $(\hat{m},\hat{n})$ of the world-line of the particle circulating in the loop around the two independent cycles of T^2. After the straightforward gaussian integration the result is

$$\begin{aligned}
\mathcal{V}A_{R^4} &= \tilde{K}\mathcal{V}\int_0^\infty d\hat{t}\hat{t}^{\frac{1}{2}} \sum_{\hat{m},\hat{n}} e^{-\mathcal{V}\hat{t}\frac{|\hat{m}+\hat{n}\Omega|^2}{\Omega_2}} \\
&= \tilde{K}\mathcal{V}C + \tilde{K}\mathcal{V}^{-\frac{1}{2}} \sum_{(\hat{m},\hat{n})\neq(0,0)} \frac{\Omega_2^{3/2}}{|\hat{m}+\hat{n}\Omega|^3} \\
&= \tilde{K}\left\{\mathcal{V}C + \mathcal{V}^{-1/2}\left[2\zeta(3)(\Omega_2)^{3/2} + \frac{2\pi^2}{3}(\Omega_2)^{-1/2} + \cdots\right]\right\},
\end{aligned} \tag{5.9}$$

where $\hat{t} = t^{-1}$. The cubic ultraviolet divergence in (5.9) is contained in the zero winding term, $\hat{m} = \hat{n} = 0$, which has the divergent coefficient C. The second equality in (5.9) gives the first two terms in the expansion for large Ω_2 (this expansion will be discussed in detail in section 6). The amplitude (5.9) comes from a term in the effective action of the form (2.21) with $f^{(0,0)}$ again of the form (2.22). As remarked earlier, in the limit $\mathcal{V} \to 0$ the regularized $\mathcal{V}C$ term vanishes and the amplitude has a finite limit in type IIB coordinates, i.e., $r_B \to \infty$ with e^{ϕ^B} held constant. Substituting Ω by $\tau = C^{(0)} + ie^{-\phi}$ leads to precisely the same expression for the $\mathcal{R}^4$ term in the type IIB theory that we deduced earlier from supersymmetry.

It is easy to argue on dimensional grounds that the finite part of the coefficient of the $\mathcal{R}^4$ term can only come from one loop in eleven dimensions. Higher loops lead to finite contributions that have extra derivatives acting on R^4 [63]. However, the systematics of the divergent parts is more obscure and the coefficient C has contributions from divergences of diagrams with arbitrary numbers of loops. These divergences must be regularized by microscopic effects in the eleven-dimensional theory that lie outside the realm of perturbation theory. Although we do not know how to fix the regularized value of C directly it is precisely determined by symmetry considerations. Keeping $\mathcal{V}$ nonzero and comparing the leading terms of the large-Ω_2 expansion in (5.9) with (2.21) shows that these expressions only agree if C has the value

$$C = \frac{2\pi^2}{3}. \tag{5.10}$$

We saw that the coefficients in (2.21) are consistent with T-duality that relates the IIA and IIB string perturbation expansions in nine dimensions. This means that only for the value of C in (5.10) is (5.9) consistent with T-duality. This leads to the same value of C as the argument based on supersymmetry that was mentioned at the end of section 2.

5.2 One loop with sixteen dilatini

In a similar manner all the interactions that are related by linearized supersymmetry to the $\mathcal{R}^4$ term can be deduced by considering one-loop processes in eleven-dimensional supergravity on T^2 [14]. As we saw in section 3 the generic terms of this type violate the $U(1)$ R-symmetry charge of the type IIB theory that is conserved classically. An extreme example of this is the interaction between sixteen dilatini. Since the dilatino λ carries a $U(1)$ charge of 3/2 the inte raction λ^{16} violates 24 units of charge.

As in the case of four-graviton scattering the calculation of the general loop amplitude that describes the scattering of sixteen external dilatini is facilitated by using the light-cone superspace description of the eleven-dimensional super-particle. However, the λ^{16} term of interest can be extracted from the zero-momentum process and this can be calculated without using the general gravitino vertex operators. The zero-momentum loop amplitude has the structure (ignoring an overall constant)

$$A_{\lambda^{16}} = \frac{1}{\mathcal{V}}\sum_{m,n}\int d^9p \int \frac{d\tau}{\tau}\tau^{16}\mathrm{tr}_S(V_\lambda^{(1)}(0)\cdots V_\lambda^{(16)}(0))e^{-\tau(p^2+\frac{1}{\mathcal{V}\Omega_2}|m+n\Omega|^2)}, \tag{5.11}$$

where $V_\lambda^{(r)}(0)$ is the zero-momentum vertex for the λ component of the rth gravitino. The trace in the above expression is over the fermionic operators in the vertices.

The fermions of the IIB theory compactified on S^1, the complex fields ψ and λ, are related to specific projections of the eleven-dimensional gravitino of M theory on T^2. These relations are described in appendix D (based on [14]). It follows that the vertex V_λ is a specific projection of the vertex operator for the eleven-dimensional gravitino, V_Ψ. The gravitino vertex can again be deduced from eleven-dimensional supersymmetry in the light-cone gauge [14]. The zero-momentum vertex is very simply expressed in terms of the 32-component eleven-dimensional supercharge, Q, by the covariant expression

$$V_\Psi = \bar{\zeta}^*_{\hat{\mu}} Q p^{\hat{\mu}}, \tag{5.12}$$

where $\zeta_{\hat{\mu}}$ is the zero-momentum wave function, $p_{\hat{\mu}}$ is the momentum of the particle circulating around the loop and $\bar{\zeta}^*_{\hat{\mu}} \equiv \zeta_{\hat{\mu}}\Gamma^0$ (with no complex conjugation). Since the zero momentum vertex is independent of the proper time around the loop the volume of integration over the proper times of the vertices gives the factor of τ^{16} in (5.11).

We now want to consider the components of $\Psi_{\hat{\mu}}$ containing λ. Inverting the relations in equation (D.12) in appendix D gives,

$$\Psi_z = \mathcal{P}_{\bar{z}}\Gamma_z\chi + \mathcal{P}_z\Gamma_{\bar{z}}\lambda, \qquad \Psi_{\bar{z}} = \mathcal{P}_z\Gamma_{\bar{z}}\chi^* + \mathcal{P}_{\bar{z}}\Gamma_z\lambda^*. \tag{5.13}$$

from which it follows that

$$V_\lambda = -\bar{\lambda}^*\Gamma_{\bar{z}}\mathcal{P}_{\bar{z}}Qp_{\bar{z}} = -\bar{\lambda}^*\Gamma_{\bar{z}}qp_{\bar{z}}, \tag{5.14}$$

where $q = \mathcal{P}_{\bar{z}}Q$ is a projected supercharge that satisfies the anticommutation relations

$$\{q^A, q^B\} = \mathcal{P}_{\bar{z}}^{AC}\mathcal{P}_{\bar{z}}^{BD}\{Q^C, Q^D\} = \Gamma_z^{AB}p_{\bar{z}}. \tag{5.15}$$

The momentum dependence can be scaled out by changing to $\hat{q}^A = q_A(p_{\bar{z}})^{-\frac{1}{2}}$, which satisfies $\{\hat{q}^A, \hat{q}^B\} = \Gamma_z^{AB}$. In the chirally projected subspace $\Gamma_z^{AB} \propto \delta^{AB}$ and the commutation relation does not depend on $\bar{z}$.

Substituting the vertex (5.14) into the loop amplitude (5.11) (contact terms are again not needed in this spin-$\frac{1}{2}$ process) and integrating over the loop momentum gives an expression of the form

$$\mathcal{V}A_{\lambda^{16}} = \hat{K}\mathcal{V}^{-\frac{1}{2}}f^{(12,-12)}(\Omega, \bar{\Omega}), \tag{5.16}$$

where the kinematic prefactor,

$$\hat{K} = \bar{\lambda}^{(1)*}_{A_1}\cdots\bar{\lambda}^{(16)*}_{A_{16}}\,\mathrm{tr}(\hat{q}^{A_1}\cdots\hat{q}^{A_{16}}), \tag{5.17}$$

is manifestly antisymmetric under permutations of the (commuting) fermion wave functions due to the anticyclic property of the trace. It can be rewritten (up to an overall constant) as

$$\hat{K} = \bar{\lambda}^*_{A_1} \cdots \bar{\lambda}^*_{A_{16}} \varepsilon_{16}^{A_1 \ldots A_{16}}, \tag{5.18}$$

where ε_{16} is the rank sixteen epsilon tensor with spinor indices.

The rest of the expression (5.16) depends on the circumference r_B through the factor $\mathcal{V}^{-\frac{1}{2}}$. The Ω and $\bar{\Omega}$ dependence comes from the loop integration and is given by

$$\begin{aligned} f^{(-12,12)}(\Omega, \bar{\Omega}) &= c \sum_{m,n} \int \frac{dt}{t} t^{23/2} \left(\frac{1}{\sqrt{\Omega_2}} (m + n\bar{\Omega}) \right)^{24} \exp\left(-t \frac{1}{\Omega_2} |m + n\Omega|^2 \right) \\ &= \frac{c}{\sqrt{\Omega_2}} \Gamma(23/2) \sum_{m,n} \frac{(m + n\bar{\Omega})^{24}}{|m + n\Omega|^{23}} \end{aligned} \tag{5.19}$$

(where c is an overall constant). This expression is superficially cubically divergent due to the sum over the Kaluza-Klein charges. However, this does not take into account the phase dependence which, for generic Ω, can lead to a cancellation between the growing terms in the sum. Indeed, for asymptotically large m and n the sum over discrete momenta p_z and $p_{\bar{z}}$ can be replaced by integrals and the result is proportional to

$$\int dp_z dp_{\bar{z}} p_{\bar{z}}^{24} (p_z p_{\bar{z}})^{-23/2}, \tag{5.20}$$

which vanishes if $|p_z|$ is regularized. Equivalently, it vanishes if the phase integration is carried out before the integration over $|p_z|$. This is the regularization prescription that will be used in the following, motivated by the fact that it leads to an $SL(2, \mathbb{Z})$ invariant answer.

In order to extract the finite result it is useful, as before, to use a double Poisson summation to transform from the discrete momentum sum to a sum over windings of the loop around the torus. Writing

$$f^{(12,-12)}(\Omega, \bar{\Omega}) = \frac{c}{\sqrt{\Omega_2}} \left(\frac{\partial}{\partial \alpha} \right)^{24} \sum_{m,n} \int \frac{dt}{t} t^{23/2} \, e^{-t|m+n\Omega|^2 + \alpha(m+n\bar{\Omega})} \Big|_{\alpha=0}, \tag{5.21}$$

the Poisson resummation equates this to the manifestly finite expression,

$$f^{(-12,12)}(\Omega,\bar{\Omega}) =$$
$$= \frac{c}{\sqrt{\Omega_2}}\left(\frac{\partial}{\partial\alpha}\right)^{24}\sum_{\hat{m},\hat{n}}\frac{\pi}{\Omega_2}\int\frac{dt}{t}t^{21/2}\exp\left(-\frac{\pi^2}{t\Omega_2^2}|\hat{n}+\hat{m}\Omega|^2+\frac{i\pi\alpha}{t\Omega_2}(\hat{n}+\hat{m}\bar{\Omega})\right)\Bigg|_{\alpha=0}$$
$$= \frac{c}{\pi^2}\Gamma(27/2)\Omega_2^{3/2}\sum_{(\hat{m},\hat{n})\neq(0,0)}\frac{(\hat{n}+\hat{m}\bar{\Omega})^{24}}{|\hat{n}+\hat{m}\Omega|^{27}}. \tag{5.22}$$

where $\hat{m}$ and $\hat{n}$ are winding numbers.

After translating from the M-theory coordinates $(\Omega, \mathcal{V})$ to the string-frame IIB coordinates (τ, r_B) the expression (5.22) is identical to (4.26) up to an overall normalization that can be absorbed in c. The amplitude $A_{\lambda^{16}}$ corresponds to the λ^{16} term in the IIB effective action, (3.10), in the limit, $r_B \to \infty$, of ten decompactified dimensions. In fact, in this case there is no divergent term at all even at finite r_B. The eleven-dimensional limit of $A_{\lambda^{16}}$ is not only finite but it vanishes. The nine-dimensional effective action in the IIB parameterization contains the term

$$S^{(3)}_{\lambda^{16}} = \frac{1}{3\cdot(4\pi)^7 l_S}\int d^9x\, r_B \det e\, e^{-\phi/2}\,\lambda^{16} f^{(12,-12)}(\tau,\bar{\tau}). \tag{5.23}$$

This expression has been written in the string frame. The transformation from Einstein frame to string frame includes the rescaling $\lambda \to \lambda e^{-\phi/8}$ in addition to the rescaling of the zehnbein, $e^a_\mu \to e^a_\mu e^{\phi/4}$.

6 Fourier series and properties of D-instantons

We have seen that the terms in the ten-dimensional type IIB effective action of order α'^{-1} have the form (in string frame) [65],

$$\frac{1}{\alpha'}\int d^{10}x \det e\, e^{-\phi/2} f^{(w,-w)}(\tau,\bar{\tau})\mathcal{O}_{\{u_r\}}, \tag{6.1}$$

where $\mathcal{O}_{\{u_r\}}$ is a function of the fields that reduces, at the linearized level to a monomial of p fields with $U(1)$ R-symmetry charges $u_1, u_2, \ldots u_p$ (see (3.13)). The total $U(1)$ charge of the interaction is $2w$ where $w = p - 4$ (see (3.14)) and the coefficient $f^{(w,-w)}$ is given by (4.27). The coefficient functions can be expressed as Fourier series,

$$\tau_2^{1/2} f^{(w,-w)} = \sum_{K=-\infty}^{\infty}\mathcal{F}_{K,w}e^{2\pi i K\tau_1}, \tag{6.2}$$

where the Fourier coefficients, $\mathcal{F}_{K,w}$, with nonzero K are the D-instanton terms when $K > 0$ and anti D-instanton terms when $K < 0$. This expansion is readily derived by a generalization of the method that leads to the analogous expansion of the Eisenstein series, E_s (see, for example, [56]). First one performs the sum over all the terms in (4.27) with $n = 0$ and $m \neq 0$. This leads to the 'tree-level' term of the form $2\zeta(3)e^{-2\phi}$. For the terms with $n \neq 0$ one may perform a Poisson resummation that converts the sum over m into a sum over a conjugate integer, $\hat{m}$. The sum over all $n \neq 0$ with $\hat{m} = 0$ gives the 'one-loop' term that is independent of τ_2 while the remaining sum (over all $n \neq 0$, $\hat{m} \neq 0$) gives the D-instanton contributions with $K = \hat{m}\, n$.

The result is [65,24]

$$\tau_2^{1/2} f^{(w,-w)} = 2\zeta(3)e^{-2\phi} + \frac{2\pi^2}{3} c_w + \sum_{K=1}^{\infty} \left(\mathcal{F}_{K,w} e^{2\pi i K \tau_1} + \mathcal{F}_{K,-w} e^{-2\pi i K \tau_1}\right), \tag{6.3}$$

where

$$c_w = (-1)^w \frac{\pi}{4} \frac{1}{\Gamma(\frac{3}{2}+w)\Gamma(\frac{3}{2}-w)}. \tag{6.4}$$

The first two terms in (6.3) have the interpretation of the tree-level and one-loop string terms while the instanton and anti-instanton terms are contained in

$$\mathcal{F}_{K,w} = 4\pi^{1/2} Z_K (2\pi K \tau_2)^{1/2} \sum_{k=0}^{\infty} \frac{c_{w,k}}{(2\pi K \tau_2)^{k-w}} e^{2\pi i K \tau}, \tag{6.5}$$

where

$$c_{w,k} = \frac{(-1)^w}{2^{k-w} k!} \frac{\Gamma(3/2)}{\Gamma(-w+3/2)} \frac{\Gamma(k-w-1/2)}{\Gamma(-k+w-1/2)} \tag{6.6}$$

and

$$Z_K = \sum_{\hat{m}|K} \frac{1}{\hat{m}^2}. \tag{6.7}$$

The infinite series in 6.6 begins with the power τ_2^w for a D-instanton ($K > 0$) while the series of corrections to the anti D-instanton ($K < 0$) starts with the power τ_2^{-w}.

A particularly simple example of this formula is the expansion of the coefficient $f^{(0,0)}$ of the $\mathcal{R}^4$ term which is given by

$$\tau_2^{1/2} f^{(0,0)}(\tau,\bar{\tau}) = 2\zeta(3)\tau_2^2 + \frac{2\pi^2}{3} + 8\pi\tau_2 \sum_{K\neq 0} |K|\, Z_K\, K_{-1}(2\pi|K|\tau_2) e^{2\pi i K \tau_1}$$

$$= 2\zeta(3)\tau_2^2 + \frac{2\pi^2}{3} + 4\sqrt{2}\pi \sum_{K=1}^{\infty} (K\tau_2)^{1/2} Z_K \tag{6.8}$$

$$\times \left(e^{2\pi i K\tau} + e^{-2\pi i K\bar{\tau}}\right) \left(1 + \sum_{k=1}^{\infty} (4\pi K\tau_2)^{-k} \frac{\Gamma(k-1/2)}{\Gamma(-k-1/2)k!}\right),$$

where the asymptotic expansion of the Bessel function $K_{-1}(2\pi|K|\tau_2)$ (which is useful for small coupling, or large τ_2) has been used in the second equality.

The expressions (6.3) and (6.8) reproduce precisely the perturbative tree-level and one-loop terms and demonstrate that there is a perturbative non-renormalization theorem beyond one loop. The absence of higher-order perturbative corrections must be related to the fact that the terms in (6.1) are given by integrals over half the on-shell superspace although a direct proof of the absence of higher-order perturbative corrections has not been given (see however [28]). The non-perturbative terms in (6.8) are contributions of D-instantons (and anti-D-instantons) of arbitrary charge K together with an infinite sequence of perturbative fluctuations around each instanton configuration.

The D-instanton sum has a simple origin from the point of view of the loop integrals considered in section 5. There, the circulating particle can be considered to be a Kaluza–Klein mode of charge n of M theory compactified on the first circle of circumference R_{11}. For $n \neq 0$ this is a charge-n D-particle. The world-line of this D-particle winds $\hat{m}$ times around the second circle of radius r_A (in IIA string units). The action for such a configuration is given by the Dirac–Born–Infeld action for the D-particle and is equal to

$$2\pi|\hat{m}n|(r_A e^{\phi^A} \mp iC^{(1)}) \tag{6.9}$$

(where the $\mp$ distinguishes instantons and anti-instantons). After performing a T-duality on the euclidean circle [11] the real part of this expression is identified with the charge-K D-instanton action,

$$S_{K=\hat{m}n} = 2\pi|K|e^{\phi^B} \tag{6.10}$$

and the $\mathcal{R}\otimes\mathcal{R}$ potential $C^{(1)}$ is identified with $C^{(0)}$.

From (6.5) we see that the leading contribution to the N D-instanton contribution in $\mathcal{F}_{K,w}(\tau,\bar{\tau})$ is

$$\mathcal{F}_{K,w} \sim Z_K S_K^{-7/2+p} e^{-(S_K - 2\pi i K C^{(0)})} \left(1 + o(Ke^{\phi})\right), \tag{6.11}$$

where $p = w + 4$ and we have factored out the measure factor Z_K (6.7), which depends on K but not on S_K and is normalized so that $Z_1 = 1$. The power

$-7/2$ in (6.11) arises from the combination of ten bosonic zero modes and sixteen fermionic zero modes while the factor of S_K for each external state comes from the normalization of the external states.

7 D-particle bound states and the Witten index

One of the central planks of the web of dualities that relate string theory and M theory is the identification of D-particles (D0-branes) of type IIA superstring theory with Kaluza–Klein modes of eleven-dimensional supergravity compactified on S^1 [1]. There is a single Kaluza–Klein mode of a given charge, N, with mass N times the mass of charge-one D-particle. This implies that there must be precisely one threshold bound state (a bound state with zero binding energy) of N D-particles. The interaction of N D-particles is described by the 'one-dimensional matrix model' — a quantum mechanical system that is isomorphic to the dimensional reduction of ten-dimensional $U(N)$ supersymmetric Yang-Mills theory to one (time) dimension [30]. It is notoriously difficult to analyze the possible threshold bound states of this theory. One approach used in [32,33] in the case $N = 2$ is to demonstrate that the Witten index is equal to one. This shows that there is at least one bound state. Here we will show how the results of the last section indicate that the Witten index is equal to one for arbitrary N (this section is based on [31]) [m].

7.1 D-particles and D-instantons

The Witten index for the system of N D-particles was defined in [32,33] by

$$
\begin{aligned}
I^{(N)} &= \int d^9x \lim_{\beta\to\infty} \operatorname{tr}(-1)^F e^{-\beta H}(x,x) \\
&= \lim_{R\to\infty}\lim_{\beta\to 0}\left\{\int_{|x|<R} d^9x \operatorname{tr}(-1)^F e^{-\beta H}\right. \\
&\quad \left. + \frac{1}{2}\int_{|x|=R} d^9x \int_\beta^\infty d\beta' \operatorname{tr}e_n(-1)^F Q e^{-\beta' H}\right\},
\end{aligned} \tag{7.1}
$$

where the first term defines a bulk contribution $I^{(N)}_{bulk}$ (which in general is not an integer) and the second term a deficit contribution $I^{(N)}_{def}$ (which corrects the bulk term). The separation of the integer $I^{(N)}$ into these two parts is dependent on the order of limits. We shall always consider the infinite volume limit to be taken before the limit $\beta \to 0$.

---

[m] See [66] for a very different approach to the threshold bound state problem

The trace in (7.1) is defined over gauge invariant states of the Hilbert space, which may be written as a trace over all states if a projector onto gauge invariant states (states satisfying the constraints of Gauss' law, $C|\rangle = 0$) is inserted. For the bulk term this gives

$$I_{bulk}^{(N)} = \frac{1}{Vol(SU(N)/Z_N)} \lim_{\beta\to 0} \int_{SU(N)} d\eta \int d^9x \, \mathrm{tr}(-1)^F e^{i\eta C} e^{-\beta H}(x,x) \quad (7.2)$$

where η^a is the gauge parameter. Using the heat kernel approximation for the propagator in the hamiltonian H, which is valid in the limit $\beta \to 0$, it was observed in [32] and [33] that the exponent of the bulk part of the index can be written in a $SO(10)$ invariant form. This is possible because the parameter η turns into a tenth bosonic coordinate, x^9.

The resulting integrals are seen to be equivalent to the $SU(N)$ integral in the partition function for the *zero*-dimensional matrix model. This is the model that is obtained by compactification of supersymmetric Yang–Mills theory from ten (euclidean) dimensions to zero dimensions that describes dynamics in the presence of N D-instantons [30,67,68]. The situation with D-instantons is somewhat distinct from that of other Dp-branes since $p = -1$ and the world-volume reduces to a space-time point. The configuration space of N D-instantons is determined by bosonic and fermionic $U(N)$ matrices A_μ and ψ_α, where μ and α denote $SO(10)$ vector and spinor indices, respectively. The reduction of the supersymmetric Yang–Mills action to a point is given by

$$S = \frac{1}{4}\mathrm{tr}([A_\mu, A_\nu]^2) + \frac{i}{2}\mathrm{tr}(\bar{\psi}\Gamma^\mu[A_\mu, \psi]). \quad (7.3)$$

The 'center of mass' degrees of freedom are associated with the element of the Cartan subalgebra of $U(N)$ proportional to the unit matrix and do not appear in the action (7.3). They do, however, enter as shift symmetries in the supersymmetry transformations,

$$\begin{aligned} \delta A_\mu &= i\bar{\eta}\Gamma_\mu\psi, \\ \delta\psi &= [A_\mu, A_\nu]\Gamma^{\mu\nu}\eta + \mathbf{1}\,\epsilon. \end{aligned} \quad (7.4)$$

The sixteen-dimensional spinor η parameterizes the dimensionally reduced supersymmetry of the $SU(N)$ YM theory and the spinor ϵ acts as a constant shift on the fermions in the 'center of mass' degrees of freedom. Hence the U(1) part of $U(N)$ fermionic fields ψ plays the role of the sixteen fermionic collective coordinates associated with the charge-N D-instanton.

After factoring out the centre of mass coordinates the N D-instanton partition function is given by

$$Z = \int d^{10}y \int d^{16}\epsilon \, Z_N, \quad (7.5)$$

where

$$Z_N = \frac{1}{Vol(SU(N)/Z_N)} \int_{SU(N)} d\psi dA \exp(-S_{SYM}[A,\psi]). \qquad (7.6)$$

The integration over y^μ and ϵ^A is the integral over the overall bosonic and fermionic collective coordinates of the collection of D-instantons of total charge N. The integration measure in (7.6) is the group invariant Haar measure which defines the appropriate normalization (as discussed in [69,70]). A non-zero value for Z_N can only arise from configurations in which there are no extra fermionic zero modes (in addition to ϵ) which is characteristic of a single multiply-charged D-instanton. According to [30] the partition function of the SU(N) zero-dimensional matrix model should be identified with the D-instanton measure, Z_N (6.7). The integral (7.6) was evaluated for the case $N = 2$ in [32,33]. The sub-integral over configurations in which the elements of the Cartan subalgebra, A^3_μ, ψ^3, are fixed was evaluated in [71] and corresponds to two D-instantons at a fixed separation. More recently, the integral has been explicitly evaluated for general N by deforming the integrand into a topological density [72] and numerical estimates have been made in [69].

The fact that the integral (7.6) is the same as the bulk integral which appears in the calculation of the Witten index in the bound state problem of D-particles is the statement of T-duality. The radius β of the euclidean circle in (7.2) shrinks to zero size and the insertion of $(-1)^F$ enforces supersymmetric (periodic) boundary conditions on the fermionic fields. The conjectured form for the measure, Z_N given by (6.7) therefore suggests that the value of the bulk integral is given, for arbitrary N, by

$$I^{(N)}_{bulk} = \sum_{\hat{m}|N} \frac{1}{\hat{m}^2} = Z_N, \qquad (7.7)$$

where an overall constant has been fixed by noting that $I^{(1)}_{bulk} = 1$. Note that for $N = 2$ the value of $I^{(2)}_{bulk} = 5/4$ in (7.7) agrees with the explicit calculation of [32,33].

7.2 The deficit term

The appearance of a deficit term is related to the presence of the boundary terms in (7.1). For the $SU(2)$ case a heuristic evaluation of this term was given in [32] and a more rigorous argument in [33]. It was argued that since the boundary term for the system of two D-particles arises from the region in which the particles are separated it is obtained by treating the particles as identical *free* particles with moduli space R^9/Z_2. Since the Witten index vanishes for free particles the deficit term, $I^{(2)}_{def}$, must cancel the bulk term for the free system, $I^{(2)}_{0\,bulk}$, where the subscript 0 indicates the free theory. But $I^{(2)}_{0\,def}$ is easily evaluated and is equal to 1/4. In the following[31] we shall assume that this prescription generalizes to $N > 2$ D-particles so that, for *prime* values of N,

$$I^{(N)}_{def} = -I^{(N)}_{0\,bulk}, \tag{7.8}$$

where $I^{(N)}_{0\,bulk}$ is the bulk index for N identical free particles moving on $R^{9(N-1)}/S_N$. For *non-prime* values, $N = \hat{m}n$, the generalization will take into account the regions of moduli space of $\hat{m}$ free charge-n particles on $R^{9(\hat{m}-1)}/S_{\hat{m}}$.

In the $N = 2$ case considered in [32,33] the only configuration that contributed to the trace over the free two-particle states was one with an odd permutation of the two particles. If the trace is expressed as a functional integral this includes only configurations in which the two D-particles are described by a single euclidean world-line that winds twice around the compact β direction. In order to generalize this to arbitrary N we need to consider the action of S_N, which is the Weyl group of $SU(N)$. This can be parameterized by matrices M_{ab} acting on the positions of the N particles modded out by the overall translation invariance, X^i_a ($i = 1, \cdots, 9$), and the fermions ψ^α_a. The index $a = 1, .., N-1$ labels the different $U(1)$'s in the Cartan subalgebra. The action of an element in S_N is given by

$$X^i_a \to M^b_a X^i_b, \qquad \psi^\alpha_a \to M^b_a \psi^\alpha_b, \tag{7.9}$$

where the vector index runs over $i = 1, \cdots, 9$ and the spinor index $\alpha = 1, \cdots 16$.

The fermion fields ψ satisfy the following anticommutation relations,

$$\{\psi^\alpha_a, \psi^\beta_b\} = \delta_{ab}\delta^{\alpha\beta}. \tag{7.10}$$

It is convenient to build up the fermionic Hilbert space by defining fermionic creation and anihilation operators $\hat{\psi}^\alpha_a, \hat{\psi}^{\dagger\alpha}_a$ by

$$\hat{\psi}^\alpha_a = \frac{1}{\sqrt{2}}\left(\psi^{2\alpha-1}_a + i\psi^{2\alpha}_a\right), \quad \hat{\psi}^{\dagger\alpha}_a = \frac{1}{\sqrt{2}}\left(\psi^{2\alpha-1}_a - i\psi^{2\alpha}_a\right), \quad \alpha = 1, \cdots, 8, \tag{7.11}$$

which satisfy $\{\hat{\psi}^{\dagger\alpha}_a, \hat{\psi}^\beta_b\} = \delta_{ab}\delta^{\alpha\beta}$. A general wave function $|\,\Psi\rangle$ can be expanded as

$$|\,\Psi\rangle = \left(\Psi^{(0)}(X_a) + \Psi^{(1)}_{a_1}(X_a)\hat{\psi}^\dagger_{a_1} + \frac{1}{2}\Psi^{(2)}_{a_1 a_2}(X_a)\hat{\psi}^\dagger_{a_1}\hat{\psi}^\dagger_{a_2} + \cdots\right)|\,0\rangle, \tag{7.12}$$

where the vector and spinor indices are suppressed and the highest term has $8 \times N$ fermionic creation operators acting on the Fock space vacuum.

The expression for the bulk contribution to the Witten index for N free D-particles (with N a prime) is given by

$$I^{(N)}_{0\,bulk} = \lim_{\beta \to 0} \mathrm{tr}(-1)^F e^{-\beta H}. \tag{7.13}$$

The trace is taken over gauge invariant states in the Hilbert space and hence one has to insert a projector on states which are invariant under the Weyl permutation group,

$$\mathcal{P} = \frac{1}{N!} \sum_{\pi \in S_N} M_\pi. \tag{7.14}$$

The matrix M_π representing $\pi \in S_N$ act as in (7.9) on the coordinates X_a and the fermions $\hat{\psi}_a$ in the wave function Ψ given in (7.12). The bulk index for the free theory is therefore given by

$$I^{(N)}_{0\,bulk} = \lim_{\beta \to 0} < \Psi | (-1)^F e^{-\beta H} \mathcal{P} | \Psi > . \tag{7.15}$$

The factor of $(-1)^F$ counts bosons (even number of $\hat{\psi}$) with $+1$ and fermions (odd number of $\hat{\psi}$) with -1. Because the fermions transform under the symmetric group matrices M, the introduction of the projector $\mathcal{P}$ can give a non-vanishing contribution.

The action of M_π on $\hat{\psi}_a$ and X_a in (7.15) factorizes into a trace over the Hilbert space built from the fermionic creation operators and a bosonic gaussian integral coming from the heat kernel approximation for the free propagator in the limit $\beta \to 0$ [32].

$$I^{(N)}_{0\,bulk} = \lim_{\beta \to 0} \frac{1}{N!} \sum_{\pi \in S_N} \mathrm{tr}_\psi \left((-1)^F M_\pi \right) \int \prod_{i=1}^{9} \prod_{a=1}^{N-1} dX^i_a \, \frac{e^{-(X - M_\pi X)^2/2\beta}}{(2\pi\beta)^{(N-1)9/2}}. \tag{7.16}$$

For all values of α the fermionic trace is given by

$$\begin{aligned} \mathrm{tr}_\psi \left((-1)^F M_\pi \right) &= \langle 0 \mid 0 \rangle - \sum_a \langle 0 \mid \hat{\psi}_a M_{\pi\,ab} \hat{\psi}^\dagger_b \mid 0 \rangle \\ &\quad + \frac{1}{2} \sum_{a,b} \langle 0 \mid \hat{\psi}_a \hat{\psi}_b M_{\pi\,ac} M_{\pi\,bd} \hat{\psi}^\dagger_c \hat{\psi}^\dagger_d \mid 0 \rangle + \cdots \\ &= 1 - \mathrm{tr}(M_\pi) + \frac{1}{2}\mathrm{tr}(M_\pi^2) - \frac{1}{2}\mathrm{tr}(M_\pi)^2 + \cdots \\ &= \det(1 - M_\pi), \end{aligned} \tag{7.17}$$

so that the trace over all eight fermion components gives a factor of $\det(1 - M_\pi)^8$. The gaussian integration over the coordinates X_a^i similarly gives a factor of $\det(1 - M_\pi)^{-9}$

The determinant is easily evaluated by using an explicit representation for the matrices M_π. Recall that these represent the action of S_N on the elements of the Cartan subalgebra. The roots of $SU(N)$ are the vectors in R^N,

$$e_i - e_j, \qquad i \neq j; \quad i,j = 1, \cdots, N, \tag{7.18}$$

where e_i the ith unit vector of R^N. All the roots lie in a $(N-1)$-dimensional subspace $\mathcal{R}_\omega$ orthogonal to the vector $\omega = e_1 + e_2 + \cdots + e_N = (1, 1, \cdots, 1)$ and an element $\pi \in S_N$ acts as a permutation, $\pi : e_i \to e_{\pi(i)}$.

Two clases of permutations need to be distinguished:
(a) *Cyclic permutations*
These can be represented up to conjugation by the $N \times N$ matrix

$$M_c = \begin{pmatrix} 0 & 1 & 0 & 0 & \cdots & 0 \\ 0 & 0 & 1 & 0 & \cdots & 0 \\ 0 & 0 & 0 & 1 & \cdots & 0 \\ \cdots & \cdots & \cdots & \cdots & \cdots & \cdots \\ 1 & 0 & 0 & 0 & \cdots & 0 \end{pmatrix}, \tag{7.19}$$

which has eigenvalues

$$\lambda_k = e^{\frac{2\pi i k}{N}}, \quad k = 0, 1, \cdots, N-1. \tag{7.20}$$

The matrix M_c does not leave any of the roots (7.18) invariant and ω is the unique eigenvector with eigenvalue $\lambda_0 = 1$. We are interested in evaluating $\det(1 - M_\pi) = \det'(1 - M_c)$, where the prime indicates the omission of the zero eigenvalue. This is the determinant in the space orthogonal to ω and is given by the product of non-zero eigenvalues,

$$\det{}'(1 - M_c) = \prod_{k=1}^{N-1} (1 - \lambda_k) = 2^{N-1} \prod_{k=1}^{N-1} \sin \frac{\pi k}{N} = N. \tag{7.21}$$

(b) *Non-cyclic permutations*
These can be represented by

$$M_{nc} = \begin{pmatrix} M_1 & 0 & 0 & \cdots & 0 \\ 0 & M_2 & 0 & \cdots & 0 \\ \cdots & \cdots & \cdots & \cdots & \cdots \\ 0 & 0 & 0 & \cdots & M_p \end{pmatrix} \tag{7.22}$$

where M_1, M_2, $\cdots$, M_p, represent cyclic permutations of subsets of elements and can be written in the form of (7.19). The matrix M_{nc} has $p > 1$ unit eigenvalues so there are $p-1$ eigenvectors $w_j \in R_\omega$ for which $(1-M_{nc})w_j = 0$ $(j = 1, \cdots, p-1)$. As a result, some elements of the Cartan subalgebra are left invariant by M_π and

$$\det(1 - M_\pi) = \det{}'(1 - M_{nc}) = 0, \tag{7.23}$$

where the prime again indicates the omission of the zero eigenvalue associated with ω.

To be more precise, a zero eigenvalue of the bosonic determinant $\det(1 - M_{nc})$ should be interpreted as the inverse volume, R^{-1} in the limit $R \to \infty$, whereas the fermionic determinant vanishes identically. This means that only the cyclic permutations contribute to the index and the expression (7.16) reduces to

$$I^{(N)}_{0\,bulk} = \sum_{c \in S_N} \frac{1}{N!} (\det{}'(1 - M_c))^{-1}. \tag{7.24}$$

The non-vanishing determinant $\det'(1 - M_c) = N$ simply counts the winding number of a fundamental D-particle world-line and the result is the obvious generalization of the $N = 2$ case. The non-cyclic permutations do not contribute to the index but have the interpretation of p disconnected D-particle world-lines winding around the euclidean circle. This is a multi D-instanton configuration and the vanishing of the determinant is due to the appearance of extra fermionic zero modes. Since each of the $(N-1)!$ cyclic elements of S_N gives the same contribution to (7.24) the result is

$$I^{(N)}_{0\,bulk} = \frac{1}{N^2} = -I^{(N)}_{def}\,. \tag{7.25}$$

Combining this with our earlier ansatz that $I^{(N)}_{bulk} = Z_N = 1 + 1/N^2$ leads to the conclusion that

$$I^{(N)} = I^{(N)}_{bulk} + I^{(N)}_{def} = 1, \tag{7.26}$$

for prime values of N. This is consistent with the presence of at least one bound state for every prime value of N.

Now consider the non-prime charge sector in which $N = \hat{m}n$ with $\hat{m}, n > 1$. It is useful to recall the interpretation of the D-instanton of charge N as a single wrapped euclidean D-particle world-line. The contributions to the instanton measure labelled by n were associated with the world-line of a charge-n D-particle winding $\hat{m} = N/n$ times. In the case when N is prime the only contribution to the deficit term comes from $n = 1$, $\hat{m} = N$, which corresponds to $\hat{m}$ windings of a charge-$n = 1$ D-particle world-line. The other possibility, $n = N$, $\hat{m} = 1$, gives the net contribution of one to the Witten index of the threshold bound state. It is therefore very compelling to *assume* that the contribution to the Witten index in the non-prime case, $N = \hat{m}n$, of a subset of $\hat{m}$ D-particles of charge n can be calculated in the same way as that of N fundamental ($n = 1$) D-particles. In other words, we assume a generalization of the property that was strongly motivated in the $N = 2$ case in [32,33] that the deficit term is determined by free supersymmetric particle dynamics with 16 supercharges. This means that the space of free-particle states that enters in the calculation of $I_{0\,bulk}$ should be enlarged to include the infinite tower of charge-n threshold bound states. In that case a cyclic permutation of $\hat{m}$ identical charge-n particles contributes a term

$$I^{(\hat{m})}_{0\,bulk} = \frac{1}{\hat{m}^2}, \tag{7.27}$$

which leads to a total deficit in the charge-N sector of

$$I^{(N)}_{def} = -\sum_{\substack{\hat{m}|N \\ \hat{m}>1}} I^{(\hat{m})}_{0\,bulk} = -\sum_{\substack{\hat{m}|N \\ \hat{m}>1}} \frac{1}{\hat{m}^2}, \tag{7.28}$$

consistent with the Witten index $I^{(N)} = 1$ for all integer N.

The success of these assumptions points to the systematics that needs to emerge from a more precise treatment of the integration over the boundary term when N is not prime. This suggests that there are regions of moduli space in which the non-abelian integration leads to $\hat{m}$ charge-n threshold bound states which behave as free particles. Related issues arise in the considerations of[19,20] concerning D-string instantons in $d = 8$ type I string theories.

The systematics of the ten-dimensional D-particle problem is consistent with the expected features of D-particle quantum mechanics with less supersymmetry. The clearest examples are in six and four dimensions. The six-dimensional case is realized in the compactification of IIA on $K3$ by wrapping D2-branes around a two-cycle in $K3$ and scaling to the limit in which the two-cycle vanishes. Using duality with the heterotic string on T^4 identifies the resulting D-particles as massless gauge particles which have no threshold bound states. The four-dimensional case is realized by wrapping the type IIB D3-brane around a three-cycle in a Calabi–Yau threefold and again scaling to the limit of zero area as in[73]. The successful resolution of the conifold singularity in terms of light states again requires the absence of threshold bound states. The absence of threshold bound states in these cases implies that the Witten index should vanish. In both cases the bulk term in the N-particle index can be evaluated explicitly (as in [72]) or extracted from the T-dual D-instanton problem (as in [65]) and has the form

$$I^{(N)\,D=4,6}_{bulk} = \frac{1}{N^2}. \tag{7.29}$$

Furthermore, it is easy to see that in these cases the deficit term is given by $I^{(N)\,D=4,6}_{def} = -1/N^2$ (assuming, as before, that it is given by the free-particle index). The extra factor of N here compared to (7.25) comes about because in these lower-dimensional cases the number of fermions is equal to the number of bosons. This systematics suggests that

$$I^{(N)\,D=4,6} = I^{(N)\,D=4,6}_{bulk} + I^{(N)\,D=4,6}_{def} = 0, \tag{7.30}$$

as expected.

8 The AdS/CFT correspondence and higher derivative terms

Up to now the discussion of the higher derivative terms has been in the context of the low energy effective action of string theory or M theory compactified on a circle or two-torus. We now turn to consider the effect of the higher derivative terms when the IIB string propagates in an $AdS_5 \times S^5$ background (much of this section and section 9 is based on [74,75]). For our purposes it will be sufficient to consider the euclidean version of AdS_5, which has a boundary with topology S^4 (or R^4). The metric will be parameterized by

$$ds^2 = \frac{L^2}{\rho^2}(dx \cdot dx + d\rho^2) + d\omega_5{}^2 = \frac{L^2}{\rho^2}(dx \cdot dx + dy \cdot dy), \tag{8.1}$$

where (x^μ, y^i) ($\mu = 0, \cdots, 3$, $i = 1, \cdots, 6$) are ten-dimensional cartesian coordinates with $\rho^2 = y^2$ and $d\omega_5^2$ is the spherically-symmetric constant curvature metric on S^5 with radius L. In these coordinates the AdS boundary is located at $\rho = 0$. The second equality in (8.1) makes it obvious that the metric is conformally equivalent to the flat ten-dimensional metric. This will prove to be important later. The only nonvanishing curvature components in this background are

$$R_{MNPQ} = -\frac{1}{L^2}(g_{MP}g_{NQ} - g_{MQ}g_{NP}) \qquad R_{mnpq} = +\frac{1}{L^2}(g_{mp}g_{nq} - g_{mq}g_{np}) \tag{8.2}$$

(upper case Latin indices, $M, N, \ldots = 0, 1, 2, 3, 5$, label the AdS_5 coordinates and lower case Latin indices, $m, n, \ldots = 1, 2, 3, 4, 5$ label the S^5 coordinates).

The non-vanishing components of the Ricci tensor are

$$R_{MN} = -\frac{4}{L^2} g_{MN} \qquad R_{mn} = +\frac{4}{L^2} g_{mn} \,, \tag{8.3}$$

so that the scalar curvature vanishes. The dilaton and $C^{(0)}$ are arbitrary constants and the only other field that is nonvanishing in the $AdS_5 \times S^5$ background is F_5, which has the solution,

$$F_{MNPQR} = \frac{1}{L}\varepsilon_{MNPQR} \,, \qquad F_{mnpqr} = \frac{1}{L}\varepsilon_{mnpqr} \,, \tag{8.4}$$

which corresponds to a configuration in which the total $\mathcal{R}\otimes\mathcal{R}$ charge associated with F_5 is equal to $\alpha'^{-2}L^4 e^{-\phi}$, which must be an integer, N. The isometry of this space is the product $SO(4,2) \times SU(4)$. These factors package together with fermionic symmetries into the supergroup $OSp(4,2|4)$ which also acts as a superconformal symmetry group on the boundary.

This background is maximally supersymmetric (just like the Minkowski vacuum) and there are 32 conserved supercharges that transform as a complex chiral spinor of the tangent-space group, $SO(4,1) \times SO(5)$. In the basis where the ten dimensional Γ_Λ matrices[n] a re given by $\Gamma_M = \sigma_1 \otimes \gamma_M \otimes I$ and $\Gamma_m = \sigma_2 \otimes I \otimes \gamma_m$, the supersymmetries are generated by the Killing spinors that satisfy

$$D_\Lambda \epsilon - \frac{1}{2L}(\sigma_1 \otimes I \otimes I)\Gamma_\Lambda \epsilon = 0, \tag{8.5}$$

which follows from the requirement that the gravitino supersymmetry transformation should vanish. In this basis the complex chiral supersymmetry parameter reads

$$\epsilon = \begin{pmatrix} 1 \\ 0 \end{pmatrix} \otimes \zeta \otimes \kappa \,, \tag{8.6}$$

where ζ is a complex four-component $SO(4,1)$ spinor and κ is a complex four-component $SO(5)$ spinor. The Killing spinor equation (8.5) has components

$$D_M \zeta_\pm - \frac{1}{2L}\gamma_M \zeta_\pm = 0, \tag{8.7}$$

$$D_m \kappa_\pm \mp i\frac{1}{2L}\gamma_m \kappa_\pm = 0. \tag{8.8}$$

The Killing spinors $\kappa_\pm$ on S^5 may be used to construct the Kaluza-Klein excitations of all the fields in the IIB gauged supergravity starting from the modes

[n] In this section and the next the ten-dimensional gamma matrices will be denoted by upper case Γ while lower case γ will be reserved for four-dimensional gamma matrices.

of the massless complex singlet scalar. We will later make use of the Euclidean continuation of the Killing spinors on AdS_5, which requires strategically placed factors of i.

The metric (8.1) can be obtained from the geometry of a stack of coincident D3-branes, where each D3-brane carries a unit charge associated with $F_5 = dC^{(4)}$. The effective world-volume theory of N coincident D3-branes is described by $U(N)$ gauge theory, where the integer N is again the total $\mathcal{R}\otimes\mathcal{R}$ charge. In terms of the low energy supergravity theory the classical D3-brane geometry interpolates from the Minkowski region far away from the horizon to the near-horizon geometry, which is $AdS_5 \times S^5$. The transition region grows with N and the $N \to \infty$ limit considered in [76,34] is one in which the horizon region fills all space. This leads to the AdS/CFT conjecture [34] which suggests that the bulk string theory is *equivalent* to superconformal $\mathcal{N} = 4$ SU(N) Yang–Mills theory on the boundary[o] (see also [35,36]). The supergroup $OSp(4,2|4)$ is that associated with the superconformal $\mathcal{N} = 4$ Yang–Mills theory.

According to this idea the effective action of type IIB supergravity, evaluated on a solution of the equations of motion with prescribed boundary conditions, is equated with the generating functional of connected gauge-invariant correlation functions in the Yang–Mills theory. The parameters of the $\mathcal{N} = 4$ Yang–Mills theory and the IIB superstring on $AdS_5 \times S^5$ are related by the dictionary,

$$g = \frac{g_{YM}^2}{4\pi} \qquad 2\pi\tilde{C}^{(0)} = \theta_{YM} \qquad \frac{L^2}{\alpha'} = \sqrt{g_{YM}^2 N}\,, \tag{8.9}$$

where $g = e^{\phi}$ and $\tilde{C}^{(0)}$ are the arbitrary constant scalar string fields, g_{YM} is the Yang–Mills coupling and θ_{YM} is the vacuum angle. The complex Yang–Mills coupling is therefore identified with the constant boundary value of the complex scalar field of the IIB superstring,

$$S = \frac{\theta_{YM}}{2\pi} + \frac{4\pi i}{g_{YM}^2} = \tilde{C}^{(0)} + \frac{i}{g}. \tag{8.10}$$

This explicit connection between the bulk theory and the boundary theory can be expressed symbolically in terms of the generating functions [36,35,78],

$$\exp(-S_{IIB}[\Phi_m(J)]) = \int DA\, \exp(-S_{YM}[A] + \mathcal{O}_\Delta[A]J). \tag{8.11}$$

The left-hand side is the generating function for the amplitudes in IIB superstring theory or its low energy supergravity limit. The effective action S_{IIB} is

[o]The fact that the theory should be $SU(N)$ and not $U(N)$ is explained in [36,77].

evaluated in terms of the 'massless' supergravity fields and their Kaluza–Klein descendents, that we have generically indicated with $\Phi(z;\omega)$, where ω are the coordinates on S^5 and $z^M \equiv (x^\mu, \rho)$ ($M = 0,1,2,3,5$ and $\mu = 0,1,2,3$) are the AdS_5 coordinates. The notation in (8.11) indicates that this action depends on the boundary data, $J(x)$, of the bulk fields. The right-hand side of (8.11) is the generating function of the correlation functions of the gauge-invariant composite operators, $\mathcal{O}(A)$, to which J couples in the boundary $\mathcal{N} = 4$ supersymmetric Yang–Mills, where the fluctuating Yang–Mills potential is denoted by A.

For general values of the dimensionless ratio, α'/L^2, an expansion in powers of α' is not necessarily a good approximation to the theory. Only for AdS_5 scales that are large compared to the string scale, i.e. $\alpha' << L^2$, is t he curvature small and the α' expansion can still be interpreted as a derivative expansion in the nontrivial background. It follows from (8.9) that in this regime classical supergravity is related to the Yang–Mills limit in which $g^2_{YM}N \to \infty$. This requires $N \to \infty$ (since $SL(2,\mathbb{Z})$ duality can always be used to map the large string coupling to a value that is not large). Since, according to 't Hooft [79] the quantity $g^2_{YM}N$ is the coupling constant of large-N Yang–Mills theory we see that the AdS/CFT conjecture relates the strong coupling limit of one theory to the weak coupling limit of the other. Several tests of this conjecture have been made at the level of the two-point and three-point correlations of currents [80] based on the semiclassical approximation to the bulk supergravity ($g << 1$) which is valid in the limit $\alpha'/L^2 << 1$, which corresponds to the large-N limit with $g^2_{YM}N$ fixed at a large value.

The spectrum of IIB supergravity compactified on $AdS_5 \times S^5$ is described in [81]. Here we are most interested in the identification of the lowest lying modes that form the supergravity supermultiplet. While the massless dilaton is associated with the constant mode on S^5, *i.e.* with the scalar spherical harmonic Y_ℓ with $\ell = 0$, the other scalars in the supermultiplet are associated with excitations on the 5-sphere. In particular the real scalars Q^{ij} ($i,j,k = 1,2,\ldots,6$), with mass $m^2 = -4/L^2$ in the $\mathbf{20}_R$ of the $SO(6)$ isometry group of S^5 result from a combination of the trace of the internal metric and the self-dual $\mathcal{R}\otimes\mathcal{R}$ five-form field, $F^{(5)}$, with $\ell = 2$ (Q^{ij} are quadrupole moments of S^5). Similarly the complex scalars E^{AB} with mass $m^2 = -3/L^2$ and their conjugates (in the $\mathbf{10}$) are associated with the pure two-form fluctuations with $\ell = 1$ of the complexified antisymmetric tensor in the internal directions. The $\mathbf{15}$ massless vectors $V^{[ij]}_M$ that gauge the $SO(6)$ isometry group are in one-to-one correspondence with the Killing vectors of S^5 and result from a linear combination with $\ell = 1$ of the mixed components of the metric and the internal three-form components of the $\mathcal{R}\otimes\mathcal{R}$ four-form potential, $C^{(4)}$. The $\mathbf{6}$ complex antisymmetric tensors $B_{MN}^{[AB]}$ with $m^2 = 1/L^2$, which have first order equations of motion, result from scalar spherical harmonics with $\ell = 1$. The analysis of the fermions is similar. The **4** dilatini[p] Λ^A with mass $m = -3/(2L)$ are proportional to the internal Killing spinors κ_+. The $\mathbf{20}_c$ spinors χ^A_{BC} with mass $m = -1/(2L)$ correspond to internal components of the gravitino with $\ell = 1$. Finally the supergravity multiplet is completed by the massless $\mathbf{4}^*$ gravitinos Ψ_{MA} which are proportional to the internal Killing spinors κ_-.

The above fields are those that act as sources for the superconformal currents in the boundary Yang–Mills theory. The higher Kaluza–Klein modes on S^5, which have higher values of ℓ, are bulk fields of dimension $\Delta = l$. The boundary value of any of these fields couples to a composite gauge invariant operator in the Yang–Mills theory that can be expres sed as a rank l symmetric traceless tensor of $SO(6)$ (the isometry group of S^5) made from a product of l superfields of the boundary theory. These operators will not concern us in the following but are a vital part of the complete story.

Whereas the duality group for the type IIB theory compactified on T^6 is $E_{6(+6),\mathbb{Z}}$ the duality group for the $AdS_5 \times S^5$ background is $SL(2,\mathbb{Z})$, which is inherited from ten dimensions. The local symmetry group is $U(4)$ where the $U(1)$ factor is a remnant of the local symmetry of flat ten-dimensional classical supergravity. However, there is no corresponding $U(1)$ symmetry in the $\mathcal{N} = 4$ Yang–Mills boundary theory where the R-symmetry is $SU(4)$. This fits with the fact that in IIB string theory the classical $SL(2,\mathbb{R})$ is replaced by (local) $SL(2,\mathbb{Z})$ and the continuous $U(1)$ symmetry is not present.

8.1 $\mathcal{N} = 4$ Yang—Mills fields and supercurrents

The $\mathcal{N} = 4$ four-dimensional Yang–Mills theory is classically invariant under superconformal transformations as well as under global $SU(4)$ transformations, which form the R-symmetry group of automorphisms of the $\mathcal{N} = 4$ supersymmetry algebra. The physical fields comprise six real scalars, four Weyl spinors, and one gauge vector potential, all in the adjoint representation of the gauge group, which may be arbitrary in general but should be $SU(N)$ for the most straightforward application of the AdS/CFT correspondence. The real scalars, φ^i $(i = 1, \ldots, 6)$, form a **6** of the $SU(4)$ R symmetry group and can be written as antisymmetric tensors in the defining representation of $SU(4)$, $\varphi^i = \frac{1}{2}\bar{t}^i{}_{AB}\,\varphi^{AB}$ (where $\bar{t}^i{}_{AB}$ is a Clebsch–Gordan coefficient and $\varphi^{AB} = -\varphi^{BA}$

[p]In this section and the next the dilatini will be indicated by the upper case Λ in order to avoid confusion with the gaugino of the Yang–Mills theory which is denoted by lower case λ.

with $\overline{\varphi}_{AB} = \frac{1}{2}\varepsilon_{ABCD}\varphi^{CD})$[q]. The spinors, λ^A, $\overline{\lambda}_B$ transform as **4** and **4***, respectively, and the vector A_μ is a singlet of $SU(4)$. The classical Minkowski signature action is given by

$$S = \int d^4x\, \mathrm{Tr}\Big\{(D_\mu\varphi^{AB})(D^\mu\overline{\varphi}_{AB}) - \frac{1}{2}i(\lambda^{\alpha A}\delta\gamma \cdot D_{\alpha\dot{\alpha}}\overline{\lambda}^{\dot{\alpha}}{}_A) - \frac{1}{4}F_{\mu\nu}F^{\mu\nu} \qquad (8.12)$$
$$- g_{YM}\lambda^{\alpha A}[\lambda_\alpha{}^B, \overline{\varphi}_{AB}] - g_{YM}\overline{\lambda}_{\dot{\alpha}A}[\overline{\lambda}^{\dot{\alpha}}{}_B, \varphi^{AB}] + 2g_{YM}^2[\varphi^{AB}, \varphi^{CD}][\overline{\varphi}_{AB}, \overline{\varphi}_{CD}]\Big\},$$

where Tr denotes a trace over the $SU(N)$ colour indices and $D_\mu^{ab} = \delta^{ab}\partial_\mu + ig_{YM}f^{abc}A_{\mu c}$ is the covariant derivative in the adjoint representation. We are here using a standard chiral basis for the spinors with the 2×2 Pauli matrices $\sigma^\mu_{\alpha\dot{\alpha}}$ coupling the complex chiral spinors.

The supersymmetry transformations on the component fields with parameters $\eta_\alpha{}^A$ and $\overline{\eta}^{\dot{\alpha}}{}_A$ are given by,

$$\begin{aligned}
\delta\varphi^{AB} &= \frac{1}{2}(\lambda^{\alpha A}\eta_\alpha{}^B - \lambda^{\alpha B}\eta_\alpha{}^A) + \frac{1}{2}\varepsilon^{ABCD}\overline{\eta}_{\dot{\alpha}C}\overline{\lambda}^{\dot{\alpha}}_D \\
\delta\lambda_\alpha{}^A &= -\frac{1}{2}F^{-}_{\mu\nu}\sigma^{\mu\nu}{}_\alpha{}^\beta\eta_\beta{}^B + 4i(\gamma \cdot D_{\alpha\dot{\alpha}}\varphi^{AB})\overline{\eta}^{\dot{\alpha}}_B - 8g_{YM}[\overline{\varphi}_{BC}, \varphi^{CA}]\eta_\alpha{}^B \\
\delta A_\mu &= -i\lambda^{\alpha A}\sigma^\mu{}_{\alpha\dot{\alpha}}\overline{\eta}^{\dot{\alpha}}{}_A - i\eta^{\alpha A}\sigma^\mu{}_{\alpha\dot{\alpha}}\overline{\lambda}^{\dot{\alpha}}{}_A.
\end{aligned} \qquad (8.13)$$

These are ordinary global supersymmetries when the 16 components of η_α^A and $\overline{\eta}^{\dot{\alpha}}{}_A$ are constant but the symmetry extends to the thirty-two component superconformal symmetry when η and $\overline{\eta}$ are linear fun ctions of x^μ,

$$\eta_\beta^A \to \frac{1}{\sqrt{\rho_0}}\left(\rho_0\eta_\beta^A + (x - x_0)_\mu\sigma^\mu_{\beta\dot{\beta}}\bar{\xi}^{\dot{\beta}A}\right) \equiv \left(\frac{1-\gamma^5}{2}\right)\frac{1}{\sqrt{\rho_0}}\gamma_M z^M\zeta^{(0)}, \quad (8.14)$$

where ξ is the Grassmann spinor associated with the special supersymmetries which is packaged together with η into a four-component spinor in the second equality where

$$\zeta^{(0)} = \begin{pmatrix} \eta_\alpha^A \\ \bar{\xi}^{\dot{\alpha}}{}_A \end{pmatrix}, \qquad (8.15)$$

$z^M \equiv (x^\mu - x_0^\mu, \rho)$ and $\gamma^M \equiv (\gamma^\mu, \gamma^5)$. Equation (8.15) is just a chiral projection on a spinor of AdS_5 (with tangent-space group $SO(4,1)$). The normalization factor of $\rho_0^{-1/2}$ in (8.14) has been chosen for later convenience

The β function for this theory vanishes which means that the quantum theory is expected to share the superconformal invariance of the classical theory.

[q]The superscripts $A, B = 1, \ldots, 4$ label **4**'s of $SU(4)$ while subscripts label the **4***'s.

The Noether currents associated with these superconformal transformations, as well as the chiral $SU(4)$ transformations, form a supermultiplet with 256 component currents, including the energy-momentum tensor [82,83]. For example, the lowest member of the supermultiplet is the scalar current that couples to Q^{ij},

$$Q^{ij} = \varphi^i \varphi^j - \delta^{ij} \varphi_k \varphi^k / 6. \tag{8.16}$$

Equation (8.16) only involves the bilinear terms in the current, but it is important that the full nonabelian currents also contain higher-order contributions [75] which are crucial in verifying the AdS/CFT correspondence [84]. The current that couples to the boundary value of the dilatino is

$$\hat{\Lambda}^A_\alpha = -\sigma^{\mu\nu}{}_\alpha{}^\beta F^-{}_{\mu\nu} \lambda_\beta{}^A \tag{8.17}$$

(again in the linearized approximation). The complete list of currents in the supermultiplet is given in [83]. These superconformal currents are in one to one correspondence with the fields of the supergravity multiplet.

From the AdS/CFT conjecture the supersymmetrized Einstein–Hilbert action gives the leading large-$g^2_{YM}N$ contributions to the correlation functions of the superconformal currents. For example, a correlation function of M stress tensors,

$$\langle T_{\mu_1\nu_1}(x_1) T_{\mu_2\nu_2}(x_2) \ldots T_{\mu_r\nu_r}(x_r) \ldots T_{\mu_M\nu_M}(x_M) \rangle, \tag{8.18}$$

corresponds to bulk superstring amplitudes in which the boundary values of the gravitons have polarizations, $h_{\mu_r\nu_r}$, which are oriented in the four-dimensional Minkowski directions, $\mu_r = M_r$, $\nu_r = N_r$ where $M_r, N_r = 0,1,2,3$. Since the coefficient of the Einstein–Hilbert term is $e^{-2\phi}\alpha'^{-4} = L^8 N^2$ the leading behaviour of the correlation functions obtained in this manner is of order N^2. In general there is little known about such Yang–Mills correlation functions at large 't Hooft coupling. However, the free-field value of the correlation functions of two superconformal currents, such as $\langle T_{\mu_1\nu_1}(x_1^{\mu_1}) T_{\mu_2\nu_2}(x_2^{\mu_2})\rangle$, is known to be exact [85,86] by virtue of the relation of this correlation function to the R-symmetry anomaly, which is not renormalized due to the Adler–Bardeen theorem. An analogous Ward identity prevents the three-point correlation function of superconformal currents (in the $\mathcal{N} = 4$ theory, but not for $\mathcal{N} < 4$) from receiving renormalizations beyond those of the free field theory [87]. These free-field correlators of two and three superconformal currents are both proportional to

$$\dim\,(SU(N)) = N^2 - 1. \tag{8.19}$$

The leading factor of N^2 corresponds to the coefficient of $\alpha'^{-4} g^{-2}$ in the Einstein–Hilbert action. This would be the exact N-dependence if the gauge group were $U(N)$ but (8.19) shows that there must be one other contribution to the Einstein–Hilbert action in the $AdS_5 \times S^5$ background that arises at order N^{-2} (order α'^4) relative to the leading term. It is easy to see that such a term would come from a on e string loop effect which has not yet been elucidated. Since the terms that concern us here are of order α'^3 relative to the leading term they dominate over this effect.

8.2 *Generalities concerning higher derivative terms*

We wish to consider the effects of the $o(\alpha'^{-1})$ terms on the AdS/CFT conjecture. Since these terms are suppressed by α'^3 relative to the leading terms they will correspond to $O(N^{-3/2})$ corrections to the leading behaviour. The interesting feature of these terms is that they manifest the $SL(2,\mathbb{Z})$ duality in a very nontrivial fashion.

We saw from the definition of the $\mathcal{R}^4$ term, (3.11), that the only components of the curvature which contribute are those of the Weyl tensor. We can expand R_{θ^4} (defined in (3.12)) in fluctuations around its classical value, $R^{(0)}_{\theta^4}$,

$$R_{\theta^4} = R^{(0)}_{\theta^4} + \hat{R}_{\theta^4}, \tag{8.20}$$

where $\hat{R}$ is the fluctuation[r]. Since $AdS_5 \times S^5$ is conformally flat the Weyl tensor vanishes and it follows that

$$R^{(0)}_{\theta^4} = 0. \tag{8.21}$$

This leads immediately to several consequences:

1. $\mathcal{R}^4 = 0$ in the $AdS_5 \times S^5$ background, which means that the dilaton equation of motion is unchanged in this background.

2. $\delta R^4/\delta g_{\mu\nu} = 0$ (since the differential is proportional to $(R^{(0)}_{\theta^4})^3$), which means that the Einstein equation is also unaffected. It is also easy to see that none of the other $o(\alpha'^3)$ terms which are related to $\mathcal{R}^4$ by supersymmetry contribute to the equations of motion. This means that the $AdS_5 \times S^5$ background is unaltered by the presence of this term, so this background is a solution of the effective equations of motion of string theory through $o(\alpha'^3)$, to all orders in g (a more general superspace argument leads to the statement that this background is exact to all orders in α' [89]).

[r]A more complete discussion, given in [88], shows that there is a field definition in which the off-shell fluctuations of the metric can be restricted to those that only give fluctuations to the Weyl com ponents of the curvature.

3. Equation (8.21) also implies that

$$\frac{\delta}{\delta g_{\mu\nu}}\frac{\delta}{\delta g_{\rho\sigma}}R^4\Big|_{AdS_5\times S^5} = 0 = \frac{\delta}{\delta g_{\mu\nu}}\frac{\delta}{\delta g_{\rho\sigma}}\frac{\delta}{\delta g_{\tau\omega}}R^4\Big|_{AdS_5\times S^5}. \tag{8.22}$$

This shows that there is no renormalization of the graviton two-point or three-point functions in the $AdS_5 \times S^5$ background at $o(\alpha'^3)$.

However, there is a non-zero four-graviton contribution from the $\mathcal{R}^4$ term which is obtained by differentiating four times with respect to the metric. This gives a contribution to the amplitude which adds to the term which arises from the Einstein–Hilbert action, so that up to $o(\alpha'^3)$ the four-graviton amplitude is proportional to

$$\alpha'^{-4}e^{-2\phi}A_4^{String\,(1)} + k\alpha'^{-1}e^{-\phi/2}f^{(0,0)}(\tau,\bar\tau)A_4^{String\,(2)}, \tag{8.23}$$

where $A_4^{String\,(1)}$ is the classical amplitude obtained from the Einstein–Hilbert action while $A_4^{String\,(2)}$ is the contribution from the $\mathcal{R}^4$ term and k is a well-defined constant.

For example, the contribution of the $\mathcal{R}^4$ term to four-graviton scattering is obtained by contracting the linearized $\mathcal{R}^4$ vertex with four graviton propagators that propagate from the interaction point at (x_0^μ, ρ_0) to the boundary at points $(x_{0^\mu\, r}, \rho_{0,r} = 0)$, where r labels the position of the rth boundary field. The linearized curvature is

$$R_{\mu_1\omega_1\nu_1\tau_1} = D_{\omega_1}D_{\tau_1}h_{\mu_1\nu_1}, \tag{8.24}$$

where $h_{\mu\nu}$ is the linearized fluctuation of the metric around its value in $AdS_5 \times S^5$ and D is the $AdS_5\times S^5$ covariant derivative. The bulk-to-boundary graviton propagator in $AdS_5 \times S^5$ is an obvious generalization [90] of the scalar and spin-1/2 propagators that will appear in later sections. The four-graviton interaction vertex can be expressed in the form (2.18) where the tensors t_8 in the kinematic factor $\tilde{K}$ have the form of the product of four inverse $AdS_5 \times S^5$ metrics summed over various permutations of their indices.

Using the AdS/CFT conjecture the $\mathcal{R}^4$ term gives a contribution to the correlation functions of stress tensors in the boundary Yang–Mills field theory that are suppressed by a factor of $N^{-3/2}$ relative to the leading free-field contribution. The vanishing of the graviton one-point function, translates into the statement that the one-point function of the stress tensor $\langle T_{\mu\nu}\rangle$ vanishes, which follows from conformal invariance. Similarly, the vanishing of the $\alpha'^{-1}R^4$ contribution to the two-graviton and three-graviton S-matrix elements in IIB supergravity translates into the statements that the correlation functions of two and of three stress tensors in $\mathcal{N}=4$ Yang–Mills theory is not renormalized from the free-field value at this order in N^{-1}. This is in accord with the exact results [85,86,87] cited earlier.

It is only when we come to the four-graviton amplitude that the $\mathcal{R}^4$ term contributes and therefore the correlation function of four stress tensors gets a new contribution. The supergravity calculations give the following expression for the momentum-space correlation function of four stress tensors,

$$\begin{aligned} A_4^{YM} &= \langle T_{\mu_1\nu_1}(1)T_{\mu_2\nu_2}(2)T_{\mu_3\nu_3}(3)T_{\mu_4\nu_4}(4)\rangle = \\ &= N^2 A_4^{(1)} + \tilde{k} N^{1/2} f^{(0,0)}(S,\bar{S}) A_4^{(2)} + \ldots, \end{aligned} \tag{8.25}$$

where $A_4^{(1)}$ is the contribution at leading order in $(g_{YM}^2 N)$ and $A_4^{(2)}$ is the correction arising from the $\mathcal{R}^4$ term (and an irrelevant constant has been absorbed into $\tilde{k}$). The term $A_4^{(1)}$ comes from the four-graviton amplitude computed from the Einstein action in $AdS_5 \times S^5$, while $A_4^{(2)}$ corresponds to the four-graviton vertex in the $\mathcal{R}^4$ term. No further terms are necessary at this order in the α' expansion. The scalar field τ is now interpreted as the complex coupling constant, S.

The second term in (8.25) has a remarkable amount of information concerning the Yang–Mills theory. It is the first non-leading term in the $1/N$ expansion but is an exactly known function of the (complex) coupling. As we have seen, the factor $f^{(0,0)}$ contains two terms that are perturbative in string theory and an infinite number of D-instanton terms. Therefore, in the limit $g_{YM} \to 0$ with $g_{YM}^2 N$ fixed and large, the expression (8.25) takes the form

$$\begin{aligned} N^{-2} A_4^{YM} &= A_4^{(1)} + \tilde{k} A_4^{(2)} \Bigg[2\zeta(3) \left(\frac{g_{YM}^2 N}{4\pi}\right)^{-3/2} + \frac{2\pi^2}{3N^2} \left(\frac{g_{YM}^2 N}{4\pi}\right)^{1/2} \\ &+ \frac{(4\pi)^{3/2}}{N^{3/2}} \sum_{K=1}^{\infty} Z_K K^{1/2} \left(e^{-K\left(8\pi^2 g_{YM}^{-2} + i\theta\right)} + e^{-K\left(8\pi^2 g_{YM}^{-2} - i\theta\right)} \right) \left(1 + o(g_{YM}^2/K)\right) \Bigg], \end{aligned} \tag{8.26}$$

which includes an infinite series of instanton corrections. We will return to a discussion of these corrections later. An important property of this expression is its invariance under $SL(2,\mathbb{Z})$ (which is interpreted as Montonen–Olive electromagnetic duality). This transformation, which maps the small coupling regime to large coupling, is manifest in (8.25) but not in (8.26), which is the 't Hooft expansion, and is only valid when g_{YM} is very small. The presence of noninteger powers of $g_{YM}^2 N$ in this expansion is one indication of how far removed the strong coupling expansion is from weakly coupled perturbation theory.

8.3 The scalar AdS Green function and the instanton profile.

In order to make use of (8.11) to perform explicit detailed calculations of Yang–Mills correlation functions we need to evaluate the bulk-to-boundary Green functions for the various fields of the bulk theory. These are defined as specific normalized limits of bulk-to-bulk Green functions [35,36,87] when one point is taken to the *AdS* boundary. The precise forms of these propagators depend on the spin and mass of the field. For example, the normalized bulk-to-boundary Green function for the *AdS* Laplace operator for a dimension Δ scalar field is given by

$$G_\Delta(x,\rho,\omega;x',0,\omega') = c_{{}_\Delta} K_\Delta(x^\mu,\rho;x'^\mu,0), \tag{8.27}$$

which is independent of ω and where $c_{{}_\Delta} = \Gamma(\Delta)/(\pi^2\Gamma(\Delta-2))$ and

$$K_\Delta(x^\mu,\rho;x'^\mu,0) = \frac{\rho^\Delta}{(\rho^2+(x-x')^2)^\Delta}. \tag{8.28}$$

The expression (8.27) is appropriate for an 'S-wave' process in which there are no excitations in the directions of the five-sphere, S^5. In terms of K_Δ the bulk field

$$\Phi_m(z;J) = c_{{}_\Delta}\int d^4x' K_\Delta(x,\rho;x',0)J_\Delta(x') \tag{8.29}$$

satisfies the boundary condition as $\rho\to 0$,

$$\Phi_m(x,\rho;J) \approx \rho^{4-\Delta}J_\Delta(x), \tag{8.30}$$

since $\rho^{\Delta-4}K_\Delta$ reduces to a δ-function on the boundary. The conformal dimension of the operator is related to the *AdS* mass of the corresponding bulk field by $(mL)^2 = \Delta(\Delta-4)$, so that $\Delta_\pm = 2\pm\sqrt{4+(mL)^2}$ and only the positive branch, $\Delta=\Delta_+$, is relevant for the lowest-'mass' supergravity multiplet. In the case of a massless scalar field ($\Delta_+ = 4$) the propagator reduces to $\delta^{(4)}(x^\mu - x'^\mu)$ in the limit $\rho\to 0$. In comparing the D-instanton contributions of the bulk theory with the Yang–Mills instanton contributions to the boundary theory it is of significance, as we will see in section 9, that the Green function K_4 defined in (8.28) has the same form (up to an additive constant) as the profile of e^ϕ in the D-instanton solution centered on the point (x^μ,ρ) in $AdS_5\times S^5$ and evaluated at the boundary point $(x'^\mu,0)$.

We now turn to the Yang–Mills instanton. Recall that that the instanton solution for the $SU(2)$ Yang–Mills potential gives a nontrivial nonabelian field strength of the form

$$F^{a-}_{(0)\mu\nu}(x^\mu;x_0^\mu,\rho_0) = -\frac{4}{g_{YM}}\frac{\eta^a_{\mu\nu}\rho_0^2}{\left(\rho_0^2+(x-x_0)^2\right)^2}, \tag{8.31}$$

where $\eta^a_{\mu\nu}$ is the standard 't Hooft symbol, ρ_0 is the arbitrary scale and ${x_0}^\mu$ the position of the instanton on S^4. Squaring F^- and using properties of η leads to the equality

$$(F^-_{(0)})^2 = \frac{4}{g^2_{YM}} K_4(x^\mu_0, \rho_0; x^\mu, 0), \tag{8.32}$$

which is again equal to the scalar Green function for the $\Delta = 4$ case where the fifth coordinate ρ is now identified with the instanton scale. This is a key observation in identifying D-instanton effects of the bulk theory with Yang–Mills instanton effects in the boundary theory. It is related to the fact that the moduli space of a Yang–Mills instanton on S^4 has an AdS_5 factor.

9 D-instantons and Yang–Mills instantons

We now want to compare the leading order one-instanton contribution to the supersymmetric Yang–Mills correlation functions with the amplitudes of the IIB superstring theory with appropriate boundary conditions. For illustrative purposes we will only consider the sixteen-Λ amplitude in the $AdS_5 \times S^5$ background and the corresponding Yang–Mills sixteen-$\hat{\Lambda}$ correlation function [75]. From either perspective the leading instanton contribution arises from the product of sixteen factors each carrying one single fermionic zero mode. Correlation functions of other superconformal currents can be calculated in analogous fashion. For certain of these it is possible to express the result very explicitly in terms of dilogarithms [75,91].

From the bulk type IIB point of view we are interested in the leading D-instanton contribution to amplitudes, which may either be extracted directly from the exactly known $o(\alpha'^{-1})$ terms in the effective action (method (a) below) or deduced by integration over the fluctuations around the classical $AdS_5 \times S^5$ D-instanton solution (method (c)). From the Yang–Mills perspective we are interested in the instanton contributions to supercurrent correlators of the $\mathcal{N} = 4$ theory (method (b)).

Method (a). Expansion of $\alpha'^{-1} f^{(12,12)} \Lambda^{16}$ term in the bulk supergravity theory

The first method for obtaining this amplitude is to expand the function $f^{(12,-12)}$ in order to extract the one-instanton term. The leading behaviour comes from the leading power of $g^{12} = \tau_2^{-12}$ in $\mathcal{F}_{1,12}$ defined by (6.3) and (6.5). We want to consider the situation in which all sixteen fermions propagate to specific values on the boundary. The Dirac operator acting on spin-1/2 fields in AdS_5 was given in [92,93] as

$$\gamma \cdot D\Lambda = e_{\hat{L}}{}^{M} \gamma^{\hat{L}} \left(\partial_M + \frac{1}{4} \omega_M^{\hat{M}\hat{N}} \gamma_{\hat{M}\hat{N}} \right) \Lambda = (\rho \gamma^{\hat{5}} \partial_5 + \rho \gamma^{\hat{\mu}} \partial_\mu - 2\gamma^{\hat{5}})\Lambda \,, \quad (9.1)$$

where $e_{\hat{L}}{}^{M}$ is the vielbein, $\omega_M^{\hat{M}\hat{N}}$ the spin connection (hatted indices refer to the tangent space) and $\gamma^{\hat{\mu}}$ are the four-dimensional Dirac matrices. Equation (9.1) leads to the normalized bulk-to-boundary propagator of the fermionic field Λ of mass $m = -3/2L$, associated with the composite operator $\hat{\Lambda}$ of dimension $\Delta = \frac{7}{2}$,

$$K^F_{7/2}(\rho_0, x_0; x) = K_4(\rho_0, x_0; x) \frac{1}{\sqrt{\rho_0}} (\rho_0 \gamma_{\hat{5}} + (x_0 - x)^\mu \gamma_{\hat{\mu}}) \,, \quad (9.2)$$

which, suppressing all spinor indices, leads to

$$\Lambda_J(x_0, \rho_0) = \int d^4 x K^F_{7/2}(\rho_0, x_0; x) J_\Lambda(x) \,, \quad (9.3)$$

where $J_\Lambda(x)$ is a left-handed boundary value of Λ and acts as the source for the composite operator $\hat{\Lambda}$ in the boundary $\mathcal{N} = 4$ Yang–Mills theory satisfying $\hat{\gamma}^5 J_\Lambda = J_\Lambda$. As a result, the classical action for the operator $(\hat{\Lambda})^{16}$ in the $AdS_5 \times S^5$ supergravity action is

$$S_\Lambda[J] = (const.)\, e^{-2\pi(\frac{1}{g} - iC^{(0)})} g^{-12} \, V_{S^5} \int \frac{d^4 x_0 d\rho_0}{\rho_0^5}$$
$$\varepsilon_{16} \prod_{p=1}^{16} \left[K_4(\rho_0, x_0; x_p) \frac{1}{\sqrt{\rho_0}} \left(\rho_0 \gamma^{\hat{5}} + (x_0 - x_p)^\mu \gamma_{\hat{\mu}} \right) J_\Lambda(x_p) \right] , (9.4)$$

where $J_\Lambda(x_p)$ is the wave-function of the dilatino evaluated at the boundary point $(x_p, 0)$ and we have set $e^\phi = g$ and $C^{(0)} = \tilde{C}^{(0)}$ (since the scalar fields are taken to be constant in the $AdS_5 \times S^5$ background) and $V_{S^5} = \pi^3$ is the S^5 volume. The spinor indices on the J's are contracted with the rank-sixteen antisymmetric tensor, $\varepsilon_{16}^{a_1 a_2 \ldots a_{16}}$. We have not kept track of the overall constant normalization in this expression. The final integration over the moduli x_0^μ, ρ_0 has not been carried out explicitly in (9.4). For certain of the other correlation functions the integration over the instanton moduli can be evaluated in terms of standard functions [75,91]. According to (8.11) the sixteen-$\hat{\Lambda}$ Yang–Mills correlation function is extracted from (9.4) by differentiating with respect to all the J_Λ's.

Method (b). Sixteen-$\hat{\Lambda}$ correlator in the boundary Yang–Mills theory

We will now consider the correlation function of sixteen fermionic superconformal current operators in a charge-one instanton background,

$$G_{\hat{\Lambda}^{16}}(x_p) = \langle \prod_{p=1}^{16} g^2_{YM} \hat{\Lambda}^{A_p}_{\alpha_p}(x_p) \rangle_{K=1} , \tag{9.5}$$

where $\hat{\Lambda}$ is defined in (8.17). The $SU(2)$ Yang–Mills instanton has sixteen supermoduli which correspond to the fermionic zero modes that are induced by the broken supersymmetries. The special feature of the product of fields inside the correlator (9.5), as well as the others that are related by supersymmetry, is that they provide precisely the sixteen fermionic zero modes that are needed to give a nonzero result in the instanton background. In particular, it is easy to see that $\hat{\Lambda}$ is linear in fermionic zero modes so a total of sixteen factors is needed for a non-zero result. To leading order in g_{YM}, $G_{\hat{\Lambda}^{16}}$ does not receive contribution from anti-instantons.

The fermionic zero modes can be deduced by applying the broken components of the supersymmetry transformations to the fields. The bilinear $\hat{\Lambda}^A_\alpha = -\sigma^{\mu\nu\,\beta}{}_{\alpha} F^-_{\mu\nu} \lambda^A_\beta$ is proportional to the zero mode of $\lambda_\alpha{}^A$ which will be denoted $\lambda_{(0)\,\alpha}{}^A$ and can be deduced from the second line of (8.13),

$$\lambda^A_{(0)\alpha} \equiv \delta\lambda^A_\alpha = \frac{1}{2} F^-_{(0)\mu\nu} \sigma^{\mu\nu\,\beta}_{\alpha} \frac{1}{\sqrt{\rho_0}} \left(\rho_0 \eta^A_\beta + (x - x_0)_\mu \sigma^\mu_{\beta\dot{\beta}} \bar{\xi}^{\dot{\beta} A} \right) . \tag{9.6}$$

The leading term in $\hat{\Lambda}$ in the one-instanton sector is simply obtained by using the instanton profile $F^-_{(0)\mu\nu}$ (8.31) and substituting each $\lambda_\alpha{}^A$ by the corresponding zero mode, $\lambda_{(0)\alpha}{}^A$. The resulting correlation function has the form

$$G_{\hat{\Lambda}^{16}}(x_p) = (const.) g^8_{YM} e^{-\frac{8\pi^2}{g^2_{YM}} + i\theta_{YM}} \int \frac{d^4x_0 d\rho_0}{\rho_0^5} \int d^8\eta d^8\bar{\xi}$$

$$\prod_{p=1}^{16} \left[K_4(x^\mu_p, \rho_0; x^\mu_0, 0) \frac{1}{\sqrt{\rho_0}} \left(\rho_0 \eta^{A_p}_{\alpha_p} + (x_p - x_0)_\mu \sigma^\mu_{\alpha_p \dot{\alpha}_p} \bar{\xi}^{\dot{\alpha}_p A_p} \right) \right] \tag{9.7}$$

Integration over the fermion zero modes leads to the sixteen-index invariant tensor ε_{16}. Converting to four-component spinor notation and using (8.11) this result coincides with (9.4), up to the undetermined overall constant.

One is entitled to ask why such a comparison of a D-instanton effect with a $SU(2)$ Yang–Mills instanton effect is being made in the first place. In method (a) we used the expression for the $o(\alpha'^{-1})$ interaction in the $AdS_5 \times S^5$ background which should only be a good approximation to strongly coupled $SU(N)$

Yang–Mills in the limit $N \to \infty$. On the other hand method (b) computed the effect of an $SU(2)$ Yang–Mills instanton. However, as will be described in the final subsection, the N-dependence of these correlation functions was shown in [39] to only affect the overall coefficient of the correlation function and not its functional form. Furthermore, the coefficient has the expected larg e-N limit, $N^{1/2}$. This will be reviewed in the last subsection along with the results in [40] which demonstrate that the large-N limit of the Yang–Mills calculation agrees perfectly with the leading D-instanton contribution even in the multi-instanton sectors.

9.1 The classical D-instanton in flat space and in $AdS_5 \times S^5$

Later in this subsection we will evaluate the single D-instanton terms in the sixteen-dilatino amplitude in $AdS_5 \times S^5$ by semi-classical quantization in the background of a D-instanton solution of type IIB supergravity (this is method (c)). First we need to recall some properties of the classical D-instanton solution.

In flat ten-dimensional euclidean space the charge-K D-instanton solution is a finite-action euclidean supersymmetric (BPS–saturated) solution in which the metric is trivial ($g_{\mu\nu} = \eta_{\mu\nu}$ in the Einstein frame) but the complex scalar $\tau = C^{(0)} + ie^{-\phi}$ has a nontrivial profile with a singularity at the position of the D-instanton. The (euclidean) $\mathcal{R} \otimes \mathcal{R}$ scalar is related to the dilaton by the BPS condition $\partial_\Sigma C^{(0)} = \pm i \partial_\Sigma e^{-\phi}$, while the dilaton solution is the harmonic function (correcting an error in [52])

$$e^{\hat{\phi}^{(10)}} = g + \frac{3K\alpha'^4}{\pi^4 |X - X_0|^8}. \tag{9.8}$$

This is the classical solution of the ten-dimensional Laplace equation, $\partial^2 e^\phi = 0$, outside an infinitesimal sphere centered on the point X_0^Λ (where X^Λ is the ten-dimensional coordinate and X_0^Λ is the location of the D-instanton[s]), g is the asymptotic value of the string coupling and the normalization of the second term has a quantized value by virtue of a condition analogous to the Dirac–Nepomechie–Teitelboim condition that quantizes the charge of an electrically charged p-brane and of its magnetically charged p'-brane dual[94,95]. It is notable that the solution in (9.8) is simply the Green function for a scalar field to propagate from X_0 to X subject to the boundary condition that $e^\phi = g$ at $|X| \to \infty$ or $|X_0| \to \infty$.

[s]Having run out of many conventional options we are here using upper case Greek letters to signify the ten-dimensional indices.

We are now interested in solving the equations of motion of the IIB theory in euclidean $AdS_5 \times S^5$. The BPS condition for a D-instanton in this background again requires $\partial_\Sigma e^{-\phi} = \pm i \partial_\Sigma C^{(0)}$ that leads to

$$g^{\Lambda\Sigma}\nabla_\Lambda \nabla_\Sigma e^{\phi} = 0, \tag{9.9}$$

and (in the Einstein frame) the Einstein equations are unaltered by the presence of the D-instanton so that $AdS_5 \times S^5$ remains a solution. Equation (9.9) is identical to the equation for the Green function of a massless scalar propagating between the location of the D-instanton (x_0^μ, y_0^i) and the point (x^μ, y^i), which is the bulk-to-bulk propagator (subject to the boundary condition that it is constant in the limits $\rho \to 0$ and $\rho \to \infty$). This is easy to solve using the conformal flatness of $AdS_5 \times S^5$ which implies that the solution for the dilaton is of the form

$$e^{\hat\phi} = g + \frac{\rho_0^4 \rho^4}{L^8}\left(e^{\hat\phi^{(10)}} - g\right), \tag{9.10}$$

where $\rho_0 = |y_0|$ and $e^{\hat\phi^{(10)}}$ is the harmonic function that appeared in the flat ten-dimensional case, (9.8). In evaluating D-instanton dominated amplitudes we will only be interested in the case in which the point (x^μ, y^i) approaches the boundary $(\rho \equiv |y| \to 0)$, in which case it is necessary to rescale the dilaton profile (just as it is necessary to rescale the scalar bulk-to-bulk propagator, [35,36]) so that the combination

$$\rho^{-4}\left(e^{\hat\phi} - g\right) = \frac{3K(\alpha')^4}{L^8 \pi^4} \frac{\rho_0^4}{((x-x_0)^2 + \rho_0^2)^4}, \tag{9.11}$$

is of relevance in the $\rho \to 0$ limit.

As mentioned earlier, the correspondence with the Yang–Mills instanton follows from the fact that $\rho_0^4/((x-x_0)^2+\rho_0^2)^4 = K_4$ is proportional to the instanton number density, $(F^-_{(0)})^2$, in the $\mathcal{N}=4$ Yang–Mills theory. Strikingly, the scale size of the Yang–Mills instanton is replaced by the distance ρ_0 of the D-instanton from the boundary. This is another indication of how the geometry of the Yang–Mills theory is encoded in the IIB superstring. Note, in particular, that as the D-instanton approaches the boundary $\rho_0 \to 0$, the expression for $\rho^{-4} e^{\hat\phi}$ reduces to a δ function that corresponds to a zero-size Yang–Mills instanton.

The BPS condition implies that we can write the solution for the $\mathcal{R}\otimes\mathcal{R}$ scalar as

$$\hat C^{(0)} = \tilde C^{(0)} + i f(x,y), \tag{9.12}$$

where $\tilde{C}^{(0)}$ is the constant real part of the field (which corresponds to $\theta_{YM}/2\pi$) and

$$f = A - \frac{1}{g} + e^{-\hat{\phi}}. \tag{9.13}$$

Since the action is independent of constant shifts of $C^{(0)}$ it does not depend on the arbitrary constant, A. In a manner that follows closely the flat ten-dimensional case considered in the appendix of [11] the action for a single D-instanton of charge K can be written as

$$S_K = -\frac{L^{10}}{\alpha'^4} \int \frac{d\rho d^4 x d^5 \omega}{\rho^5} g^{\Lambda\Sigma} \nabla_\Lambda (e^{2\hat{\phi}} f \partial_\Sigma f), \tag{9.14}$$

which reduces to an integral over the boundaries of $AdS_5 \times S^5$ and the surface of an infinitesimal sphere centered on the D-instanton at $x = x_0$, $y = y_0$. With the choice $A = 0$ in (9.13) the entire D-instanton action comes from the boundary of the infinitesimal sphere. Substituting for f from (9.13 gives

$$S_K = \frac{2\pi |K|}{g}, \tag{9.15}$$

which is the same answer as in the flat ten-dimensional case. On the other hand, with the choice $A = 1/g$ in (9.13) the expression (9.14) reduces to an integral over the boundary at $\rho = 0$ but the total action remains the same as S_K in (9.15). Remarkably, in this case the boundary integrand is *identical* to the action density of the standard four-dimensional Yang–Mills instanton. It is crucial for this agreement that the integral (9.14) includes an integral over the S^5 factor even though the classical moduli space of the Yang–Mills instanton does not include an S^5. This is a concrete manifestation of the holographic principle whereby the physics of the bulk is encoded on the boundary [96,97].

Whereas the $AdS_5 \times S^5$ metric remains unchanged by the presence of the D-instanton in the Einstein frame it is radically altered in the string frame where the instanton is manifested as a space-time wormhole (as in the flat ten-dimensional case [52]). For finite values of K the dilaton becomes large in the Planck-scale neck and the classical solution is not reliable in that region. However, for very large instanton number, the neck region becomes much larger than the Planck scale so, by analogy with the D-brane examples studied in [34], it should be very interesting to study the implications of the modified $AdS_5 \times S^5$ geometry in the large-K limit of the large-N theory.

Method (c). Fluctuations around a classical D-instanton

The D-instanton contribution to the amplitude with sixteen external dilatini, Λ^A_α, may now be obtained directly by semi-classical quantization around the classical D-instanton solution in $AdS_5 \times S^5$. The leading instanton contribution can be determined by applying supersymmetry transformations to the scalar field which has an instanton profile given by (9.11). Since the D-instanton background breaks half the supersymmetries the relevant transformations are those in which the supersymmetry parameter corresponds to the Killing spinors for the sixteen broken supersymmetries. These Killing spinors have $U(1)$ charge 1/2 and are defined by a modified version of (8.5) that includes the non-trivial composite $U(1)$ connection, Q_M [58], that is made from the IIB scalar field [52],

$$\mathcal{D}_M\zeta \equiv (D_M - \frac{i}{2}Q_M)\zeta = \frac{1}{2L}\gamma_M\zeta. \tag{9.16}$$

Substituting the euclidean D-instanton solution into the expression for the composite connection gives

$$Q_M = \frac{i}{2}e^{-\hat{\phi}}\partial_M e^{\hat{\phi}} \tag{9.17}$$

with $\hat{\phi}$ defined by (9.11). The solution of (9.16) is

$$\zeta_\pm = e^{-\hat{\phi}/4}\frac{z_M\gamma^{\hat{M}}}{\sqrt{\rho_0}}\zeta_\pm^{(0)}, \tag{9.18}$$

where $\zeta_\pm^{(0)}$ is a constant spinor satisfying $\gamma_5\zeta_\pm^{(0)} = \pm\zeta_\pm^{(0)}$.

The sixteen broken supersymmetry transformations associated with $\zeta_-^{(0)}$ give rise to the dilatino zero-modes,

$$\Lambda_{(0)} = \delta\Lambda = (\gamma^M\hat{P}_M)\zeta_-, \tag{9.19}$$

where $\hat{P}_M$ is the expression for $P_M \equiv i\partial_M\tau^*/2\tau_2$ in the D-instanton background [11],

$$\hat{P}_M = e^{-\hat{\phi}}\partial_M e^{\hat{\phi}}. \tag{9.20}$$

Using the Killing spinor equation and the D-instanton equation $\mathcal{D}^M\hat{P}_M = 0$ it is easy to check (recalling that P_M has $U(1)$ charge 2) that

$$\gamma^M\mathcal{D}_M\Lambda_{(0)} = -\frac{3}{2L}\Lambda_{(0)}, \tag{9.21}$$

so that $\Lambda_{(0)}$ is a solution of the appropriate massive Dirac equation. We will be interested in amplitudes with external states located on the boundary, in which case we may use the fact that for $\rho \sim 0$,

$$\hat{P}_M \sim \frac{1}{g}\partial_M e^{\hat{\phi}} \tag{9.22}$$

in (9.19), which leads to

$$\Lambda_{(0)} \sim \frac{4}{g}(e^{\hat{\phi}} - g)\zeta_- \,. \tag{9.23}$$

This means that near $\rho = 0$ the dilatino profile in the D-instanton background is proportional to $\rho^4 K_4(x_0, \rho_0; x, 0)$.

As a result the leading contribution to the sixteen-dilatino amplitude again reproduces the corresponding sixteen-current correlator in $\mathcal{N} = 4$ supersymmetric Yang–Mills theory. Explicitly, the D-instanton approximation to the amplitude with sixteen external dilatini, ${\Lambda_\alpha}^A$, at points on the $\rho = 0$ boundary is (up to an overall constant factor)

$$\langle \prod_{p=1}^{16} \Lambda_{\alpha_p}^{A_p}(x_p, 0) \rangle_J = (const.)\, g^{-12} e^{-2\pi K(\frac{1}{g} + iC^{(0)})} V_{S^5} \int \frac{d^4 x_0 d\rho_0}{\rho_0^5} \int d^{16}\zeta_-^{(0)}$$
$$\prod_{p=1}^{16} \left[K_4(x_0, \rho_0; x_p) \frac{1}{\sqrt{\rho_0}} \left(\rho_0 \eta_{\alpha_p}^{A_p} + (x_p - x_0)_\mu \sigma^\mu_{\alpha_p \dot{\alpha}_p} \bar{\xi}^{\dot{\alpha}_p A_p} \right) J_\Lambda(x_p) \right] . \tag{9.24}$$

The overall factor of $(const.)\, g^{-12}$ in this expression should be derived from the normalization of the bosonic and fermionic zero modes but it has been inserted by hand. Up to this overall constant factor, the amplitude (9.24) agrees with (9.7) and therefore with (9.4).

In similar manner the instanton profiles of all the fields in the supergravity multiplet follow by applying the broken supersymmetries to P_M any number of times, just as they do in the flat ten-dimensional case [11]. The single D-instanton contributions to any correlation function can then be determined.

9.2 Instantons in the limit of large N

We now return briefly to the issue of comparing the D-instanton and Yang–Mills instanton effects for gauge groups $SU(N)$ in the limit $N \to \infty$. The string computation was valid only in the region of large $g^2 N$ since we kept only the first term in the α' expansion that contributes to $\mathcal{R}^4$. In the classical $\mathcal{N} = 4$ Yang–Mills theory with a gauge group $SU(N)$ an instanton background has $8N$ fermion zero modes, $2N$ for each of the 4 adjoint Weyl fermions. It is a special feature of the $N = 2$ case that the number of fermionic zero modes coincides with the number of broken supersymmetries but for larger values of N there are more fermionic zero modes in the classical D-instanton background than the number of broken supersymmetries. However, most of these modes receive perturbative corrections that gives them nonzero energy and the only zero modes which are protected are the sixteen which correspond to broken supersymmetries. In other words, most of the classical zero modes are not protected by symmetries and do not cause the correlation functions that we have been discussing (such as the $\hat{\Lambda}^{16}$ correlator) to vanish.

The measure for a Yang–Mills instanton for the group $SU(N)$ was considered a long time ago in nonsupersymmetric theories. The measure is defined as the jacobian for the transformation from bosonic zero-modes to the collective coordinates and for pure $SU(N)$ Yang–Mills it has the form [98]

$$\frac{(2\pi)^{4N}}{(N-1)!(N-2)!}\left(\frac{\rho_0}{g_{YM}}\right)^{4N}\frac{d\rho_0 d^4x_0}{\rho_0^5} \tag{9.25}$$

(where μ_0 is an arbitrary scale), which accounts for the number of ways of embedding the $SU(2)$ instanton solution in $SU(N)$ and the coupling constant dependence is due to the eight bosonic zero modes [99]. In supersymmetric theories this coefficient is modified due to the gaussian fluctuations of the other fields. In [39] the measure for a single Yang–Mills instanton in $SU(N)$ $\mathcal{N}=4$ supersymmetric Yang–Mills was computed, including the integration over fermionic zero modes. The result is that the factor (9.25) is replaced by (dropping irrelevant constants)

$$g_{YM}^8 2^{-2N}\frac{(2N-2)!}{(N-1)!(N-2)!}, \tag{9.26}$$

so that the overall power of the coupling is the same as in the $SU(2)$ case with $\mathcal{N}=4$ supersymmetry that we considered earlier but there is now an explicit N-dependent coefficient. The large-N limit of this expression can be extracted by using Stirling's approximation for the gamma functions with the result that the overall measure at large N is proportional to

$$g_{YM}^8 N^{1/2} \sim \alpha'^{-1}\tau_2^{-7/2}, \tag{9.27}$$

which agrees with the $K=1$ case of the expression deduced from the D-instanton measure in (6.11).

The large-N limit of the contribution of K instantons to the protected processes that we have been considering was derived in an interesting recent paper [40]. Making extensive use of the ADHM construction this paper showed that the large-N limit is dominated by a saddle point which constrains the K $SU(2)$ instantons to a very special point in the multi-instanton moduli space. This is a point in the subspace of the full moduli space at which the instantons are in K commuting SU(2) subgroups of SU(N). Clearly, this can only be precisely true for $N \to \infty$. For finite N such configurations may dominate for $K << N$ but when $K > N/2$ no such configurations exist. Furthermore, it was shown in [40] that at the dominant saddle point the instantons coincide in space and have equal scale sizes — they behave as a single charge-K instanton. Even more impressively, the evaluation of the large-N limit of the fluctuations of the fields around the coincident instantons induces five new effective moduli that parameterize the five-sphere. Finally, the overall coefficient is identified in [40] with the partition function of the zero-dimensional matrix model, Z_K. The result is that the effective moduli space of the leading contribution of K Yang–Mills instantons to the protected processes is $AdS_5 \times S^5$. The resulting measure coincides precisely with the leading term in the D-instanton measure in (6.11) for general $K \geq 1$.

These results apply to the instanton corrections to the protected processes at leading power of the Yang–Mills coupling constant. However, the full $SL(2, \mathbb{Z})$-invariant structure of the Yang–Mills correlation functions suggested by the AdS/CFT correspondence in [74] is very much richer. We have seen that in the limit $g^2_{YM} N \to \infty$ $(\alpha' << L^2)$ not only must each instanton contribution to a correlation function be accompanied by an infinite number of perturbative corrections but there are vital non-instanton terms which correspond to the tree-level and one-loop contributions in the string theory. These are the terms in (8.26) that have fractional powers of g_{YM} and are not visible in weakly-coupled Yang–Mills theory and therefore cannot be obtained by matching the semiclassical approximations of the bulk and boundary theories. Explaining the nature of these terms from the Yang–Mills perspective would appear to be of some interest.

Acknowledgments:

I am very grateful to Pierre Vanhove and Constantin Bachas for useful comments concerning the material in this article.

Appendices

A Derivative expansion of the string tree amplitude

The tree amplitude for four-graviton scattering in either of the type II superstring theories is given by

$$A_4^{tree} = \alpha'^{-4} e^{-2\phi} \frac{\tilde{K}}{stu} \frac{\Gamma(1-\alpha' s)\Gamma(1-\alpha' t)\Gamma(1-\alpha' u)}{\Gamma(1+\alpha' s)\Gamma(1+\alpha' t)\Gamma(1+\alpha' u)} \tag{A.1}$$

$$= \alpha'^{-4} e^{-2\phi} \frac{\tilde{K}}{stu} \exp\left(\sum_{n=1}^{\infty} \frac{\alpha'^{2n+1} 2\zeta(2n+1)}{2n+1} (s^{2n+1} + t^{2n+1} + u^{2n+1}) \right),$$

where the second equality follows from elementary properties of Γ functions. The overall kinematic factor $\tilde{K}$, defined in (2.18), is the linearized expansion of $\mathcal{R}^4$ and s, t, u are the Mandelstam invariants in string frame (and satisfy $s + t + u = 0$). These are related to the Mandelstam invariants in M-theory coordinates (which we will denote S, T and U) by

$$s = \frac{S}{R_{11}}, \qquad t = \frac{T}{R_{11}}, \qquad u = \frac{U}{R_{11}}. \tag{A.2}$$

Every term in the exponent of (A.1) can be expressed as a polynomial in s and t multiplied by stu, as can be seen from the identity,

$$s^{2n+1} + t^{2n+1} + u^{2n+1} = stu \left[\sum_{r=1}^{n} \sum_{q=0}^{2n-2r} \frac{(2n+1)!}{r!(2n+1-r)!} (-1)^q s^{2n-1-r-q} t^{r+q-1} \right], \tag{A.3}$$

where $n \geq 1$. This means that the massless poles only contribute to the first term in the expansion of the exponential.

When expressed in terms of the Mandelstam invariants in the M-theory metric the expression (A.1) has the low-energy expansion,

$$\begin{aligned} A_4^{tree} \sim \tilde{K} \Bigg(& \frac{1}{STU} + \frac{2\zeta(3)}{R_{11}^3} + \frac{2\zeta(5)}{R_{11}^5}(S^2 + ST + T^2) + \frac{2\zeta(3)^2}{R_{11}^6} STU \\ & + \frac{2\zeta(7)}{R_{11}^7}(S^4 + 2S^3T + 3S^2T^2 + 2ST^3 + T^4) + \frac{2\zeta(3)\zeta(5)}{R_{11}^8} STU(S^2 + ST + T^2) \\ & + \frac{2\zeta(9)}{R_{11}^9}(S^6 + 4S^5T + \cdots + T^6) + \frac{4}{3R_{11}^9}\zeta(3)^3 S^2T^2U^2 + \cdots \Bigg). \end{aligned} \tag{A.4}$$

The first term in this expansion combines with the kinematic factor, $\tilde{K}$, to give the tree-level amplitude that is described by the Einstein–Hilbert action of eleven-dimensional supergravity. The term with coefficient $\zeta(3)$ can be identified (section 5 of these notes) with a one-loop effect in four-graviton scattering in eleven-dimensional supergravity compactified on a circle of radius $R_{11} = e^{2\phi/3}$. This explains the corresponding term in (A.1) when A_4^{tree} is interpreted as a type IIA tree amplitude. When (A.1) is interpreted as a IIB tree amplitude the $\zeta(3)$ term comes from the zero-volume limit of M theory on T^2. Similarly, there is some evidence [63] that the term with coefficient $\zeta(5)$ in (A.4) can be identified with a two-loop effect in eleven-dimensional supergravity. The higher-order terms in (A.4) give terms in the effective action with derivatives acting on $\mathcal{R}^4$ and other higher-derivative terms.

B Modular covariant derivatives.

The various coefficient functions in the effective action are $(w, \hat{w})$ forms, where w refers to the holomorphic modular weight and $\hat{w}$ to the anti-holomorphic modular weight. A nonholomorphic modular form $F^{(w,\hat{w})}$ transforms as,

$$F^{(w,\hat{w})} \to F^{(w,\hat{w})} (c\tau + d)^w (c\bar{\tau} + d)^{\hat{w}}, \tag{B.1}$$

under the $SL(2,\mathbb{Z})$ transformation 1.6. Equation (B.1) describes a $U(1)$ transformation when $\hat{w} = -w$.

The modular covariant derivative,

$$\mathcal{D}_w = i\left(\frac{\partial}{\partial\tau} - i\frac{w}{2\tau_2}\right), \tag{B.2}$$

maps $F^{(w,\hat{w})}$ into $F^{(w+2,\hat{w})}$ while the anti-holomorphic covariant derivative, $\bar{\mathcal{D}}_{\hat{w}} = \mathcal{D}^*_{\hat{w}}$, maps $F^{(w,\hat{w})}$ into $F^{(w,\hat{w}+2)}$. It is more convenient for our purposes to define the covariant derivatives,

$$D_w = \tau_2 \mathcal{D} = i\left(\tau_2\frac{\partial}{\partial\tau} - i\frac{w}{2}\right), \qquad \bar{D}_{\hat{w}} = \tau_2 \bar{\mathcal{D}} = -i\left(\tau_2\frac{\partial}{\partial\bar{\tau}} + i\frac{\hat{w}}{2}\right) \tag{B.3}$$

which change the $U(1)$ charge of F by two units,

$$D_w F^{(w,\hat{w})} = F^{(w+1,\hat{w}-1)}, \qquad \bar{D}_{\hat{w}} F^{(w,\hat{w})} = F^{(w-1,\hat{w}+1)}. \tag{B.4}$$

The Laplace operator on the fundamental domain of $SL(2,\mathbb{Z})$ is defined to be,

$$\nabla_0^2 \equiv \nabla^2 = 4\tau_2^2 \frac{\partial}{\partial\tau}\frac{\partial}{\partial\bar{\tau}}, \tag{B.5}$$

when acting on $(0,0)$ forms. More generally, we shall be interested in the Laplacian acting on $(w, -w)$ forms. There are two such Laplacians which are defined by,

$$\nabla^2_{(-)w} = 4D_{w-1}\bar{D}_{-w} = 4\tau_2^2\frac{\partial}{\partial\tau}\frac{\partial}{\partial\bar{\tau}} - 2iw\tau_2\left(\frac{\partial}{\partial\tau} + \frac{\partial}{\partial\bar{\tau}}\right) - w(w-1), \tag{B.6}$$

and

$$\begin{aligned}\nabla^2_{(+)w} &= 4\bar{D}_{-w-1}D_w = 4\tau_2^2\frac{\partial}{\partial\tau}\frac{\partial}{\partial\bar{\tau}} - 2iw\tau_2\left(\frac{\partial}{\partial\tau} + \frac{\partial}{\partial\bar{\tau}}\right) - w(w+1), \\ &= \nabla^2_{(-)w} - 2w.\end{aligned} \tag{B.7}$$

Now consider a $(w, -w)$ form that is an eigenfunction of the Laplace operator $\nabla^2_{(-)\,w}$ with eigenvalue σ_w,

$$\nabla^2_{(-)\,w}\, F^{(w,-w)} = 4D_{w-1}\,\bar{D}_{-w}\, F^{(w,-w)} = \sigma_w\, F^{(w,-w)}. \tag{B.8}$$

Applying $\bar{D}_{-w}$ to this equation gives,

$$\nabla^2_{(+)w-1}\, F^{(w-1,-w+1)} = \sigma_w\, F^{(w-1,-w+1)}. \tag{B.9}$$

It is also easy to see that,

$$\begin{aligned}\nabla^2_{(-)\,w-1}\, F^{(w-1,-w+1)} &= 4D_{w-2}\,\bar{D}_{-w+1}\, F^{(w-1,-w+1)},\\ &= (\sigma_w + 2w - 2)\, F^{(w-1,-w+1)}\end{aligned} \tag{B.10}$$

where $F^{(w-1,-w+1)} = \bar{D}_{-w}\, F^{(w,-w)}$. Repeating this for m steps gives

$$\begin{aligned}\nabla^2_{(-)\,w-m}\, F^{(w-m,-w+m)} &= 4D_{w-m-1}\,\bar{D}_{-w+m+1}\, F^{(w-m,-w+m)},\\ &= (\sigma_w + 2mw - m^2 - m)\, F^{(w-m,-w+m)}.\end{aligned} \tag{B.11}$$

Similarly,

$$\nabla^2_{(+)\,w-m}\, F^{(w-m,-w+m)} = (\sigma_w + 2mw - 2w - m^2 + m)\, F^{(w-m,-w+m)}. \tag{B.12}$$

This relation between eigenvalue equations will be useful in analyzing the equations that are satisfied by the modular forms that enter in $S^{(3)}$.

In section 4 it is proved that $f^{(12,-12)}$ satisfies

$$\nabla^2_{(-)\,12} f^{(12,-12)} = \left(-132 + \frac{3}{4}\right) f^{(12,-12)}. \tag{B.13}$$

This is (B.11) with $w = 12$, $m = 0$ and $\sigma_{12} = -132 + 3/4$. It is also argued in section 3 that $f^{(12,-12)}$ is related to $f^{(0,0)}$ by

$$f^{(12,-12)} = D^{12} f^{(0,0)} \equiv D_{11} \cdots D_1 D_0 f^{(0,0)}. \tag{B.14}$$

This equation together with (B.13) imply that $f^{(0,0)}$ satisfies (4.25).

More generally, let us denote a solution of the scalar Laplace equation with eigenvalue $\sigma = s(s-1) > 1/4$ by $E_s(\tau)$ [56],

$$\nabla^2 E_s = s(s-1)\, E_s. \tag{B.15}$$

We can express $E_s(\tau)$ in terms of the nonholomorphic Eisenstein series,

$$E_s(\tau) = \frac{1}{2}\tau_2^s \sum_{(m,n)=1} |m\tau + n|^{-2s}, \tag{B.16}$$

where $(m,n) = 1$ indicates that m and n are coprime. The eigenfunctions $E_s(\tau)$ are singled out by their power law behavior near the boundary of the moduli space, which agrees with the known tree-level and perturbative contributions to the interactions that we are considering. The solution to the Laplace equation (B.15) is unique for a given s assuming that E_s is a modular function. Therefore, if we assume that $f^{(0,0)}$ is well-defined on the fundamental domain of $SL(2,\mathbb{Z})$ the solution is unique and its behaviour near the $\tau_2 \to \infty$ boundary agrees with the known tree-level and one-loop contributions. It follows from (B.14) that $f^{(12,-12)}$ is also determined uniquely by its Laplace equation (B.13). In this case there are no string calculations with which to compare the perturbative terms.qyn

C Some properties of Type IIB Supergravity

C.1 Spinors and gamma matrices

The spinors that enter into the IIB theory are complex Weyl spinors. The gravitino ψ_μ and dilatino λ have opposite chiralities and the supersymmetry parameter ϵ has the same chirality as the gravitino. The complex conjugate of the product of a pair of spinors is defined by

$$(\lambda_a\, \rho_b)^* = -\lambda_a^*\, \rho_b^*. \tag{C.1}$$

The conjugate of any spinor is defined by, $\bar\lambda = \lambda^* \gamma^0$. We will choose our metric to be space-like and the γ matrices to be real and satisfy the Clifford algebra,

$$\{\gamma^\mu, \gamma^\nu\} = 2\eta^{\mu\nu}. \tag{C.2}$$

Noting that,

$$\gamma^0\gamma^\mu = -(\gamma^\mu)^T\gamma^0, \tag{C.3}$$

it follows that two complex chiral spinors of the same chirality, λ_1 and λ_2, satisfy the relations,

$$\begin{aligned} \bar\lambda_1\gamma^\mu\lambda_2 &= -\bar\lambda_2^*\gamma^\mu\lambda_1^*, \\ \bar\lambda_1\gamma^{\mu\nu\rho}\lambda_2 &= \bar\lambda_2^*\gamma^{\mu\nu\rho}\lambda_1^*, \\ \bar\lambda_1\gamma^{\rho_1\cdots\rho_5}\lambda_2 &= -\bar\lambda_2^*\gamma^{\rho_1\cdots\rho_5}\lambda_1^*, \end{aligned} \tag{C.4}$$

while two chiral spinors of opposite chiralities, λ and ϵ, satisfy,

$$\begin{aligned} \bar\lambda\,\epsilon &= \bar\epsilon^*\,\lambda^*, \\ \bar\lambda\gamma^{\rho_1\rho_2}\epsilon &= -\bar\epsilon^*\gamma^{\rho_1\rho_2}\lambda^*, \\ \bar\lambda\gamma^{\rho_1\rho_2\rho_3\rho_4}\epsilon &= \bar\epsilon^*\gamma^{\rho_1\rho_2\rho_3\rho_4}\lambda^*. \end{aligned} \tag{C.5}$$

The Fierz identity for ten-dimensional complex Weyl spinors can be expressed as,

$$\lambda_1^a \bar{\lambda}_2^b = -\frac{1}{16}\bar{\lambda}_2\gamma_\mu\lambda_1\,\gamma^\mu_{ab} + \frac{1}{96}\bar{\lambda}_2\gamma_{\mu\nu\rho}\lambda_1\,\gamma^{\mu\nu\rho}_{ab} - \frac{1}{3840}\bar{\lambda}_2\gamma_{\rho_1\cdots\rho_5}\lambda_1\,\gamma^{\rho_1\cdots\rho_5}_{ab}, \quad (C.6)$$

where λ_1 and λ_2 are two chiral spinors of the same chirality.

An additional useful identity is,

$$\gamma^{\rho_1\ldots\rho_5}\lambda_1\,\bar{\lambda}_2\gamma_{\rho_1\ldots\rho_5}\lambda_3 = 0, \quad (C.7)$$

where λ_1, λ_2 and λ_3 are three chiral spinors of the same chirality.

Some gamma matrix identities that are useful in proving the various relationships in the text are,

$$\begin{aligned}\mathrm{tr}(\gamma_{\mu\nu\rho}\gamma^{\rho_1\rho_2\rho_3}) = -16\,\big(&\delta^{\rho_1}_\mu\delta^{\rho_2}_\nu\delta^{\rho_3}_\rho - \delta^{\rho_2}_\mu\delta^{\rho_1}_\nu\delta^{\rho_3}_\rho + \delta^{\rho_2}_\mu\delta^{\rho_3}_\nu\delta^{\rho_1}_\rho \\ &-\delta^{\rho_3}_\mu\delta^{\rho_2}_\nu\delta^{\rho_1}_\rho + \delta^{\rho_3}_\mu\delta^{\rho_1}_\nu\delta^{\rho_2}_\rho - \delta^{\rho_1}_\mu\delta^{\rho_3}_\nu\delta^{\rho_2}_\rho\big)\,.\end{aligned} \quad (C.8)$$

$$\begin{aligned}\gamma^\mu\,\gamma_\sigma\,\gamma_\mu &= -8\gamma_\sigma\,,\\ \gamma^\mu\,\gamma_{\sigma_1\sigma_2\sigma_3}\,\gamma_\mu &= -4\gamma_{\sigma_1\sigma_2\sigma_3}\,,\\ \gamma^\mu\gamma_{\sigma_1\ldots\sigma_5}\gamma_\mu &= 0,\\ \gamma^{\mu\nu\rho}\,\gamma_\sigma\,\gamma_{\mu\nu\rho} &= -288\gamma_\sigma\,,\\ \gamma^{\mu\nu\rho}\,\gamma_{\sigma_1\sigma_2\sigma_3}\,\gamma_{\mu\nu\rho} &= -48\gamma_{\sigma_1\sigma_2\sigma_3}\,,\\ \gamma^{\mu\nu\rho}\,\gamma_{\sigma_1\ldots\sigma_5}\,\gamma_{\mu\nu\rho} &= -14\gamma_{\sigma_1\ldots\sigma_5}\,.\end{aligned} \quad (C.9)$$

In section 4 the first two terms in parentheses on the right-hand-side of (3.10) are considered in detail. The precise notation used in those terms uses the following definitions

$$(\lambda^r)_{a_{r+1}\cdots a_{16}} \equiv \frac{1}{r!}\epsilon_{a_1\cdots a_{16}}\lambda^{a_1}\ldots\lambda^{a_r}, \quad (C.10)$$

so that,

$$\lambda^{16} = \frac{1}{16!}\epsilon_{a_1\ldots a_{16}}\lambda^{a_1}\ldots\lambda^{a_{16}}, \quad (C.11)$$

and

$$\begin{aligned}\hat{G}\lambda^{14} &\equiv \hat{G}_{\mu\nu\rho}(\gamma^{\mu\nu\rho}\gamma^0)_{a_{15}a_{16}}(\lambda^{14})_{a_{15}a_{16}}\,,\\ &= \frac{1}{14!}\hat{G}_{\mu\nu\rho}(\gamma^{\mu\nu\rho}\gamma^0)_{a_{15}a_{16}}\,\epsilon_{a_1\ldots a_{16}}\lambda^{a_1}\ldots\lambda^{a_{14}}.\end{aligned} \quad (C.12)$$

The following identities follow very simply from (C.10),

$$(\lambda^{14})_{ab}\,\lambda^c = (\lambda^{15})_b\,\delta^c_a - (\lambda^{15})_a\,\delta^c_b,$$
$$(\lambda^{15})_a\,\lambda^b = \delta^b_a\,\lambda^{16},$$
$$(\lambda)^{15}_a\,\lambda^a = 16\lambda^{16}, \tag{C.13}$$

and

$$(\lambda^{14})_{ab}\,\lambda_c\lambda_d = \lambda^{16}\,(\delta_{ac}\delta_{bd} - \delta_{ad}\delta_{bc}). \tag{C.14}$$

C.2 The fields and their supersymmetry transformations

Here we will review various features of type IIB supergravity that are useful in the body of the paper. Most of this material can be found in [57,58] in a form that is adapted to the field definitions in which the global symmetry is $SU(1,1)$ and the scalar fields parameterize the coset space $SU(1,1)/U(1)$, which is the Poincaré disk. It is simple to transform this to our parameterization in which the global symmetry is $SL(2,\mathbb{R})$ and the scalars parameterize the coset space $SL(2,\mathbb{R})/U(1)$, or the upper half plane.

The theory is then defined in terms of the following fields: the scalar fields can be parameterized by the frame field,

$$V \equiv \begin{pmatrix} V^1_- & V^1_+ \\ V^2_- & V^2_+ \end{pmatrix} = \frac{1}{\sqrt{-2i\tau_2}} \begin{pmatrix} \bar\tau e^{-i\phi} & \tau e^{i\phi} \\ e^{-i\phi} & e^{i\phi} \end{pmatrix}, \tag{C.15}$$

where $V^\alpha_\pm$ ($\alpha = 1,2$) is a $SL(2,\mathbb{Z})$ matrix that transforms from the left by the global $SL(2,\mathbb{R})$ and from the right by the local $U(1)$. Note that we are using a complex basis for convenience. A general transformation is then written as,

$$(V^\alpha_+, V^\alpha_-) \to U^\alpha_{\ \beta}\left(V^\beta_+ e^{i\Sigma}, V^\beta_- e^{-i\Sigma}\right), \tag{C.16}$$

where U is a $SL(2,\mathbb{R})$ matrix and Σ is the $U(1)$ phase. An appropriate choice of Σ fixes the gauge and eliminates the scalar field ϕ. We will make the gauge choice $\phi = 0$. Since this gauge is not maintained by generic symmetry transformations, it is necessary to compensate a symmetry transformation with an appropriate local $U(1)$ tranformation to maintain the gauge. In particular, the local supersymmetry transformations require compensating local $U(1)$ transformations. The supersymmetry and $U(1)$ transformations of V^α_- are given by,

$$\delta^{(0)} V^\alpha_- = i V^\alpha_+ \bar\epsilon \lambda^* - i\Sigma V^\alpha_-. \tag{C.17}$$

This choice ensures that the gauge $\phi = 0$ is maintained if a local supersymmetry transformation is accompanied by a $U(1)$ transformation with parameter,

$$\Sigma = \frac{1}{2}(\bar{\epsilon}\lambda^* - \bar{\epsilon}^*\lambda). \tag{C.18}$$

The $SL(2,\mathbb{R})$ singlet expression,

$$Q_\mu = -i\epsilon_{\alpha\beta}\, V_+^\alpha \partial_\mu V_-^\beta, \tag{C.19}$$

is the composite $U(1)$ connection and transforms as $Q \to Q + \partial_\mu \Sigma$ under infinitesimal local $U(1)$ transformations, while the $SL(2,\mathbb{R})$ singlet expression

$$P_\mu = -\epsilon_{\alpha\beta} V_+^\alpha \partial_\mu V_-^\beta, \tag{C.20}$$

transforms with $U(1)$ charge $u_P = 2$. In the gauge $\phi = 0$, the expression for P_μ takes the simple form,

$$P_\mu = \frac{i}{2}\frac{\partial_\mu \tau}{\tau_2}, \tag{C.21}$$

while

$$Q_\mu = -i\epsilon_{ab} V_+^a \partial_\mu V_-^b = -\frac{1}{2}\frac{\partial_\mu \rho_1}{\rho_2}. \tag{C.22}$$

The fermions comprise the complex chiral gravitino, ψ_μ^a, which has $U(1)$ charge $u_\psi = 1/2$, and the dilatino, λ^a, with $U(1)$ charge $u_\lambda = 3/2$. These two fields have opposite chiralities. The graviton is a $U(1)$ and $SL(2,\mathbb{R})$ singlet as is the antisymmetric fourth-rank potential, $C^{(4)}$, which has a field strength $F_5 = dC^{(4)}$. As is well known, this field strength has an equation of motion that is expressed by the self-duality condition $F_5 = *F_5$, which cannot be obtained from a globally well-defined Lagrangian. For this reason, our considerations are restricted to statements concerning the on-shell properties of the theory where the fields satisfy the equations of motion.

The two antisymmetric second-rank potentials, $B_{\mu\nu}$ and $C_{\mu\nu}^{(2)}$, have field strengths F^1 ($NS \otimes NS$) and F^2 ($R \otimes R$) that form an $SL(2,\mathbb{R})$ doublet, F^α. It is very natural to package them into the $SL(2,\mathbb{R})$ singlet fields,

$$G = -\epsilon_{\alpha\beta} V_+^\alpha F^\beta, \qquad G^* = -\epsilon_{\alpha\beta} V_-^\alpha F^\beta, \tag{C.23}$$

which carry $U(1)$ charges $u_G = +1$ and $u_{G^*} = -1$, respectively.

In a fixed $U(1)$ gauge, a global $SL(2,\mathbb{R})$ transformation which acts on τ by

$$\tau \to \frac{a\tau + b}{c\tau + d}, \tag{C.24}$$

with $ad - bc = 1$, induces a $U(1)$ transformation on the fields that depends on their charge. Thus, a field Φ with $U(1)$ charge u_Φ transforms as,

$$\Phi \to \Phi \left(\frac{c\bar{\tau} + d}{c\tau + d}\right)^{u_\Phi/2} . \tag{C.25}$$

The higher derivative terms of interest to us only repect the $SL(2, \mathbb{Z})$ subgroup of $SL(2, \mathbb{R})$ for which a, b, c, d are integers and the continuous $U(1)$ symmetry is broken.

The supersymmetry of the action is naturally described in terms of combinations of bosonic fields and fermion bilinears which are 'supercovariant', which means that they do not contain derivatives of the supersymmetry parameter ϵ in their transformations. These combinations are,

$$\begin{aligned}\hat{G}_{\mu\nu\rho} &= G_{\mu\nu\rho} - 3\bar{\psi}_{[\mu}\gamma_{\nu\rho]}\lambda - 6i\bar{\psi}^*_{[\mu}\gamma_\nu\psi_{\rho]},\\ \hat{P}_\mu &= P_\mu - \bar{\psi}^*_\mu\lambda,\\ \hat{F}_{5\,\mu_1\ldots,\mu_5} &= F_{5\,\mu_1\ldots,\mu_5} - 5\bar{\psi}_{[\mu_1}\gamma_{\mu_2\mu_3\mu_4}\psi_{\mu_5]} - \frac{1}{16}\bar{\lambda}\gamma_{\mu_1\ldots\mu_5}\lambda.\end{aligned} \tag{C.26}$$

We will now present the lowest-order supersymmetry transformations, suitably adapted from those given in [58] to the $SL(2, \mathbb{R})$ parameterization. From (C.17) and (C.18), it follows that

$$\delta^{(0)}\tau = 2\tau_2\bar{\epsilon}^*\lambda, \qquad \delta^{(0)}\bar{\tau} = -2\tau_2\bar{\epsilon}\lambda^*. \tag{C.27}$$

It follows from the definition of Q_μ and the transformations of τ and $\bar{\tau}$ that

$$\delta^{(0)}Q_\mu = -\bar{\epsilon}\lambda^*\, P_\mu \,+\, c.c. \tag{C.28}$$

Also, the supersymmetry transformation of the zehnbein is given by,

$$\delta^{(0)}e^m_\mu = i(\bar{\epsilon}\gamma^m\psi_\mu + \bar{\epsilon}^*\gamma^m\psi^*_\mu). \tag{C.29}$$

The transformation of the dilatino is given, in the fixed $U(1)$ gauge, by

$$\begin{aligned}\delta^{(0)}\lambda &= i\gamma^\mu\epsilon^*\,\hat{P}_\mu - \frac{1}{24}i\gamma^{\mu\nu\rho}\epsilon\hat{G}_{\mu\nu\rho} + \delta^{(0)}_\Sigma\lambda\\ &= i\gamma^\mu\epsilon^*\,\hat{P}_\mu + \frac{i}{8}\gamma^{\mu\nu\tau}\epsilon\,(\bar{\psi}_{[\mu}\gamma_{\nu\tau]}\lambda) - i\gamma^\mu\epsilon^*\,(\bar{\psi}^*_\mu\lambda) + \delta^{(0)}_\Sigma\lambda + \ldots\end{aligned} \tag{C.30}$$

where we have only kept the terms that are needed in the body of this paper in the second line. The δ_Σ arises from the compensating $U(1)$ gauge transformation,

$$\delta^{(0)}_\Sigma\lambda_a = \frac{3}{2}i\Sigma\,\lambda_a = \frac{3}{4}i\lambda_a(\bar{\epsilon}\lambda^*) - \frac{3}{4}i\lambda_a(\bar{\epsilon}^*\lambda). \tag{C.31}$$

The gravitino transformation is given by,

$$\begin{aligned}\delta^{(0)}\psi_\mu = D_\mu\epsilon &+ \frac{1}{480}i\gamma^{\rho_1\dots\rho_5}\gamma_\mu\epsilon\hat{F}_{\rho_1\dots\rho_5} + \frac{1}{96}\left(\gamma_\mu{}^{\nu\rho\lambda}\hat{G}_{\nu\rho\lambda} - 9\gamma^{\rho\lambda}\hat{G}_{\mu\rho\lambda}\right)\epsilon^* \\ &-\frac{7}{16}\left(\gamma_\rho\lambda\,\bar{\psi}_\mu\gamma^\rho\epsilon^* - \frac{1}{1680}\gamma_{\rho_1\dots\rho_5}\lambda\,\bar{\psi}_\mu\gamma^{\rho_1\dots\rho_5}\epsilon^*\right) \\ &+\frac{1}{32}i\left[\left(\frac{9}{4}\gamma_\mu\gamma^\rho + 3\gamma^\rho\gamma_\mu\right)\epsilon\,\bar{\lambda}\gamma_\rho\lambda\right. \\ &\left.-\left(\frac{1}{24}\gamma_\mu\gamma^{\rho_1\rho_2\rho_3} + \frac{1}{6}\gamma^{\rho_1\rho_2\rho_3}\gamma_\mu\right)\epsilon\,\bar{\lambda}\gamma_{\rho_1\rho_2\rho_3}\lambda + \frac{1}{960}\gamma_\mu\gamma^{\rho_1\dots\rho_5}\epsilon\,\bar{\lambda}\gamma_{\rho_1\dots\rho_5}\lambda\right] \\ &+\delta^{(0)}_\Sigma(\psi_\mu),\end{aligned} \tag{C.32}$$

where the compensating $U(1)$ transformation is given by

$$\delta^{(0)}_\Sigma\psi_\mu = \frac{1}{2}i\Sigma = \frac{1}{4}i\psi_\mu(\bar{\epsilon}\lambda^*) - \frac{1}{4}i\psi_\mu(\bar{\epsilon}^*\lambda). \tag{C.33}$$

By using (C.6) and (C.32) extensively we may manipulate the variation of $\gamma^\mu\psi^*_\mu$ into the form,

$$\delta^{(0)}(\gamma^\mu\psi^*_\mu)_a = -\frac{3}{4}i\lambda^*_a(\bar{\epsilon}\lambda) + \frac{1}{1920}i(\gamma^{\rho_1\dots\rho_5}\epsilon^*)_a\left(\bar{\lambda}\gamma_{\rho_1\dots\rho_5}\lambda\right) + \dots, \tag{C.34}$$

where we have only kept the terms bilinear in λ, λ^*. This implies the relation,

$$(\lambda)^{15}_a\,\delta^{(0)}(\gamma^\mu\psi^*_\mu)_a = -15i\lambda^{16}(\bar{\lambda}\epsilon^*) + \dots, \tag{C.35}$$

which we use in the body of the text.

D Fermions in type IIB on S^1 and M theory on T^2

The massless fermions of the type II string theories on S^1 are identified with particular projections of the eleven-dimensional gravitino compactified on T^2. In order to calculate the λ^{16} loop amplitude we need to identify the particular projection that isolates the spin-half dilatino of the type IIB theory.

Compactification of the IIB theory on a circle S^1 of circumference r_B (in string frame) in the direction x^9 breaks the $SO(9,1)$ Lorentz symmetry to $SO(8,1)$. The complex chiral spin-$\frac{1}{2}$ fermion λ simply becomes a complex spinor of $SO(8,1)$. The gravitino decomposes into a nine-dimensional gravitino $\hat{\psi}_\alpha$ (where the $SO(8,1)$ vector index $\alpha = 0, 1, \cdots, 8$) together with a second complex spin-$\frac{1}{2}$ fermion,

$$\chi^A = r_B\psi^A_9, \tag{D.1}$$

where the factor of r_B comes from the component e_{99} of the zehnbein. The nine-dimensional gravitino is defined by shifting ψ_a,

$$\hat{\psi}_a = \psi_a + \frac{1}{7}\Gamma_a\Gamma^9\chi \tag{D.2}$$

(where a labels the nine-dimensional tangent space, $a = 0, 1 \cdots, 8$) so that the kinetic term is diagonal.

We will now identify the components of the T^2 compactification of eleven-dimensional gravitino, $\Psi_{\hat{\mu}}$ ($\hat{\mu} = 0, \cdots, 9, 11$), that correspond to the IIB fermions on S^1. Compactification on T^2 breaks the local Lorentz symmetry from $SO(10,1)$ to $SO(8,1) \times SO(2)$. The nine-dimensional fermions can then be organized into eigenstates of $SO(2) \equiv U(1)$, which is related to the $U(1)$ in the denominator of the coset space of the IIB theory. The world indices split into the compact directions $\sigma = 9, 11$ and the noncompact directions $\alpha = 0, \cdots, 8$. Making a block diagonal ansatz, the elfbein $e_{\hat{\mu}}^{\ \hat{m}}$, is written as

$$e_{\hat{\mu}}^{\ \hat{m}} = \begin{pmatrix} e_\sigma^{\ s} & 0 \\ 0 & e_\alpha^{\ a} \end{pmatrix} \tag{D.3}$$

where a again labels the nine-dimensional tangent space and s the two-dimensional tangent space ($s = 1, 2$). The zweibein, $e_\sigma^{\ s}$, of T^2 may be chosen, in a special Lorentz frame, to be

$$e_\sigma^{\ s} = \sqrt{\frac{\mathcal{V}}{\Omega_2}} \begin{pmatrix} \Omega_2 & \Omega_1 \\ 0 & 1 \end{pmatrix} \tag{D.4}$$

Symplectic reparametrizations of the torus act as $SL(2, \mathbb{R})$ matrices from the left and local Lorentz transformations act as $SO(2)$ transformations from the right. The condition that the zweibein remains in the frame (D.4) leads to the standard $SL(2, \mathbb{Z})$ transformation of the complex structure of the torus (which is (C.24) with τ replaced by Ω) and induces a specific Ω-dependent $U(1)$ transformation on the fermions.

The spin-$\frac{1}{2}$ components of the compactified gravitino are written in a complex basis $z = x_{11} + ix_9$, $\bar{z} = x_{11} - ix_9$, as $\Psi_z = \Psi_{11} + i\Psi_9$ and $\Psi_{\bar{z}} = \Psi_{11} - i\Psi_9$. To relate these 32-component spinors to λ and χ of the IIB theory we shall first organize them into eigenstates of the $U(1)$ rotations of the compact T^2 generated by $i\Gamma^{11}\Gamma^9/2 + j_{vec}$, where j_{vec} acts on the vector index so that $j_{vec}\Psi_z = \Psi_z$ and $j_{vec}\Psi_{\bar{z}} = -\Psi_{\bar{z}}$. Those components of the spin connection that have tangent-space indices in the compact directions are given by

$$\omega_{z,\alpha z} = \frac{i}{2}\frac{\partial_\alpha \Omega}{\Omega_2}, \qquad \omega_{\bar{z},\alpha\bar{z}} = \frac{i}{2}\frac{\partial_\alpha \bar{\Omega}}{\Omega_2}, \qquad \omega_{z,\alpha\bar{z}} = \omega_{\bar{z},\alpha z} = \frac{1}{\mathcal{V}}\partial_\alpha \mathcal{V}. \tag{D.5}$$

The type IIB complex scalar field is identified with Ω so the quantities $P_\alpha = \omega_{z,\alpha z}$ and $\bar{P}_\alpha = \omega_{\bar{z},\alpha\bar{z}}$ correspond to the fields with the same symbol defined in appendix C. The third real scalar field in the nine-dimensional theory is $\omega_{z,\alpha\bar{z}} = -4\partial_\alpha \ln r_B^E/3$.

We shall make use of projectors made out of the eleven-dimensional Γ matrices. The complex combinations, $\Gamma_z = \frac{1}{2}(\Gamma^{11}+i\Gamma^9)$ and $\Gamma_{\bar{z}} = \frac{1}{2}(\Gamma^{11}-i\Gamma^9)$, obey

$$\Gamma_z^2 = \Gamma_{\bar{z}}^2 = 0, \qquad \{\Gamma_z, \Gamma_{\bar{z}}\} = 1. \tag{D.6}$$

so that the products

$$\mathcal{P}_z = (\mathcal{P}_{\bar{z}})^T = \Gamma_{\bar{z}}\Gamma_z = \frac{1}{2}(1 - i\Gamma^9\Gamma^{11}), \qquad \mathcal{P}_{\bar{z}} = (\mathcal{P}_z)^T = \Gamma_z\Gamma_{\bar{z}} = \frac{1}{2}(1 + i\Gamma^9\Gamma^{11}) \tag{D.7}$$

(where the superscript T indicates the transpose) are projectors, satisfying

$$\mathcal{P}_z^2 = \mathcal{P}_z, \qquad \mathcal{P}_{\bar{z}}^2 = \mathcal{P}_{\bar{z}}, \qquad \mathcal{P}_z\mathcal{P}_{\bar{z}} = 0. \tag{D.8}$$

The action of these projectors on a 32-component real spinor is to replace it with a complex sixteen component spinor which is an eigenstate of the $U(1)$ spin, $i\Gamma^{11}\Gamma^9/2$. Therefore, we may identify the four spin-1/2 fermions of the nine-dimensional compactified theory with the components,

$$\mathcal{P}_z\Psi_z, \qquad \mathcal{P}_{\bar{z}}\Psi_{\bar{z}}, \tag{D.9}$$

$$\mathcal{P}_z\Psi_{\bar{z}}, \qquad \mathcal{P}_{\bar{z}}\Psi_z. \tag{D.10}$$

The components (D.9) have $U(1)$ charges $+\frac{3}{2}$ and $-\frac{3}{2}$, respectively, while the components (D.10) are a mixture of the states with $U(1)$ charges $\pm\frac{1}{2}$. These fields are simply related to the spin-$\frac{1}{2}$ fermions of the IIB theory by converting λ and χ (and their complex conjugates) from eigenstates of Γ^{11} to eigenstates of $i\Gamma^{11}\Gamma^9$. This leads to the identifications,

$$\mathcal{P}_z\Psi_z = \Gamma_{\bar{z}}\lambda, \qquad \mathcal{P}_{\bar{z}}\Psi_z = \Gamma_z\chi \tag{D.11}$$

which gives the eigenstates of Γ^{11},

$$\lambda = (1 + \Gamma_z)\mathcal{P}_z\Psi_z, \qquad \chi = (1 + \Gamma_{\bar{z}})\mathcal{P}_{\bar{z}}\Psi_z. \tag{D.12}$$

In terms of the real components

$$\lambda_1 = \psi_{11}^+ - \Gamma^9\psi_9^-, \qquad \lambda_2 = \psi_9^+ + \Gamma^9\psi_{11}^-. \tag{D.13}$$

where the superscripts $^{\pm}$ indicate the chirality (the value of Γ^{11}). Similarly, writing $\chi = \chi_1 + i\chi_2$ and multiplying (D.11) by $(1+\Gamma^{11})$ leads to

$$\chi_1 = \Gamma^9 \psi_9^+ - \psi_{11}^-, \qquad \chi_2 = -\psi_9^- - \Gamma^9 \psi_{11}^+. \tag{D.14}$$

The remaining components, Ψ_a form the nine-dimensional gravitino after a shift similar to (D.2),

$$\hat{\Psi}_a = \Psi_a + \frac{1}{7}\Gamma_a\left(\Gamma_z \Psi_{\bar{z}} + \Gamma_{\bar{z}} \Psi_z\right) = \Psi_a + \frac{i}{7}\Gamma_a \Gamma^9 (\mathcal{P}_z \chi - \mathcal{P}_{\bar{z}} \chi^*). \tag{D.15}$$

The relations between the M-theory fields and the IIB fields can be confirmed by comparing the way they behave under supersymmetry transformations.

For completeness the IIB fermions can be related to the IIA fermions in a straightforward manner. The spin-1/2 fermions in the IIA theory follow by direct dimensional reduction of eleven-dimensional supergravity [100]. The nine-dimensional fermions that arise from the spin-$\frac{1}{2}$ fields in ten dimensions are

$$\lambda_A = \psi_{11} + \Gamma^9 \Gamma^{11} \psi_9, \tag{D.16}$$

which decomposes into the chiral components,

$$\lambda_A^+ = \psi_{11}^+ - \Gamma^9 \psi_9^-, \qquad \lambda_A^- = \psi_{11}^- + \Gamma^9 \psi_9^+. \tag{D.17}$$

Similarly,

$$\chi_A = \psi_9 + \Gamma^9 \Gamma^{11} \psi_{11}, \tag{D.18}$$

which has chiral components

$$\chi_A^- = \psi_9^- + \Gamma^9 \psi_{11}^+, \qquad \chi_A^+ = \psi_9^+ - \Gamma^9 \psi_{11}^-. \tag{D.19}$$

Comparing (D.13) with (D.17) and (D.14) with (D.19) gives the identification of fields of IIA and IIB in nine dimensions,

$$\lambda_1 = \lambda_A^+, \qquad \lambda_2 = \Gamma^9 \lambda_A^-, \qquad \chi_2 = -\chi_A^-, \qquad \chi_1 = \Gamma^9 \chi_A^+, \tag{D.20}$$

which are in agreement with the world sheet T-duality rules.

References

1. E. Witten, *String theory dynamics in various dimensions*, hep-th/9503124; Nucl. Phys. **B443** (1995) 85.
2. P. Aspinwall, *Some Relationships Between Dualities in String Theory*, in Proceedings of 'S-duality and mirror symmetry', Trieste 1995, hep-th/9508154; Nucl. Phys. Proc. **46** (1996) 30.
3. J.H. Schwarz, *Lectures on Superstring and M-theory dualities*, hep-th/9607201; J.H. Schwarz, *An $SL(2,\mathbb{Z})$ multiplet of type IIb superstrings*, hep-th/9508143; Phys. Lett. **360B** (1995) 13.
4. E. Cremmer, B. Julia and J. Scherk, *Supergravity theory in eleven dimensions*, Phys. Lett. **76B** (1978) 409.
5. C.M. Hull and P.K. Townsend, *Unity of superstring dualities*, hep-th/9410167; Nucl. Phys. **B438** (1995) 109; P.K. Townsend, *The eleven-dimensional supermembrane revisited*, hep-th/9501068; Phys. Lett. **B350** (1995) 184.
6. G. Dall'Agata G.K. Lechner, Dmitri Sorokin, *Covariant actions for the bosonic sector of d = 10 IIB supergravity*, hep-th/9707044; Class.Quant.Grav. **14** (1997) L195.
7. G. Dall'Agatai, K. Lechner, M. Tonin, *Action for IIB supergravity in 10-dimensions*, 2nd Conference on Quantum Aspects of Gauge Theories, Supersymmetry and Unification, Corfu, Greece, 21-26 Sep 1998; hep-th/9812170; *D=10, N = IIB supergravity: Lorentz invariant actions and duality*, hep-th/9806140; JHEP **9807** (1998).
8. C. Vafa and E. Witten, *A one-loop test of string-string duality*, hep-th/9505053; Nucl. Phys. **B447** (1995) 261.
9. C.G. Callan and J.A. Harvey, *Anomalies and fermion zero modes on strings and domain walls*, Nucl. Phys. **B250** (1985) 427.
10. M.J. Duff, J.T. Liu and R. Minasian, *Eleven-dimensional origin of string-string duality: a one loop test*, hep-th/9506126; Nucl. Phys. **452B** (1995)261.
11. M.B. Green and M. Gutperle, *Effects of D-instantons*, hep-th/9701093; Nucl. Phys. **B498** (1997)195.
12. M.B. Green and P. Vanhove, *D-instantons, Strings and M-theory*, hep-th/9704145; Phys. Lett. **B408** (1997) 122.
13. P.S. Howe and P.C. West, *The complete $N = 2$, $d = 10$ supergravity*, Nucl. Phys. **B238** (1984) 181.
14. M.B. Green, M. Gutperle and H. Kwon, *Sixteen Fermion and Related Terms in M theory on T^2*, hep-th/9710151; Phys. Lett. **B421** (1998) 149.

15. M.B. Green and S. Sethi, *Supersymmetry constraints on type IIB supergravity*, hep-th/9808061.
16. M.B. Green, M. Gutperle and P. Vanhove, *One Loop in Eleven-Dimensions*, hep-th/9706175; Phys. Lett. **409B** (1997) 177.
17. E. Kiritsis and B. Pioline, *On R^4 threshold corrections in IIB string theory and (p,q) string instantons*, hep-th/9707018; Nucl. Phys. **B508** (1997)509.
18. E. Kiritsis and B. Pioline, *U-duality and D-brane Combinatorics*, hep-th/9710078; Phys. Lett. **418B**(1998)61.
19. C. Bachas, C. Fabre, E. Kiritsis, N.A. Obers and P. Vanhove, *Heterotic / type I duality and D-brane instantons*, hep-th/9707126; Nucl. Phys. **B509**(1998)33.
20. C. Bachas, *Heterotic versus Type I*, Talk at Strings'97 (Amsterdam, 1997), hep-th/9710102.
21. E. Kiritsis and N.A. Obers, *Heterotic/Type-I Duality in D<10 Dimensions, Threshold Corrections and D-Instantons*, hep-th/9709058; JHEP **9710** (1997) 004.
22. J. Russo and A.A. Tseytlin, *One-loop four-graviton amplitude in eleven-dimensional supergravity*, hep-th/970713; Nucl. Phys. **B508**(1997)245.
23. J.G. Russo, *An ansatz for a non-perturbative four-graviton amplitude in type IIB superstring theory*, hep-th/9707241; Phys. Lett. **B417** (1998) 253.
24. A. Kehagias and H. Partouche, *D-Instanton Corrections as (p,q)-String Effects and Non-Renormalization Theorems*, hep-th/9712164; Int. J. Mod. Phys. **A13** (1998) 5075.
25. A. Kehagias and H. Partouche, *The exact quartic effective action for the type IIB superstring*, hep-th/9710023; Phys. Lett. **B422** (1998) 109.
26. N. Berkovits and C. Vafa, *Type IIB R^4H^{4g-4} Conjectures*, hep-th/9803145; Nucl. Phys. **B533** (1998).
27. I. Antoniadis, B. Pioline and T.R. Taylor, *Calculable $e^{-1/\lambda}$ effects*, hep-th/9707222; Nucl. Phys. **B512**(1998)61.
28. N. Berkovits, *Construction of R^4 terms in N=2 D=8 superspace*, hep-th/9709116; Nucl. Phys. **B514** (1998) 191.
29. B. Pioline, *A note on non-perturbative R^4 couplings*, hep-th/9804023; Phys. Lett. **B431** (1998) 73.
30. E. Witten, *Bound States of Strings and p-branes*, hep-th/9510135; Nucl. Phys. **B460** (1996) 335.
31. M.B. Green and M. Gutperle, *D-particle bound states and the D-instanton measure*, hep-th/9711107; JHEP **9801** (1998) 005.
32. P. Yi, *Witten index and threshold bound states of D-branes*, hep-

th/9704098; Nucl. Phys. **B505** (1997) 307.
33. S. Sethi and M. Stern, *D-brane bound states redux*, hep-th/9705046; Comm. Math. Phys. **194** (1998) 675.
34. J. Maldacena, *The large N limit of superconformal field theories and supergravity*, hep-th/9711200; Adv. Theor. Math. Phys. **2** (1998) 231.
35. S.S. Gubser, I.R. Klebanov and A.M. Polyakov, *Gauge theory correlators from non-critical string theory*, hep-th/9802109; Phys. Lett.**B428** (1998) 105.
36. E. Witten, *Anti de Sitter space and holography*, hep-th/9802150; Adv. Theor. Math. Phys. **2** (1998) 253.
37. D.I. Olive and C. Montonen, *Magnetic monopoles as gauge particles?*, Phys. Lett. **72B** (1977) 117.
38. A. Sen, *Dyon - monopole bound states, selfdual harmonic forms on the multi - monopole moduli space, and SL(2,Z) invariance in string theory*,hep-th/9402032;
39. N. Dorey, T.J. Hollowood, V.V. Khoze and M.P. Mattis, *Multiinstantons and Maldacena's conjecture*, hep-th/9810243.
40. N. Dorey, T.J. Hollowood, V.V. Khoze, M.P. Mattis and S. Vandoren, *Yang-Mills instantons in the large N limit and the AdS / CFT correspondence*, hep-th/9808157. Phys. Lett. **B329** (1994) 217.
41. J. Polchinski, *Tasi lectures on D-branes*, hep-th/9611050.
42. N. Marcus and J.H. Schwarz, *Field theories that have no manifestly Lorentz invariant formulation*, Phys. Lett. **115B** (1982) 111.
43. E. Witten, *Five-brane effective action in M theory*, hep-th/9610234; J. Geom. Phys. **22** (1997) 103.
44. M.B. Green and J.H. Schwarz, *Supersymmetrical string theories*, Phys. Lett. **109B** (1982) 444.
45. M.B. Green and J.H. Schwarz, *Supersymmetric dual string theory (III). Loops and renormalization*, Nucl. Phys. **198B** (1982) 441.
46. D.J. Gross and E. Witten, *Superstring modifications of Einstein's equations*, Nucl. Phys. **B277** (1986) 1.
47. M.T. Grisaru, A.E.M Van de Ven and D. Zanon, *Two-dimensional supersymmetric sigma models on Ricci flat Kähler manifolds are not finite*, Nucl. Phys. **B277** (1986) 388; *Four loop divergences for the N=1 supersymmetric nonlinear sigma model in two-dimensions*, Nucl. Phys. **B277** (1986) 409.
48. I. Antoniadis, S. Ferrara, R. Minasian and K.S. Narain, R^4 *couplings in M and type II theories on Calabi-Yau spaces*, hep-th/9707013; Nucl. Phys. **B507** (1997) 571.
49. A. Strominger, *Loop corrections to the universal hypermultiplet*, hep-

th/9706195; Phys. Lett. **B421** (1998) 139.
50. S.S. Gubser, I.R. Klebanov and A.A. Tseytlin, *String theory and classical absorption by three-branes*, hep-th/9703040.
51. Pawelczyk and S. Theisen, *$AdS_5 \times S^5$ black hole metric at $O(\alpha'^3)$*, hep-th/9808126; JHEP **9809** (1998) 010.
52. G.W. Gibbons, M.B. Green and M.V. Perry, *Instantons and seven-branes in type IIB superstring theory*, Phys. Lett. **B370** (1996) 37, hep-th/9511080.
53. J. Polchinski, *Combinatorics of boundaries in string theory*, hep-th/9407031; Phys. Rev. **D50** (1994) 6041.
54. M.B. Green *A Gas of D instantons*, hep-th/9504108; Phys. Lett. **B354** (1995) 271.
55. M.B. Green, *Pointlike structure and off-shell dual strings*, Nucl.Phys. **B124** (1977) 461.
56. A. Terras, *Harmonic Analysis on Symmetric Spaces and Applications I*, Springer-Verlag (New York) 1985.
57. J.H. Schwarz and P.C. West, *Symmetries and transformations of chiral $N = 2$ $D = 10$ supergravity*, Phys. Lett. **126B** (1983) 301.
58. J.H. Schwarz, *Covariant field equations of chiral $N = 10$, $D = 10$ supergravity*, Nucl. Phys. **B226** (1983) 269.
59. B.E.W. Nilsson and A. Tollsten, *Supersymmetrization of $\zeta(3)R^4_{\mu\nu\rho\sigma}$ in superstring theories*, Phys. Lett **181B** (1986) 63.
60. N. Berkovits and C. Vafa, *N=4 Topological Strings*, hep-th/9407190, Nucl. Phys.4331995123.
61. H. Ooguri and C. Vafa, *All Loop N=2 String Amplitudes*, hep-th/9505183; Nucl. Phys. **B451** (1995) 121.
62. N.A. Obers and B. Pioline, *U duality and M theory*, hep-th/9809039; Phys. Rep. (to be published).
63. M.B. Green, *Connections between M-theory and superstrings*, Talk at Strings'97, hep-th/9712195; Nucl. Phys. B, Proc. Suppl. 68 (1998) 242-251.
64. M.B. Green, M. Gutperle and H. Kwon, (in preparation).
65. M.B. Green and M. Gutperle, *D-instanton partition functions*, hep-th/9804123; Phys.Rev. **D58** (1998) 46007.
66. M. Porrati and Rozenberg, *Bound states at threshold in supersymmetric quantum mechanics*, hep-th/9708119; Nucl. Phys. **B515** (1998) 184.
67. N. Ishibashi, H. Kawai, Y. Kitazawa and A. Tsuchiya, *A large N reduced model as superstring*, hep-th/9612115; Nucl. Phys. **B498** (1997) 467.
68. V. Periwal, *Matrices on a point as the theory of everything*, hep-th/9611103; Phys. Rev. **D55** (1977) 1711.

69. W. Krauth, H. Nicolai, M. Staudacher, *Monte Carlo approach to M theory*, hep-th/9803117; Phys. Lett. **B431** (1998) 31.
70. I.K. Kostov and P. Vanhove, *Matrix string partition functions*, hep-th/9809130; Phys. Lett. **B444** (1998) 196.
71. M.B. Green and M. Gutperle, *Configurations of two D-instantons*, hep-th/9612127; Phys. Lett. **398** (1997)69.
72. G. Moore, N. Nekrasov, S. Shatashvili, *D particle bound states and generalized instantons*, hep-th/9803265.
73. A. Strominger, *Massless black holes and conifolds in string theory*, hep-th/9504090; Nucl. Phys. **B451** (1995) 96.
74. T. Banks and M.B. Green, *Nonperturbative effects in* $AdS_5 \times S^5$ *string theory and* $d = 4$ *SUSY Yang-Mills*, hep-th/9804170; JHEP **9805** (1998) 2.
75. M. Bianchi, M.B. Green, S. Kovacs and G.C. Rossi, *Instantons in supersymmetric Yang–Mills and D-instantons in IIB superstring theory*, hep-th/9807033; JHEP **9808** (1998) 13.
76. S.S. Gubser, I.R. Klebanov, and A.W. Peet, *Entropy and temperature of black 3-branes*, hep-th/9602135; Phys. Rev. **D54** (1996) 3915.
77. O. Aharony and E. Witten, *Anti-de Sitter space and the center of the gauge group*, hep-th/9807205.
78. S. Ferrara, C. Fronsdal and A. Zaffaroni, *On* $N = 8$ *supergravity in* AdS_5 *and* $N = 4$ *superconformal Yang–Mills theory*, hep-th/9802203.
79. G. 't Hooft, *A planar diagram theory for strong interactions*, Nucl. Phys. **B72** (1974) 461.
80. W. Mück and K.S. Viswanathan, *Conformal field theory correlators from classical field theory on anti-de Sitter space, I and II*, hep-th/9804035 and hep-th/9805145; G. Chalmers, H. Nastase, K. Schalm and R. Siebelink, *R-current correlators in N=4 Super Yang–Mills theory from anti-de Sitter supergravity*, hep-th/9805105.
81. H.J. Kim, L.J. Romans, and P. van Nieuwenhuizen, *Mass spectrum of chiral ten-dimensional* $N = 2$ *supergravity on* S^2, Phys. Rev. **D32** (1985) 389; M. Gunaydin and N. Marcus, *The unitarity supermultiplet of N=8 conformal superalgebra involving fields of spin* ≤ 2, Class. Quant. Grav. **2** (1985) L11-17.
82. P. Howe, K.S. Stelle and P.K. Townsend, *Supercurrents*, Nucl. Phys. **B192** (1981) 332.
83. E. Bergshoeff, M. de Roo and B. de Wit, *Extended conformal supergravity*, Nucl. Phys. **B182** (1981) 173.
84. D.Z. Freedman, S.D. Mathur, A. Matusis, L. Rastelli, *Comments on 4 point functions in the CFT / AdS correspondence* hep-th/9808006.

85. J. Erdmenger and H. Osborn, *Conformally covariant differential operators: Symmetric tensor fields*, gr-qc/9708040; Class. Quantum Grav. **15** (1998) 273.
86. S.S. Gubser and I.R. Klebanov, *Absorption by branes and Schwinger terms in the world volume theory*, hep-th/9708005; Phys. Lett. **B413** (1997) 41.
87. D.Z. Freedman, S.D. Mathur, A. Matusis and L. Rastelli, *Correlation Functions in the CFT(D)/ADS(D+1) Correspondence*, hep-th/9804058.
88. S.S. Gubser, I.R. Klebanov and A.A. Tseytlin, *Coupling Constant Dependence in the Thermodynamics of N=4 Supersymmetric Yang-Mills Theory*, hep-th/9805156; Nucl. Phys. **B534** (1998) 202.
89. R. Kallosh and A. Rajaraman, *Vacua of M-theory and string theory*, hep-th/9805041; Phys. Rev. **D58** (1998) 125003.
90. H. Liu and A.A. Tseytlin, *D = 4 superYang-Mills, D = 5 gauged supergravity, and D = 4 conformal supergravity*, hep-th/9804083.
91. J.H. Brodie and M. Gutperle, *String corrections to four point functions in the AdS / CFT correspondence*, hep-th/9809067.
92. M. Henningson and K. Sfetsos, *Spinors in the AdS/CFT correspondence*, hep-th/9803251; Phys. Lett. **B431** (1998) 63.
93. A.M. Ghezelbash, K. Kaviani, S. Parvizi and A.H. Fatollahi, *Interacting spinors-scalars and AdS/CFT correspondence*, hep-th/9805162; Phys. Lett. **B435** (1998) 291.
94. R.I. Nepomechie, *Magnetic monopoles from antisymmetric tensor gauge fields*, Phys. Rev. **D31** (1985) 1921.
95. C. Teitelboim, *Gauge invariance for extended objects*, Phys. Lett. **167B** (1986) 63; *Monopoles of higher rank*, Phys. Lett. **B167** (1986) 69.
96. G. 't Hooft, *Dimensional reduction in quantum gravity*, Essay dedicated to Abdus Salam. Salamfest 1993:284-296; hep-th/9310026.
97. L. Susskind, *The world as a hologram*, hep-th/9409089; J. Math. Phys. **36** (1995) 6377.
98. C. Bernard, *Gauge zero modes, instanton determinants, and QCD calculations*, Phys. Rev. **D19** (1979) 3013.
99. G. 't Hooft, *Computation of quantum effects due to a four dimensional pseudoparticle*, Phys. Rev. **D14** (1976) 3432; erratum: ibid. **D18** (1978) 2199.
100. I.C.G. Campbell and P.C. West, *N=2 D = 10 Nonchiral supergravity and its spontaneous compactification*, Nucl. Phys. **B243** (1984)112.

AdS$_3$ $\times$ S^3 (Un)twisted and Squashed, and An $O(2,2;\mathbf{Z})$ Multiplet of Dyonic Strings[a]

M.J. Duff[b] and C.N. Pope [⋆c]

Center for Theoretical Physics, Texas A&M University, College Station, Texas 77843, USA
⋆SISSA, Via Beirut No. 2 − 4, 34013 Trieste, Italy[d]

H. Lü

Laboratoire de Physique Théorique de l'École Normale Supérieure[de]
24 Rue Lhomond - 75231 Paris CEDEX 05, France

We consider type IIB configurations carrying both NS-NS and R-R electric and magnetic 3-form charges, and whose near horizon geometry contains AdS$_3$ $\times$ S^3. Noting that S^3 is a $U(1)$ bundle over $CP^1 \sim S^2$, we construct the dual type IIA configurations by a Hopf T-duality along the $U(1)$ fibre. In the case where there are only R-R charges, the S^3 is untwisted to $S^2 \times S^1$ (in analogy with a previous treatment of AdS$_5 \times S^5$). However, in the case where there are only NS-NS charges, the S^3 becomes the cyclic lens space S^3/Z_p with its round metric (and is hence invariant when $p = 1$), where p is the magnetic NS-NS charge. In the generic case with NS-NS and R-R charges, the S^3 not only becomes S^3/Z_p but is also squashed, with a squashing parameter that is related to the values of the charges. Similar results apply if we regard AdS$_3$ as a bundle over AdS$_2$ and T-dualise along the fibre. We show that Hopf T-dualities relate different black holes, and that they preserve the entropy. The AdS$_3$ $\times$ S^3 solutions arise as the near-horizon limits of dyonic strings. We construct an $O(2,2;\mathbf{Z})$ multiplet of such dyonic strings, where $O(2,2;\mathbf{Z})$ is a subgroup of the $O(5,5)$ or $O(5,21)$ six-dimensional duality groups, which captures the essence of the NS-NS/R-R and electric/magnetic dualities.

[a]Talk delivered by MJD

[b]Research supported in part by NSF Grant PHY-9722090.

[c]Research supported in part by DOE grant DE-FG03-95ER40917.

[d]Supported in part by EC under TMR contract ERBFMRX-CT96-0045

[e]Unité Propre du Centre National de la Recherche Scientifique, associée à l'École Normale Supérieure et à l'Université de Paris-Sud.

1 Introduction

The six-dimensional space $AdS_3 \times S^3$ emerges as the near horizon geometry [1,2] of the self-dual string [3,4] or, more generally, the dyonic string [5,4,6,7]. The dyonic string admits the ten-dimensional interpretation [5] of an intersecting NS-NS 1-brane and 5-brane, which in a type II context is in turn related by U-duality to the D1-D5 brane system [8,9,10,11,12]. This geometry plays a part in recent studies of black holes [14,15,16,17,18,19,10,11,12,20] and has attracted a good deal of attention lately [21,22,23,24,12,25,26,27,28,29,30,31,32,33,34,35,36,13] following the conjectured duality [21] between physics in the bulk of the anti-de Sitter spacetimes and conformal field theories on their boundary. AdS_3 is particularly interesting in this regard because the conformal field theory is then of the familiar and well-understood $1+1$ dimensional variety.

In a previous paper [37], devoted mainly to $AdS_5 \times S^5$, we noted that odd-dimensional spheres S^{2n+1} may be regarded as $U(1)$ bundles over CP^n and that this permits an unconventional type of T-duality along the $U(1)$ fibre that we called "Hopf" duality. This Hopf duality has the effect of untwisting S^{2n+1} to $CP^n \times S^1$. Applying this to S^5, we were able to construct the duality chain: $n = 4$ Yang-Mills $\rightarrow$ type IIB string on $AdS_5 \times S^5$ $\rightarrow$ type IIA string on $AdS_5 \times CP^2 \times S^1$ $\rightarrow$ M-theory on $AdS_5 \times CP^2 \times T^2$. In an earlier paper [38], devoted mainly to $AdS_4 \times S^7$, we exhibited the duality: M-theory on $AdS_4 \times S^7$ $\rightarrow$ type IIA string on $AdS_4 \times CP^3$. In both contexts, similar techniques were also applied to more general spaces $AdS \times M$ where M are Einstein spaces that are not necessarily spheres. These emerge as the near horizon geometries of supermembranes with fewer Killing spinors [39,40,41] and whose boundary conformal field theories have correspondingly less supersymmetry [42,37,40,43,44].

In the present paper, we wish to apply these techniques to find type IIA and M-theory duals of six-dimensional type IIB $AdS_3 \times S^3$ configurations obtained by either T^4 or K3 compactifications. The novel ingredient is that these can be supported by both NS-NS and R-R 3-forms, in contrast to the $AdS_5 \times S^5$ example where the 5-form was strictly R-R. This has some interesting and unexpected consequences. Noting that S^3 is a $U(1)$ bundle over $CP^1 \sim S^2$, we construct the dual type IIA configurations by a Hopf T-duality along the $U(1)$ fibre. In the case where there are only R-R charges, the S^3 is untwisted to $S^2 \times S^1$ (in analogy with a previous treatment of $AdS_5 \times S^5$). However, in the case where there are only NS-NS charges, the S^3 becomes the cyclic lens space S^3/Z_p with its round metric (and is hence invariant when $p = 1$), where p is the magnetic NS-NS charge. In the generic case with NS-NS and R-R charges, the S^3 not only becomes S^3/Z_p but is also squashed, with a squashing parameter that is related to the values of the charges. Similar results apply if we regard

AdS3 as a bundle over AdS2 and T-dualise along the fibre. We note that these Hopf dualities preserve the area of the horizons, and hence they preserve the black hole entropies. In particular, we show in appendix D that a Hopf reduction of a single 3-charge isotropic black hole in $D = 5$ gives a 4-charge black hole in $D = 4$, where the fourth charge is of unit strength, and comes from the magnetic charge carried by the Kaluza-Klein vector. The reduction coordinate in this case is the $U(1)$ fibre coordinate of the foliating 3-spheres in the transverse space of the $D = 5$ black hole. In general Hopf reduction and Hopf T-duality not only relate the near-horizon limits of black holes, but also relate the full solutions themselves. We show also that this statement extends to all p-branes, including non-extremal ones, that have 4-dimensional overall transverse spaces. In all cases, the Hopf reduction maps an N-charge solution in $D+1$ dimensions to an $(N+1)$-charge solution in D dimensions.

Studying the T-duality and the U-duality multiplets of BPS solitons in the full theories obtained from the T^4 or K3 reductions of ten-dimensional supergravities is a complicated matter, owing to the large number of fields, and the size of the global symmetry groups in $D = 6$. For this reason, it is helpful to make truncations of the six-dimensional theories, to more manageable subsectors that capture the essential features that we wish to study. We therefore begin, in section 2, by making a consistent truncation of six-dimensional maximal supergravity, to a subsector of bosonic fields that includes two 2-form potentials, one NS-NS, and the other R-R. We show that this theory has an $O(2,2) \sim SL(2,\mathbb{R})_1 \times SL(2,\mathbb{R})_2$ global symmetry. The $SL(2,\mathbb{R})_1$ describes an S-duality symmetry that interchanges the NS-NS and R-R 2-form potentials, while the $SL(2,\mathbb{R})_2$ is an electric/magnetic duality symmetry of the 3-form field strengths, which acts locally only at the level of the equations of motion. This consistent truncation is most conveniently constructed from the toroidal reduction in the type IIB field variables. We then consider a different consistent truncation, which is most conveniently obtained from the toroidal reduction in the type IIA fields variables. In fact the two truncations are characterised by the feature that the six-dimensional fields that are retained are precisely the original ten-dimensional ones, with the spacetime indices simply restricted to run over the six-dimensional range, and in addition the breathing-mode scalar parameterising the volume of the 4-torus. For this reason, the truncated theories can equally well be obtained by compactifying the type IIB and type IIA theories on K3, and following the identical prescription for which fields to retain.

The two truncated theories in $D = 6$ are related by a T-duality transformation upon reduction on S^1 to $D = 5$. In section 2, supplemented by appendix A, we obtain the above truncations and perform the S^1 reductions on the two theories. We then derive the explicit T-duality transformations relating them. In appendix B, we study the $O(2,2)$ symmetry in detail, and give the explicit transformation rules. In appendix C, we use these transformation rules to construct an $O(2,2;\mathbb{Z})$ multiplet of dyonic strings carrying four independent charges, namely electric and magnetic charges for each of the NS-NS and R-R 3-forms. We then extend these results to boosted and twisted dyonic strings. Since all the solutions are obtained as solutions in a consistent truncation, the dyonic strings are therefore solutions of the original supersymmetric theories, and in fact they are, as usual, BPS states preserving some fraction of the supersymmetry.

In section 3, we study the near-horizon $\text{AdS}_3 \times S^3$ limits of the dyonic strings. Noting that S^3 can be described as a $U(1)$ bundle over S^2, we perform a Hopf T-duality transformation on the $U(1)$ fibres, and show that the S^3 can be untwisted or squashed, as described previously. In the case of solutions supported purely by NS-NS fields, we also supply a CFT proof that strings on S^3/Z_m with 3-form flux n are dual to strings on S^3/Z_n with 3-form flux m.

In section 4, we perform a similar Hopf T-duality transformation on the AdS_3 instead, exploiting the fact that AdS_3 can analogously be written in the form of a bundle over AdS_2. In section 5 we perform simultaneous Hopf T-duality transformations on the fibres of S^3 and AdS_3. Section 6 addresses the issue of supersymmetry and the Hopf T-duality transformations. We construct the Killing spinors on AdS_3 and S^3 explicitly, in coordinates appropriate to the bundle descriptions, and show that Hopf T-duality on AdS_3 or S^3 either preserves all or none of the supersymmetry, at the level of the massless Kaluza-Klein modes in supergravity, depending on the orientation of the fibration. We also discuss the supersymmetry in the context of the full string theory.

In section 7, we list all the non-dilatonic black holes in $D = 5$ and $D = 4$, and study their near-horizon limits when they are oxidised to $D = 6$. We show that all the near-horizon limits can be obtained by Hopf T-duality on $\text{AdS}_3 \times S^3$.

2 $O(2,2)$ truncation of maximal supergravity in $D = 6$

We begin from the Lagrangian in $D = 6$ obtained by dimensional reduction of type IIB on a 4-torus. Since we want to consider $\text{AdS}_3 \times S^3$ solutions that carry both NS-NS and R-R 3-form charges, we first make a consistent truncation to a subset of the fields that includes the necessary pair of 3-forms. We can do this by just retaining the subset of fields comprising the reductions of the original $F^{\text{NS}}_{(3)}$ and $F^{\text{RR}}_{(3)}$ fields themselves, together with the axion χ_1 and the dilaton $\hat{\phi}$ of $D = 10$ type IIB, and the axion χ_2 coming from the dualisation of the potential $B_{(4)}$. (All details of the full reduction are given in [45].) In $D=6$, we obtain

$$
\begin{aligned}
e^{-1}\,\mathcal{L}_6 = R &- \tfrac{1}{2}(\partial\hat\phi)^2 - \tfrac{1}{2}(\partial\vec\varphi)^2 - \tfrac{1}{2}e^{2\hat\phi}\,(\partial\chi_1)^2 - \tfrac{1}{2}e^{-2\vec a\cdot\vec\varphi}\,(\partial\chi_2)^2 \\
&- \tfrac{1}{12}e^{-\hat\phi+\vec a\cdot\vec\varphi}\,(F^{\rm NS}_{(3)})^2 - \tfrac{1}{12}e^{\hat\phi+\vec a\cdot\vec\varphi}\,(F^{\rm RR}_{(3)})^2 + \chi_2\, dA^{\rm NS}_{(2)}\wedge dA^{\rm RR}_{(2)}\,, \qquad (2.1)
\end{aligned}
$$

where $F^{\rm NS}_{(3)} = dA^{\rm NS}_{(2)}$ and $F^{\rm RR}_{(3)} = dA^{\rm RR}_{(2)} + \chi_1\, dA^{\rm NS}_{(2)}$. Here $\vec\varphi$ denotes the set of 4 dilatonic scalars coming from the reduction on T^4, and $\vec a$ is a constant vector that can be found in [45], characterising the couplings of the field strengths to the dilatonic scalars $\vec\varphi$. The combinations of the dilatons $\vec\varphi$ that are perpendicular to $\vec a$ can also be consistently truncated, resulting in the Lagrangian

$$
\begin{aligned}
e^{-1}\,\mathcal{L}_{6B} = R &- \tfrac{1}{2}(\partial\phi_1)^2 - \tfrac{1}{2}(\partial\phi_2)^2 - \tfrac{1}{2}e^{2\phi_1}\,(\partial\chi_1)^2 - \tfrac{1}{2}e^{2\phi_2}\,(\partial\chi_2)^2 \\
&- \tfrac{1}{12}e^{-\phi_1-\phi_2}\,(F^{\rm NS}_{(3)})^2 - \tfrac{1}{12}e^{\phi_1-\phi_2}\,(F^{\rm RR}_{(3)})^2 + \chi_2\, dA^{\rm NS}_{(2)}\wedge dA^{\rm RR}_{(2)} \qquad (2.2)
\end{aligned}
$$

where $\phi_1 = \phi$ and $\phi_2 = -\vec a\cdot\vec\varphi$. This truncated Lagrangian is characterised by the fact that it follows from the truncation of the T^4 reduction of the type IIB theory in which all the original fields are retained, but with their indices now running only over the remaining six dimensions. (Note that the potential for the self-dual 5-form is now dualised to give the axion χ_2.) In addition, the breathing-mode scalar ϕ_2 that parameterises the volume of T^4 is also included, but all other fields with indices internal to the 4-torus are set to zero. For this reason, (2.2) can also be obtained by making a consistent truncation of the $N=2$ theory obtained by compactifying type IIB on K3, following the identical prescription for which fields are to be retained. (One cannot tell what it is that is "breathing" if all other modes are truncated.)

The Lagrangian (2.2) has an $O(2,2)\sim SL(2,\mathbb{R})_1\times SL(2,\mathbb{R})_2$ global symmetry which is a subgroup of the original $O(5,5)$ Cremmer-Julia symmetry of maximal supergravity in $D=6$. Note that $SL(2,\mathbb{R})_1$, realised by (ϕ_1,χ_1) in the scalar sector, is a symmetry of the Lagrangian, whilst $SL(2,\mathbb{R})_2$, realised by (ϕ_2,χ_2), is a symmetry of the equations of motion. We give the explicit transformation rules in appendix B. It should emphasised that this $O(2,2)$-invariant theory is not itself the bosonic sector of any supergravity theory; rather, it is a convenient *consistent* truncation of $D=6$ maximal supergravity that contains all the fields necessary for describing the (supersymmetric) solutions of $D=6$ maximal supergravity that we are going to discuss in this paper.[g] Because of the consistency of the truncation, all the solutions in the

---

[g] A similar use of a non-supersymmetric truncation was made in [46], where the 5-form field strength of the type IIB theory was set to zero in order to simplify the discussion of the multiplet of supersymmetric NS-NS and R-R string solitons. In that case the truncation was

theory (2.2) are solutions of the original untruncated supersymmetric theory. This provides a powerful tool for studying the BPS states of the original theory, since it reduces the original $O(5,5)$ or $O(5,21)$ global symmetry to a more manageable $O(2,2)$ global symmetry that nevertheless captures the essence of the NS-NS/R-R and the electric/magnetic dualities.

The six-dimensional Lagrangian (2.2) is related by T-duality in $D=5$ to a different six-dimensional theory that is also a consistent truncation of maximal six-dimensional supergravity. This theory is most conveniently obtained by truncating maximal six-dimensional supergravity described in terms of the T^4 reduction in the type IIA field variables. In this description, it corresponds again to retaining the original ten-dimensional fields, with their indices running only over the remaining six dimensions, and including in addition the breathing-mode scalar parameterising the volume of the 4-torus. Again, as with (2.2), all other six-dimensional fields, with indices internal to the 4-torus, are set to zero. This truncated Lagrangian can therefore also be obtained as a consistent truncation of the $N=2$ theory obtained by K3 compactification of type IIA, where the same truncation prescription is applied. It is given by

$$\begin{aligned} e^{-1}\mathcal{L}_{6A} = R &- \tfrac{1}{2}(\partial\phi_1)^2 - \tfrac{1}{2}(\partial\phi_2)^2 - \tfrac{1}{48}e^{\frac{1}{2}\phi_1-\frac{3}{2}\phi_2}(F_{(4)})^2 \\ &-\tfrac{1}{12}e^{-\phi_1-\phi_2}(F_{(3)})^2 - \tfrac{1}{4}e^{\frac{3}{2}\phi_1-\frac{1}{2}\phi_2}(F_{(2)})^2\,, \end{aligned} \tag{2.3}$$

where, in the notation of [48,49], $F_{(3)}$ means the NS-NS 3-form $F_{(3)1}$, and $F_{(2)}$ means the R-R 2-form $\mathcal{F}^1_{(2)}$, with the index "1" here denoting the reduction step from $D=11$ to $D=10$. (In the rest of the paper, an index "1" will be used exclusively to denote a reduction step from 6 to 5 dimensions.) We have again performed a consistent truncation and orthogonal transformation on the dilatons. To be precise, ϕ_1 is the original ten-dimensional dilaton, and ϕ_2 is the breathing mode of the 4-torus or K3. Note that $F_{(4)} = dA_{(3)} - dA_{(2)}\wedge A_{(1)}$, while $F_{(3)} = dA_{(2)}$ and $F_{(2)} = dA_{(1)}$. The $D=6$ string coupling constant in both (2.2) and (2.3) is given by $g_6 = e^{\frac{1}{2}(\phi_1+\phi_2)}$. The Lagrangian (2.3) is in fact simply the dimensional reduction of

$$e^{-1}\mathcal{L}_7 = R - \tfrac{1}{2}(\partial\phi)^2 - \tfrac{1}{48}e^{\sqrt{\frac{8}{5}}\phi}F^2_{(4)}\,. \tag{2.4}$$

This provides another way of verifying a statement we made previously, that the six-dimensional Lagrangians (2.3) and (2.2) are consistent truncations of

---

actually *inconsistent*, since generic solutions of the truncated theory would be configurations which, in the full type IIB theory, would provide sources that would force the 5-form field strength to be non-zero. However, these source terms are actually zero for the class of solutions considered in [46], and so the truncation there was consistent only in this restricted sense.

six-dimensional maximal supergravity. We do this by noting that (2.4) is itself a consistent truncation of seven-dimensional maximal supergravity. This can be easily seen from the form of the $D = 7$ theory given in [48], and by noting that $F_{(4)}$ cannot act as a source for any of the other fields that are being truncated. (Recall that for a truncation to be consistent, every solution of the truncated theory must be a solution of the untruncated theory.) Having established that (2.4) is a consistent truncation of seven-dimensional maximal supergravity, it follows that (2.3) and (2.2) are consistent truncations of six-dimensional maximal supergravity. (For the latter case, one has to invoke the T-duality relating the two theories in $D = 5$.)

To see the T-duality relating (2.2) and (2.3), we dimensionally reduce the two six-dimensional theories to $D = 5$; the details are given in appendix A. The T-duality relations between the field strengths in the five-dimensional theories are indicated in Table 1 below, which also defines our notation for the dimensional reductions of the fields. Note that we present the identifications at the level of the field strengths because it is necessary to perform some dualisations in $D = 5$ in order to implement the identifications.[h] (The precise statement of the identifications is given in appendix A.) Note also that we are now, and henceforth, using a "1" subscript on a field strength to denote the reduction step from $D = 6$ to $D = 5$.

	IIA			IIB	
	$D = 6$	$D = 5$	T-duality	$D = 5$	$D = 6$
R-R fields	$F_{(4)}$	$F_{(4)}$	$\longleftrightarrow$	$d\chi_2$	$d\chi_2$
		$F_{(3)1}$	$\longleftrightarrow$	$F^{\rm RR}_{(3)}$	$F^{\rm RR}_{(3)}$
	$F_{(2)}$	$F_{(2)}$	$\longleftrightarrow$	$F^{\rm RR}_{(2)1}$	
		$F_{(1)1}$	$\longleftrightarrow$	$d\chi_1$	$d\chi_1$
NS-NS fields	$G_{\mu\nu}$	$\mathcal{F}_{(2)}$	$\longleftrightarrow$	$F^{\rm NS}_{(2)1}$	$F^{\rm NS}_{(3)}$
	$F_{(3)}$	$F_{(3)}$	$\longleftrightarrow$	$F^{\rm NS}_{(3)}$	
		$F_{(2)1}$	$\longleftrightarrow$	$\mathcal{F}_{(2)}$	$G_{\mu\nu}$

Table 1: Fields of the truncated type II theories in $D = 6$ and $D = 5$

Our ansatz for the reduction of the metric from $D = 6$ to $D = 5$ is

$$ds_6^2 = e^{-\varphi/\sqrt{6}}\, ds_5^2 + e^{\sqrt{\frac{3}{2}}\varphi}\, (dz + \mathcal{A}_{(1)})^2 \,. \tag{2.5}$$

[h]Note that perturbative T-duality does not require dualisations in order to relate two theories. The reason why it is necessary to make a dualisation here is that we have already dualised the 4-form potential in the type IIB picture to an axion. Had we not done so, then the identification of five-dimensional fields would not have required any dualisation.

We find that the dimensional reductions of the two Lagrangians (2.2) and (2.3) become equivalent, after making the identifications given in Table 1, provided that the dilatons of the two theories are related by the orthogonal transformation

$$\begin{pmatrix} \phi_1 \\ \phi_2 \\ \varphi \end{pmatrix}_{IIA} = \Lambda \begin{pmatrix} \phi_1 \\ \phi_2 \\ \varphi \end{pmatrix}_{IIB} = \begin{pmatrix} \frac{3}{4} & -\frac{1}{4} & -\sqrt{\frac{3}{8}} \\ -\frac{1}{4} & \frac{3}{4} & -\sqrt{\frac{3}{8}} \\ -\sqrt{\frac{3}{8}} & -\sqrt{\frac{3}{8}} & -\frac{1}{2} \end{pmatrix} \begin{pmatrix} \phi_1 \\ \phi_2 \\ \varphi \end{pmatrix}_{IIB} . \quad (2.6)$$

Note that this matrix satisfies $\Lambda = \Lambda^{-1}$. In terms of the string metrics, the radius of the compactifying circle is given by $R = e^{\frac{1}{4}\phi_1 + \frac{1}{4}\phi_2 + \sqrt{\frac{3}{8}}\varphi}$. It is easily seen that under the transformation (2.6) we have $R_{IIA} = 1/R_{IIB}$.

It is worth remarking that the Lagrangians (2.2) and (2.3) are both consistent truncations of maximal supergravity. It follows that their respective solutions that are related by T-duality transformation are all solutions of the untruncated maximal supergravity. If instead we consider the two Lagrangians (2.2) and (2.3) as consistent truncations of the K3 compactifications of the type IIB and type IIA supergravities, then their solutions that are related by T-duality remain as distinct solutions of the original type IIB and type IIA supergravities.

In the rest of the paper, we shall refer to the two Lagrangians (2.2) and (2.3) as the type IIB and the type IIA descriptions of the six-dimensional truncated theories.

3 S^3 (un)twisted and squashed

In the previous section, we obtained truncated six-dimensional type IIB and type IIA Lagrangians, and their T-duality relation. We are now in a position to consider the $\mathrm{AdS}_3 \times S^3$ solution. The Lagrangian (2.2) admits dyonic string solutions supported either by the NS-NS 3-form $F_{(3)}^{\mathrm{NS}}$ or the R-R 3-form $F_{(3)}^{\mathrm{RR}}$. More general solutions can be obtained by acting with the $O(2,2)$ symmetry of the theory, allowing us, in particular, to find solutions for dyonic strings carrying both NS-NS and R-R charges. We do this in detail in appendix C, obtaining an $O(2,2;\mathbf{Z})$ multiplet of dyonic strings.

Near the horizon, even though the above dyonic solutions carry four independent charges, the 3-forms $F_{(3)}^{\mathrm{NS}}$ and $F_{(3)}^{\mathrm{RR}}$ become self-dual, and the metric approaches that of $\mathrm{AdS}_3 \times S^3$. In fact it is more convenient to construct these solutions directly. The dilatons ϕ_1 and ϕ_2 and the axions χ_1 and χ_2 are constant in the solution, and for simplicity we shall take them to be zero. The remaining equations are solved by taking the metric and 3-forms to be

$$
\begin{aligned}
ds_6^2 &= ds^2(\text{AdS}) + ds^2(S^3) \ , \\
F_{(3)}^{\text{NS}} &= \lambda\, \epsilon(\text{AdS}) + \lambda\, \epsilon(S^3) \ , \\
F_{(3)}^{\text{RR}} &= \mu\, \epsilon(\text{AdS}) + \mu\, \epsilon(S^3) \ ,
\end{aligned} \tag{3.1}
$$

where λ and μ are constants, and the metrics on the AdS_3 and S^3 have Ricci tensors given by

$$
R_{\mu\nu} = -\tfrac{1}{2}(\lambda^2 + \mu^2)\, g_{\mu\nu} \ , \qquad R_{mn} = \tfrac{1}{2}(\lambda^2 + \mu^2)\, g_{mn} \tag{3.2}
$$

respectively. (It is easy to see from (2.2) that the equations of motion will only be satisfied by taking the dilatons to be constant if the coefficients in front of the volume forms $\epsilon(\text{AdS})$ and $\epsilon(S^3)$ for the AdS_3 and S^3 metrics are equal, and hence the 3-form field strengths are self-dual.) The constants λ and μ are related to the magnetic charges as follows:

$$
Q_{\text{NS}} \equiv \frac{1}{16\pi^2} \int F_{(3)}^{\text{NS}} = \frac{\lambda}{(\lambda^2 + \mu^2)^{3/2}} \ , \qquad Q_{\text{RR}} \equiv \frac{1}{16\pi^2} \int F_{(3)}^{\text{RR}} = \frac{\mu}{(\lambda^2 + \mu^2)^{3/2}} \ . \tag{3.3}
$$

In calculating the charges, we have made use of the fact that a 3-sphere whose Ricci tensor is given by R_{mn} in (3.2) has volume $16\pi^2(\lambda^2 + \mu^2)^{-3/2}$, and in fact its metric can be written as

$$
ds^2(S^3) = \frac{4}{\lambda^2 + \mu^2}\, d\Omega_3^2 \ , \tag{3.4}
$$

where $d\Omega_3^2$ is the metric on a unit 3-sphere. Note that since we are taking the constant values of the two dilatons ϕ_1 and ϕ_2 to be zero for simplicity, this means that the electric charges (whose calculation we have not explicitly presented above) are equal to the magnetic charges. If one chooses non-zero values for the constants ϕ_1 and ϕ_2 then the field strengths are still self-dual, but the electric and magnetic charges will be unequal. Note that the charges will all be integers.

We now make use of the fact that the metric $d\Omega_3^2$ can be written as a $U(1)$ bundle over $CP^1 \sim S^2$ as follows:

$$
d\Omega_3^2 = \tfrac{1}{4} d\Omega_2^2 + \tfrac{1}{4}(dz + B)^2 \ , \tag{3.5}
$$

where $d\Omega_2^2$ is the metric on the unit 2-sphere, whose volume form $\Omega_{(2)}$ is given by $\Omega_{(2)} = dB$. (If $d\Omega_2^2$ is written in spherical polar coordinates as $d\Omega_2^2 = d\theta^2 + \sin^2\theta\, d\phi^2$, then we can write B as $B = \cos\theta\, d\phi$.) The fibre coordinate z

has period 4π. Thus the six-dimensional metric given in (3.1) can be written as

$$ds_6^2 = ds^2(\text{AdS}) + \frac{1}{\lambda^2+\mu^2}\, d\Omega_2^2 + \frac{1}{\lambda^2+\mu^2}\,(dz+B)^2\ . \tag{3.6}$$

The four-dimensional area of the horizon is given by

$$A \sim L\,(\lambda^2+\mu^2)^{-3/2}\ , \tag{3.7}$$

where L is the contribution from $ds^2(\text{AdS})$ at the boundary at constant time. The field strengths in (3.1) can now be written as

$$\begin{aligned} F_{(3)}^{\text{NS}} &= \lambda\,\epsilon(\text{AdS}) + \frac{\lambda}{(\lambda^2+\mu^2)^{3/2}}\,\Omega_{(2)}\wedge(dz+B)\ ,\\ F_{(3)}^{\text{RR}} &= \mu\,\epsilon(\text{AdS}) + \frac{\mu}{(\lambda^2+\mu^2)^{3/2}}\,\Omega_{(2)}\wedge(dz+B)\ . \end{aligned} \tag{3.8}$$

Comparing (3.6) with the general reduction ansatz (2.5), we see that if we dimensionally reduced on the fibre coordinate we obtain the 5-dimensional metric

$$ds_5^2 = (\lambda^2+\mu^2)^{-1/3}\, ds^2(\text{AdS}) + (\lambda^2+\mu^2)^{-4/3}\, d\Omega_2^2\ , \tag{3.9}$$

while the new dilaton φ is a constant, given by

$$e^{\varphi/\sqrt{6}} = (\lambda^2+\mu^2)^{-1/3}\ . \tag{3.10}$$

Comparing (3.8) with the reduction ansätze $F_{(n)} \to F_{(n)} + F_{(n-1)}\wedge(dz+B)$ for the field strengths, we find that in $D=5$ we have

$$\begin{aligned} F_{(3)}^{\text{NS}} &= \lambda\,\epsilon(\text{AdS})\ , & F_{(2)1}^{\text{NS}} &= \frac{\lambda}{(\lambda^2+\mu^2)^{3/2}}\,\Omega_{(2)}\ ,\\ F_{(3)}^{\text{RR}} &= \mu\,\epsilon(\text{AdS})\ , & F_{(2)1}^{\text{RR}} &= \frac{\mu}{(\lambda^2+\mu^2)^{3/2}}\,\Omega_{(2)}\ ,\\ \mathcal{F}_{(2)} &= dB = \Omega_{(2)}\ . \end{aligned} \tag{3.11}$$

We are now in a position to implement the T-duality transformation from the type IIB description to the type IIA description in $D=5$. From appendix A, and (3.11), we see that the field strengths in the type IIA picture will be

$$\begin{aligned} F_{(3)} &= \lambda\,\epsilon(\text{AdS})\ , & \mathcal{F}_{(2)} &= \frac{\lambda}{(\lambda^2+\mu^2)^{3/2}}\,\Omega_{(2)}\ ,\\ F_{(3)1} &= -\mu\,\epsilon(\text{AdS})\ , & F_{(2)} &= \frac{\mu}{(\lambda^2+\mu^2)^{3/2}}\,\Omega_{(2)}\ ,\\ F_{(2)1} &= \Omega_{(2)}\ . \end{aligned} \tag{3.12}$$

From (2.6) and (3.10), together with the fact that we are taking $\phi_1 = \phi_2 = 0$ in the original type IIB solution, it follows that the dilatons in the type IIA picture will be given by

$$e^{\varphi} = (\lambda^2 + \mu^2)^{1/\sqrt{6}}\,, \qquad e^{\phi_1} = e^{\phi_2} = (\lambda^2 + \mu^2)^{1/2}\,. \tag{3.13}$$

Finally, we can oxidise the type IIA solution that we have just obtained back to $D = 6$, by retracing the standard Kaluza-Klein reduction steps. Doing so, we find that the six-dimensional metric in the type IIA picture is

$$ds_6^2 = (\lambda^2 + \mu^2)^{-1/2}\, ds^2(\mathrm{AdS}) + (\lambda^2 + \mu^2)^{-3/2} \left[d\Omega_2^2 + \frac{\lambda^2}{\lambda^2 + \mu^2}\,(dz' + B)^2\right]\,, \tag{3.14}$$

where B is a potential such that $\Omega_{(2)} = dB$, and the coordinate z' is related to z by

$$z = \frac{\lambda}{(\lambda^2 + \mu^2)^{3/2}}\, z' = Q_{\rm NS}\, z'\,. \tag{3.15}$$

(The last equality follows from (3.3).) It is straightforward to verify that the area of the horizon of the metric (3.14) is the same as that before the Hopf T-duality transformation, given by (3.7). The type IIA field strengths in $D = 6$ are given by

$$F_{(4)} = -\mu\, \epsilon(\mathrm{AdS}) \wedge (dz + \mathcal{A}_{(1)})\,, \qquad F_{(3)} = \lambda\, \epsilon(\mathrm{AdS}) + \Omega_{(2)} \wedge (dz + \mathcal{A}_{(1)})\,,$$
$$F_{(2)} = \frac{\mu}{(\lambda^2 + \mu^2)^{3/2}}\, \Omega_{(2)}\,, \tag{3.16}$$

where

$$\mathcal{A}_{(1)} = \frac{\lambda}{(\lambda^2 + \mu^2)^{3/2}}\, B = Q_{\rm NS}\, B\,. \tag{3.17}$$

Note that in the above T-duality transformation the Buscher rules[50] are insufficient, since we have R-R fields as well as NS-NS fields involved in the solution. For this reason, we have constructed the two low-energy effective actions and explicitly derived the T-duality transformations that relate them.

We find that the charges carried by these field strengths are as follows:

$$Q^{(3)}_{\rm elec} \equiv \frac{1}{16\pi^2} \int_{S^3} e^{-\phi_1 - \phi_2} {*F_{(3)}} = Q_{\rm NS}\,,$$
$$Q^{(3)}_{\rm mag} \equiv \frac{1}{16\pi^2} \int_{S^3} F_{(3)} = 1\,,$$
$$Q^{(4)}_{\rm elec} \equiv \frac{1}{4\pi} \int_{S^2} e^{\frac{1}{2}\phi_1 - \frac{3}{2}\phi_2} {*F_{(4)}} = -Q_{\rm RR}\,,$$
$$Q^{(2)}_{\rm mag} \equiv \frac{1}{4\pi} \int_{S^2} F_{(2)} = Q_{\rm RR}\,. \tag{3.18}$$

If the fibre coordinate z' in (3.14) had had the period 4π, then the topology of the compact 3-space would have been S^3. Since it is related to z as given in (3.15), and z has period 4π, it follows that z' has period $4\pi/Q_{\rm NS}$, and hence the topology of the compact 3-space is $S^3/Z_{Q_{\rm NS}}$, the cyclic lens space of order $Q_{\rm NS}$.

On the other hand the magnetic charge carried by the field strength $F_{(3)}$ is equal to 1, having started, in the original solution, as $Q_{\rm NS}$. This is because the T-duality has exchanged the original charge $Q_{\rm NS}$ with the unit charge corresponding to the unit topological charge of the $U(1)$ bundle over S^2 that describes the 3-sphere. Furthermore, we can see from (3.14) that the metric on the lens space is not in general the "round" one, but is instead squashed along the $U(1)$ fibre direction, with a squashing factor ν given by

$$\nu = \frac{\lambda}{\sqrt{\lambda^2+\mu^2}} = \frac{Q_{\rm NS}}{\sqrt{Q_{\rm NS}^2+Q_{\rm RR}^2}} \,. \tag{3.19}$$

In other words, the metric on the compact 3-space is proportional to

$$ds^2 = \tfrac14 d\Omega_2^2 + \tfrac14 \nu^2\,(dz+B)^2 \,, \tag{3.20}$$

whose Ricci tensor, in the natural orthonormal basis, is given by

$$R_{ab} = \mathrm{diag}\,(4-2\nu^2, 4-2\nu^2, 2\nu^2) \,. \tag{3.21}$$

When $\nu=1$, the metric is Einstein and the 3-sphere or lens space is "round."

As we mentioned earlier, we could have considered original solutions in which the constant dilatons ϕ_1 and ϕ_2 were non-zero, in which case the original electric and magnetic charges need not have been equal. The lens space after the Hopf T-duality transformation will then be $S^3/Z_{Q^{\rm mag}_{\rm NS}}$. Also, we can generalise the starting point further by consider a solution on the product of AdS_3 and the lens space S^3/Z_n, rather than simply $\mathrm{AdS}_3\times S^3$. (From the lower-dimensional point of view, this corresponds to giving the Kaluza-Klein vector a magnetic charge n rather than 1.) If we do this, then we find that a solution $\mathrm{AdS}_3\times S^3/Z_n$ for (2.2), with charges $Q^{\rm elec}_{\rm NS}$, $Q^{\rm mag}_{\rm NS}$, $Q^{\rm elec}_{\rm RR}$ and $Q^{\rm mag}_{\rm RR}$ will result, after the T-duality transformation, in a solution $\mathrm{AdS}_3\times S^3/Z_{Q^{\rm NS}_{\rm mag}}$ for (2.3) with charges

$$Q^{(3)}_{\rm elec} = Q^{\rm elec}_{\rm NS}\,,\qquad Q^{(3)}_{\rm mag} = n\,,\qquad Q^{(4)}_{\rm elec} = -Q^{\rm elec}_{\rm RR}\,,\qquad Q^{(2)}_{\rm mag} = Q^{\rm mag}_{\rm RR}\,. \tag{3.22}$$

To summarise, we see that if we consider the case where the original $\mathrm{AdS}_3\times S^3/Z_n$ solution in the type IIB description carries only an NS-NS charge whose magnetic component is $Q^{NS}_{mag} = m$, and so $Q_{RR} = 0$, then after the Hopf T-duality transformation the solution for (2.3) will be of the form $AdS_3 \times S^3/Z_m$, where the metric on the cyclic lens space is still the "round" one, and the magnetic charge becomes $Q^{NS}_{mag} = n$. In the special case where $n = m$ and $Q_{RR} = 0$, the $AdS_3 \times S^3/Z_n$ solution is invariant under the Hopf T-duality. If, on the other hand, we start with an $AdS_3 \times S^3/Z_n$ solution with only R-R charges, then after the Hopf T-duality transformation the S^3/Z_n is completely "untwisted," and the solution will be of the form $AdS_3 \times S^1 \times S^2$. (This is analogous to the untwisting of $AdS_5 \times S^5$ discussed in [37] and the untwisting of $AdS_4 \times S^7$ discussed in [38]. The untwisting of S^3 to $S^2 \times S^1$ was also discussed in [47].) In the generic case where the original $AdS_3 \times S^3/Z_n$ solution carries both NS-NS and R-R charges, then after the T-duality transformation the solution will be of the form $AdS_3 \times S^3/Z_m$, where the metric on the compact 3-space space is not only factored now by Z_m, but it is also squashed.

Although the construction of conformal field theories with background R-R charges is problematical, there is an exact CFT duality statement[i] in the case of pure NS-NS charge, *i.e.* when $\mu = 0$. This can be seen by looking at the original solution (3.1), and the final Hopf-dualised solution (3.14, 3.16), in the string-frame metric $ds_6^2(\text{string})$. This is related to the six-dimensional Einstein-frame metric ds_6^2 by $ds_6^2(\text{string}) = e^{\frac{1}{2}(\phi_1+\phi_2)}\, ds_6^2$. Thus we find that the original solution can be written as

$$\begin{aligned} ds_6^2(\text{string}) &= ds^2(AdS) + \lambda^{-2}\left[d\Omega_2^2 + (dz + B)^2\right] , \\ F_{(3)} &= \lambda^{-2}\,\Sigma_{(3)} + \lambda^{-2}\,\Omega_2 \wedge (dz + B) , \end{aligned} \tag{3.23}$$

where $\Sigma_{(3)}$ is the volume form of the "unit" AdS_3, and that correspondingly in the final Hopf-dualised solution we have

$$\begin{aligned} ds_6^2(\text{string}) &= ds^2(AdS) + \lambda^{-2}\left[d\Omega_2^2 + (dz' + B)^2\right] , \\ F_{(3)} &= \lambda^{-2}\,\Sigma_{(3)} + \lambda^{-2}\,\Omega_2 \wedge (dz' + B) , \end{aligned} \tag{3.24}$$

where $z' = \lambda^2\, z = Q_{NS}^{-1}\, z$. The T-duality can be understood in this case from the standard Buscher rules [50] applied to the $SU(2)$ WZW model. In general, one has the statement that a solution on S^3/Z_m with 3-form flux n is Hopf dual to a solution on S^3/Z_n with 3-form flux m. This is because the Kaluza-Klein vector in the S^3/Z_m solution carries m units of charge, whereas the winding mode vector coming from the 3-form carries n units of charge. The invariance under T-duality in the special case $n = m = 1$ is discussed in [70].

[i] We are grateful to Costas Bachas for urging us to provide a CFT proof. A CFT discussion of the $D = 10$ superstring compactification on $S^3 \times S^3$ down to $AdS_2 \times S^2$ may be found in [71]

4 AdS₃ (un)twisted and squashed

In the same way as odd-dimensional spheres can be viewed as $U(1)$ bundles over complex projective spaces, so we can view odd-dimensional AdS spacetimes as bundles over certain Lorentzianisations of the complex projective spaces. The case of AdS_3 is particularly simple, since in this case the base space is nothing but AdS_2.

Let us begin by considering the unit S^3, written as a $U(1)$ bundle over S^2:

$$ds^2 = \tfrac{1}{4}d\theta^2 + \tfrac{1}{4}\sin^2\theta\, d\phi^2 + \tfrac{1}{4}(d\psi + \cos\theta d\phi)^2\ . \tag{4.1}$$

Now perform an analytic continuation to a Lorentzian signature, by sending:

$$\theta \longrightarrow \tfrac{1}{2}\pi - i\rho\ , \qquad \psi \longrightarrow i\,x\ , \qquad \phi \longrightarrow t\ . \tag{4.2}$$

This gives us the metric (after making an overall sign change to go from east-coast to west-coast notation)

$$ds^2 = -\tfrac{1}{4}\cosh^2\rho\, dt^2 + \tfrac{1}{4}d\rho^2 + \tfrac{1}{4}(dx + \sinh\rho\, dt)^2\ . \tag{4.3}$$

It is not hard to calculate the curvature for this metric, and to verify that it has Ricci tensor $R_{\mu\nu} = -2\,g_{\mu\nu}$. Thus it is AdS_3, since the cosmological constant is negative. Note that t is periodic, $0 \le t \le 2\pi$, but ρ and x both range over the entire real line. The parameterisation of points in AdS_3, viewed as the $O(2,2)$-invariant hyperboloid $X_1^2 + X_2^2 - X_3^2 - X_4^2 = 1$ in $\mathbb{R}^4$, can be given in terms of t, ρ and x as follows:

$$\begin{aligned}
\begin{pmatrix} X_1 \\ X_2 \end{pmatrix} &= \frac{1}{\sqrt{2}} \begin{pmatrix} \cos\frac{1}{2}t & \sin\frac{1}{2}t \\ -\sin\frac{1}{2}t & \cos\frac{1}{2}t \end{pmatrix} \begin{pmatrix} \cosh x_- \\ \cosh x_+ \end{pmatrix}\ , \\
\begin{pmatrix} X_3 \\ X_4 \end{pmatrix} &= \frac{1}{\sqrt{2}} \begin{pmatrix} \cos\frac{1}{2}t & \sin\frac{1}{2}t \\ -\sin\frac{1}{2}t & \cos\frac{1}{2}t \end{pmatrix} \begin{pmatrix} \sinh x_- \\ \sinh x_+ \end{pmatrix}\ ,
\end{aligned} \tag{4.4}$$

where $x_\pm \equiv \frac{1}{2}(x \pm \rho)$. It can be shown that the coordinates t, ρ and x give a 1-1 mapping to points on the hyperboloid.

It is evident that we can view (4.3) as a bundle over a base space which is AdS_2, for which the unit-radius metric is $d\Sigma_2^2 = -\cosh^2\rho\, dt^2 + d\rho^2$. (We shall in general use $d\Sigma_n^2$ to denote the "unit" AdS_n metric, which is defined to be the one whose Ricci tensor has the form $R_{\mu\nu} = -(n-1)\,g_{\mu\nu}$. This is analogous to the definition of the unit n-sphere.) Thus we have

$$d\Sigma_3^2 = \tfrac{1}{4}d\Sigma_2^2 + \tfrac{1}{4}(dx + \widetilde{B})^2\ , \tag{4.5}$$

where $d\widetilde{B} = \Sigma_{(2)}$. (We are defining $\Sigma_{(n)}$ to be the volume form on the unit AdS_n.)

Let us note here that there is also another closely related metric on AdS_3, which also allows one to do a reduction on the fibre coordinate x, namely

$$ds^2 = -\tfrac{1}{4}e^{2\rho}\, dt^2 + \tfrac{1}{4}d\rho^2 + \tfrac{1}{4}(dx + e^{\rho}\, dt)^2 \ . \tag{4.6}$$

This can be shown to be an AdS_3 metric with $R_{\mu\nu} = -2\, g_{\mu\nu}$. It is interesting because it arises by taking the near-horizon limit of the boosted dyonic string in $D = 6$ (see appendix C). (In other words, the intersection of the dyonic string with a wave. If reduced to $D = 5$, this boosted dyonic string becomes a 3-charge black hole; *i.e.* the Reissner-Nordstrøm black hole.)

The coordinate transformation that relates the two AdS_3 metrics (4.3) and (4.6) is the following:

$$\begin{aligned} e^{\tilde\rho} &= \sinh\rho + \cosh\rho\, \cos t\ , \\ \tilde t\, e^{\tilde\rho} &= \cosh\rho\, \sin t\ , \\ e^{\tilde x} &= \frac{(e^{\rho}\, \cot\tfrac{1}{2}t - 1)\, e^{x}}{(e^{\rho}\, \cot\tfrac{1}{2}t + 1)}\ , \end{aligned} \tag{4.7}$$

where the tilded coordinates here denote the coordinates in the metric (4.6). An important feature of both (4.3) and (4.6) is that there is no coordinate-dependent function multiplying the vielbein in the fibre direction, and so the circle in the $U(1)$ reduction will have a constant length.

This Hopf T-duality on the fibres of AdS_3 should be contrasted with the T-duality discussed in [52,27], or with a T-duality on one of the horospherical coordinates of AdS. (See also [53].) In these two cases the metric is already diagonal, and the size of the compactifying circle is not constant, but instead depends on other coordinates of the AdS. It follows that after T-dualisation, the dual theory has a dilaton that is singular on the horizon, and hence so also is the metric. By contrast, in the Hopf dualisation of AdS_3 discussed above, the constant radius of the circle implies that the dilaton is non-singular, and the metric has no local curvature singularity. The difference is further highlighted by the analysis of the supersymmetry. In the case of the horospherical T-duality or the T-duality in [52,27], supersymmetry is always (at least partially) broken. In the case of Hopf T-duality on AdS_3, however, the supersymmetry is either all preserved or all broken, depending on an orientation choice in the Hopf fibering. We shall discuss this, and make detailed comparisons between the various T-dualities, in section 6.

Note that we can also have the concept of "squashing" for AdS_3 where the length of the fibres is rescaled relative to the size of the base AdS_2. As in the case of the sphere, this will be an homogeneous squashing. Thus we may consider a squashed AdS_3 metric

$$ds^2 = \tfrac{1}{4} d\Sigma_2^2 + \tfrac{1}{4}\,\nu^2\,(dx + \widetilde{B})^2\,, \tag{4.8}$$

where ν is a constant squashing parameter. The vielbein components of the Ricci tensor in the natural orthonormal basis are

$$R_{ab} = \mathrm{diag}\,(4 - 2\nu^2, 2\nu^2 - 4, -2\nu^2)\,, \tag{4.9}$$

where the first entry is the R_{00} component. More generally, there also exist squashed versions of AdS_{2n+1} for any n, of the form $ds^2(\mathrm{AdS}_{2n+1}) = ds^2(\widetilde{CP}^n) + \nu^2\,(dx + \widetilde{B})^2$, where $\widetilde{CP}^n$ denotes a Lorentianisation of CP^n, and $d\widetilde{B}$ is the volume form on $\widetilde{CP}^n$.

All the previous steps of dimensional reduction on the fibres, which we did for the S^3 factor in $\mathrm{AdS}_3 \times S^3$, can now be repeated for the AdS_3 itself. The computations are essentially identical, since the dimension of the AdS_3 is the same as that of the S^3, with the exception of the details of the field strengths. The final result is that the solution $\mathrm{AdS}_3 \times S^3$ given by (3.1) becomes, after performing a T-duality transformation on the AdS_3 fibre coordinate x,

$$ds_6^2 = (\lambda^2{+}\mu^2)^{-3/2}\left[d\Sigma_2^2{+}\frac{\lambda^2}{\lambda^2+\mu^2}\,(dx'{+}\widetilde{B})^2\right]{+}(\lambda^2{+}\mu^2)^{-1/2}\,ds^2(S^3)\,, \tag{4.10}$$

where $\widetilde{B}$ is a potential such that $\Sigma_{(2)} = d\widetilde{B}$, and the coordinate x' is related to x by

$$x = \frac{\lambda}{(\lambda^2+\mu^2)^{3/2}}\,x' = Q_{\rm NS}\,z'\,. \tag{4.11}$$

The type IIA field strengths in $D = 6$ are given by

$$F_{(4)} = -\mu\,Q_{\rm NS}\,\epsilon(S^3)\wedge(dx' + \widetilde{B})\,, \qquad F_{(3)} = \lambda\,\epsilon(S^3) + Q_{\rm NS}\,\Sigma_{(2)}\wedge(dx' + \widetilde{B})\,,$$
$$F_{(2)} = Q_{\rm RR}\,\Sigma_{(2)}\,. \tag{4.12}$$

In particular, we see that in the case where the solution is supported purely by R-R 3-form charges, then after doing the T-duality transformation we arrive at an "untwisted" solution $\mathrm{AdS}_2 \times S^1 \times S^3$. If the solution instead carries only NS-NS charges, then the structure of the T-duality transformed solution is essentially unchanged, in the sense that there are just overall rescalings $1/\lambda$ and $1/\lambda^3$ of the 3-sphere and AdS_3 factors in the metric. Since the fibre coordinate x here ranges over the entire real line, we do not really have the notion of a "lens space" for AdS. We could consider instead the spacetime that one obtains by taking the fibre coordinate x to be periodic. This will no longer be globally AdS$_3$, but instead only a part of it. If we define a spacetime U_n to be AdS$_3$ with x having a period L/n, for some specified L, then one could say that the T-dual of the solution $U_1 \times S^3$ carrying purely NS-NS charge $Q_{\rm NS}$ is the solution $U_{Q_{\rm NS}} \times S^3$. As in the case of the Hopf T-duality for S^3, all of the above discussion generalises straightforwardly to the case where we allow the dilaton moduli to be non-zero, so that there can be independent electric and magnetic charges for each field strength.

Note that the area of the horizon is preserved under the Hopf T-duality on the fibre coordinate x of the AdS$_3$. It is given by $A \sim (\lambda^2 + \mu^2)^{-2}$.

5 AdS$_3$ and S^3 (un)twisted and squashed

We may now put together the results of the previous two sections, by starting with an AdS$_3 \times S^3$ solution (3.1), performing a T-duality transformation first using the $U(1)$ fibres of S^3, and then performing a second T-duality transformation using the fibres of AdS$_3$. After doing this, we find that the final metric is given by

$$ds_6^2 = \frac{1}{(\lambda^2+\mu^2)^2}\left[d\Sigma_2^2 + \frac{\lambda^2}{\lambda^2+\mu^2}(dx'+\tilde{B})^2 + d\Omega_2^2 + \frac{\lambda^2}{\lambda^2+\mu^2}(dz'+B)^2\right], \tag{5.1}$$

where z' is given by (3.15) and x' is given by (4.11). The field strengths in the final solution are given by

$$\begin{aligned} F_{(3)}^{\rm NS} &= \Omega_{(2)} \wedge (dz + \mathcal{A}_{(1)}) + \Sigma_{(2)} \wedge (dx + \widetilde{\mathcal{A}}_{(1)})\,, \\ F_{(3)}^{\rm RR} &= Q_{\rm RR}\, \Omega_{(2)} \wedge (dx + \widetilde{\mathcal{A}}_{(1)}) - Q_{\rm RR}\, \Sigma_{(2)} \wedge (dz + \mathcal{A}_{(1)})\,, \end{aligned} \tag{5.2}$$

where $\mathcal{A}_{(1)}$ is given by (3.17) and

$$\widetilde{\mathcal{A}}_{(1)} = \frac{\lambda}{(\lambda^2+\mu^2)^{3/2}}\,\widetilde{B} = Q_{\rm NS}\,\widetilde{B}\,. \tag{5.3}$$

If the NS-NS charge is zero, then in this doubly-transformed solution both the S^3 and the AdS$_3$ are untwisted, giving (AdS$_2 \times S^1) \times (S^2 \times S^1)$. If instead the R-R charge is zero, the solution becomes twisted to $U_{Q_{\rm NS}} \times (S^3/Z_{Q_{\rm NS}})$. When the NS-NS and R-R charges are both present, the solution is squashed and twisted in both the AdS$_3$ and S^3 factors. Note that the squashing parameter is the same for both the AdS$_3$ and the S^3.

Again in this case, the Hopf T-duality transformation preserves the area of horizon, given by $A \sim (\lambda^2+\mu^2)^{-2}$. It follows that the black hole entropy, which is a quarter of the area, is also invariant under the Hopf T-duality. In the case of a 5-dimensional black hole that oxidises to a boosted dyonic NS-NS string in $D = 6$, this invariance can be easily understood: It follows from (C.17) in this case that the black hole entropy is given by $S \sim \sqrt{Q_w\, q\, p} = \sqrt{n\, Q_e^{\rm NS}\, Q_m^{\rm NS}}$. As we have seen earlier, Hopf T-duality has the effect of interchanging n and $Q_e^{\rm NS}$, and hence the entropy is left invariant. In this paper, we have shown that in general the entropy of the black hole is preserved under Hopf T-duality, even when it is supported by R-R as well as NS-NS charges. Thus even though the metric on the AdS_3 or S^3 may be (un)twisted and squashed by the transformation, the area of the horizon is preserved.

6 Supersymmetry and Killing spinors

It is well known that T-duality transformations can break supersymmetries of p-brane solutions, at the level of the low-energy effective supergravity. For example, the near-horizon limit of the D3-brane in ten dimensions, which is of the form $AdS_5 \times S^5$ and hence preserves all the supersymmetry, is T-dual to the near-horizon limit of a D4-brane, which is a product of a domain wall and a sphere (rather than an AdS and a sphere), and which breaks half of the supersymmetry. The phenomenon of supersymmetry breaking, at the level of supergravity, has also been seen in the Hopf T-duality transformations discussed in [37]. For example, the Hopf dualisation of $AdS_5 \times S^5$ to $AdS_5 \times CP^2 \times S^1$ breaks all the supersymmetry at the supergravity level [37]. In another example, the Hopf reduction of the $AdS_4 \times S^7$ solution of $D = 11$ supergravity to the $AdS_4 \times CP^3$ of $D = 10$ type IIA supergravity breaks either all of the eight supersymmetries, or two out of the eight, depending on the orientation of the internal manifold [51,38].

Of course all these statements about supersymmetry breaking are made at the level of the massless Kaluza-Klein modes in the supergravity theory. In some cases, the supersymmetry will be restored when one includes the Kaluza-Klein massive modes or the string winding modes. To analyse the behaviour of supersymmetry under T-duality, one should separate the discussion into two parts: firstly, compactification on a circle, and secondly, the T-duality transformation itself. Supersymmetry breaking, if it occurs at all, is a consequence of the compactification on the circle. The T-duality transformation in the full string theory always preserves whatever supersymmetry has survived the circle compactification. An example where compactification breaks supersymmetry is an AdS spacetime, written in horospherical coordinates $ds^2 = d\rho^2 + e^{2\rho}\, \eta_{\mu\nu}\, dx^\mu\, dx^\nu$, and compactified on one of the spatial x^μ coordinates. Although there is a translational isometry, thus allowing a circle compactification, one finds that half of the Killing spinors on the AdS spacetime depend linearly on x^μ, and thus the supersymmetries associated with these Killing spinors will be broken once the chosen x^μ coordinate is taken to be periodic [54]. The other half of the Killing spinors are independent of x^μ, and so half of the supersymmetries survive the compactification. This statement is true even after including all the massive Kaluza-Klein modes. A T-duality transformation will not result in any further breaking (or restoring) of supersymmetry.

A contrasting example is provided by the Hopf reduction on the $U(1)$ fibres of S^5 in the $\mathrm{AdS}_5 \times S^5$ solution of the type IIB theory. The fibre coordinate here is naturally periodic. Although at the level of the massless Kaluza-Klein modes the compactification on the $U(1)$ fibres breaks all the supersymmetry, it is restored once the massive Kaluza-Klein modes are included. If we perform a T-duality transformation on the fibre coordinate, the $\mathrm{AdS}_5 \times S^5$ solution is mapped to an $\mathrm{AdS}_5 \times CP^2 \times S^1$ solution in the type IIA theory. At the level of supergravity, even when the massive Kaluza-Klein modes are included, this type IIA solution breaks all the supersymmetries. However in the full string theory, where the string winding modes are also included, the full supersymmetry is reinstated [37].

Thus, for example, we see that there are two different kinds of T-duality transformations that can be performed on a D3-brane. One of these maps the D3-brane to a D4-brane. In this case, the near-horizon $\mathrm{AdS}_5 \times S^5$ limit of the D3-brane is mapped to a product of a domain wall and a 4-sphere, which is the near-horizon limit of the D4-brane. In this case it is one of the horospherical coordinates x^μ in AdS_5 that is periodically identified and used in the T-duality transformation. (Since the x^μ coordinates are in fact the world-volume coordinates of the D3-brane.) Thus half of the supersymmetry is already broken in the process of making the identification. In other words, the apparent discrepancy between the "enhanced supersymmetry" seen in the near horizon limit of an unwrapped D3-brane and the usual 1/2 supersymmetry of the near-horizon limit of the D4-brane is not the result of any breaking of supersymmetry by T-duality. Rather, it is simply a consequence of the fact that there *is* no "supersymmetry enhancement" in the near-horizon limit of a wrapped D3-brane [54]. The other kind of T-duality that can be performed on a D3-brane is on the fibre coordinate of the foliating 5-spheres in its 6-dimensional transverse space. In this case, the fibre coordinate is naturally periodic, and the "supersymmetry breaking" is a pure artifact of the low-energy supergravity approximation. In the full string theory, when the Kaluza-Klein and winding modes are included, there is no supersymmetry breaking at all [37].

At the level of the massless modes in the reduction of the supergravity theory, the Hopf T-duality transformations on $\mathrm{AdS}_3 \times S^3$ that we have been considering in this paper either preserve all the supersymmetry, or they break all the supersymmetry, depending upon an orientation associated with the Hopf reduction. In this section, we shall demonstrate this by giving an explicit construction of the Killing spinors on AdS_3 and S^3. These have been given previously [54,55,56], but in coordinate systems that are not convenient for our present purposes. The relation between the coordinates used in [56], in which the metric on the unit 3-sphere is $d\Omega_3^2 = d\theta_3^2 + \sin^2\theta_3\,(d\theta_2^2 + \sin^2\theta_2\,d\theta_1^2)$, and the coordinates in (4.1), is $\theta_1 = \frac{1}{2}(\psi+\phi)$, $\cot\theta_2 = \tan\frac{1}{2}\theta\,\cos\frac{1}{2}(\psi-\phi)$, $\cos\theta_3 = \sin\frac{1}{2}\theta\,\sin\frac{1}{2}(\psi-\phi)$. In principle, the Killing spinors obtained in [56] can be re-expressed in terms of the coordinates of the metric (4.1) using these relations, but the result will be in an inconvenient local Lorentz frame.

We begin by constructing the Killing spinors for the unit S^3, using the metric given in (4.1). The vielbein and spin connection are given by

$$\begin{aligned} &e^1 = \tfrac{1}{2}d\theta\ , \qquad e^2 = \tfrac{1}{2}\sin\theta\,d\phi\ , \qquad e^3 = \tfrac{1}{2}(d\psi + \cos\theta\,d\phi)\ ,\\ &\omega_{23} = -e^1\ , \qquad \omega_{31} = -e^2\ , \qquad \omega_{12} = -2\cot\theta\,e^2 + e^3\ . \end{aligned} \tag{6.1}$$

The Killing spinor equation is $D_\mu\,\epsilon^\pm = \pm\frac{\mathrm{i}}{2}\Gamma_\mu\,\epsilon^\pm$. In a given choice of conventions in the $AdS_3\times S^3$ supergravity solution, either the ϵ^+ Killing spinors or the ϵ^- Killing spinors will be the ones that are associated with unbroken supersymmetries. Since the round S^3 is invariant under orientation reversal, equal numbers of Killing spinors ϵ^+ and ϵ^- exist (namely two of each). Substituting (6.1) into the Killing spinor equation, we find, in the basis where $\Gamma_1=\sigma_3$, $\Gamma_2=\sigma_1$ and $\Gamma_3=\sigma_2$, with σ_i denoting the standard Pauli matrices, that the two sets of Killing spinors are

$$\epsilon_1^+ = \begin{pmatrix} e^{\frac{\mathrm{i}}{2}(\phi+\theta)} \\ \mathrm{i}\,e^{\frac{\mathrm{i}}{2}(\phi-\theta)} \end{pmatrix}\ , \qquad \epsilon_2^+ = \begin{pmatrix} e^{\frac{\mathrm{i}}{2}(\theta-\phi)} \\ -\mathrm{i}\,e^{-\frac{\mathrm{i}}{2}(\theta+\phi)} \end{pmatrix}\ , \tag{6.2}$$

$$\epsilon_1^- = \begin{pmatrix} e^{\frac{\mathrm{i}}{2}\psi} \\ -\mathrm{i}\,e^{\frac{\mathrm{i}}{2}\psi} \end{pmatrix}\ , \qquad \epsilon_2^- = \begin{pmatrix} e^{-\frac{\mathrm{i}}{2}\psi} \\ \mathrm{i}\,e^{-\frac{\mathrm{i}}{2}\psi} \end{pmatrix}\ , \tag{6.3}$$

Thus we see that if the conventions have been chosen so that the ϵ^+ Killing spinors are associated with unbroken supersymmetries, then they survive the dimensional reduction to $D=5$ since they are independent of ψ. Under these circumstances, the supergravity solution after the Hopf T-duality transformation on the ψ coordinate will still be fully supersymmetric. If, on the other hand, the conventions have been chosen so that the ϵ^- Killing spinors are associated with unbroken supersymmetries, then they will not survive the dimensional reduction process since they both depend on ψ. In this case, all the supersymmetry will be broken in the supergravity solution after the Hopf T-duality transformation. Note that in our discussion here, we have made a fixed orientation choice for our Hopf reduction (implicit in our definition of the vielbeins in (6.1), and so the two supersymmetry possibilities arise from two possible convention choices for the original supergravity solution. Equivalently, one could think of making a fixed convention choice in the original solution, and the two supersymmetry possibilities would then arise from the two possible orientations in the Hopf reduction.

In order to discuss the supersymmetry in the case of the T-duality transformations on the Hopf fibres of AdS_3, we need first to give an analogous construction of the Killing spinors in AdS_3. On the unit AdS_3, these satisfy $D_\mu\,\epsilon^\pm = \pm\frac{1}{2}\Gamma_\mu\,\epsilon^\pm$. Let us first consider the metric (4.3), for which the vielbein and spin connection will be

$$\begin{aligned} &e^0 = \tfrac{1}{2}\cosh\rho\,dt\,, \qquad e^1 = \tfrac{1}{2}d\rho\,, \qquad e^2 = \tfrac{1}{2}(dx + \sinh\rho\,dt)\,,\\ &\omega_{01} = -2\tanh\rho\,e^0 + e^2\,, \qquad \omega_{02} = e^1\,, \qquad \omega_{12} = -e^0\,. \end{aligned} \tag{6.4}$$

Taking the Dirac matrices to be $\Gamma_0 = -\mathrm{i}\,\sigma_1$, $\Gamma_1 = \sigma_3$ and $\Gamma_2 = \sigma_2$, we find that the solutions for the Killing spinors are

$$\epsilon_1^+ = \begin{pmatrix} e^{\frac{1}{2}(\rho+\mathrm{i}t)} \\ -e^{\frac{1}{2}(-\rho+\mathrm{i}t)} \end{pmatrix}\,, \qquad \epsilon_2^+ = \begin{pmatrix} e^{\frac{1}{2}(\rho-\mathrm{i}t)} \\ e^{-\frac{1}{2}(\rho+\mathrm{i}t)} \end{pmatrix}\,, \tag{6.5}$$

$$\epsilon_1^- = \begin{pmatrix} e^{\frac{1}{2}x} \\ -\mathrm{i}\,e^{\frac{1}{2}x} \end{pmatrix}\,, \qquad \epsilon_2^- = \begin{pmatrix} e^{-\frac{1}{2}x} \\ \mathrm{i}\,e^{-\frac{1}{2}x} \end{pmatrix}\,. \tag{6.6}$$

Again we see that the Hopf T-duality transformation on the fibre coordinate x will preserve either all or none of the supersymmetry, depending upon the orientation.

It is also instructive to construct the Killing spinors for the AdS_3 metric (4.6), since this is the one that arises as the near-horizon limit of the boosted dyonic string. For this, the vielbein and spin connection will be:

$$\begin{aligned} &e^0 = \tfrac{1}{2}e^\rho\,dt\,, \qquad e^1 = \tfrac{1}{2}d\rho\,, \qquad e^2 = \tfrac{1}{2}(dx + e^\rho\,dt)\,,\\ &\omega_{01} = -2\,e^0 + e^2\,, \qquad \omega_{02} = e^1\,, \qquad \omega_{12} = -e^0\,. \end{aligned} \tag{6.7}$$

Elementary calculations then show that the Killing spinors $\epsilon^\pm$, satisfying $D_\mu\,\epsilon^\pm = \pm\frac{1}{2}\,\Gamma_\mu\,\epsilon^\pm$, are given by

$$\epsilon_1^+ = \begin{pmatrix} t\,e^{\frac{1}{2}\rho} \\ \mathrm{i}\,e^{-\frac{1}{2}\rho} \end{pmatrix}\,, \qquad \epsilon_2^+ = \begin{pmatrix} e^{\frac{1}{2}\rho} \\ 0 \end{pmatrix}\,, \tag{6.8}$$

$$\epsilon_1^- = \begin{pmatrix} e^{\frac{1}{2}x} \\ -\mathrm{i}\, e^{\frac{1}{2}x} \end{pmatrix}, \qquad \epsilon_2^- = \begin{pmatrix} e^{-\frac{1}{2}x} \\ \mathrm{i}\, e^{-\frac{1}{2}x} \end{pmatrix}. \tag{6.9}$$

Here also, we see that the two ϵ^+ Killing spinors are independent of the fibre coordinate x, while the two ϵ^- Killing spinors depend on x. It is worth remarking that the metric (4.6) is in some ways reminiscent of the horospherical metric on AdS_3, and indeed this reflects itself in certain similarities in the form of the solutions for the Killing spinors [54].

The orientation dependence of the supersymmetry is echoed in $D = 4$, where 4-charge black hole solutions, can either preserve 1/8 of the supersymmetry or break it entirely, depending on the relative signs of the charges [18,57]. These black holes are all related by U-duality to a black hole in which two of the charges, one electric and one magnetic, are carried by the two Kaluza-Klein vectors coming from the reduction from $D = 6$. Upon oxidation to $D = 6$, this gives a metric whose near-horizon limit is $AdS_3 \times S^3$, with the two Kaluza-Klein vectors providing the twisting of the S^3 and the AdS_3 fibres. Thus the sign of the fibre orientations is precisely related to the signs of the charges in $D = 4$.

The above discussion is at the level of the massless Kaluza-Klein modes in the supergravity theory. If the orientation for the Hopf fibration is such that the relevant Killing spinors do not depend on the fibre coordinate, then statement of preservation of full supersymmetry under the Hopf T-duality extends to the full string theory. If the orientation is opposite, so that all the relevant Killing spinors depend on the fibre coordinate, then the discussion bifurcates into two categories in the full string theory. If the Hopf reduction is on the (naturally periodic) $U(1)$ fibres of the S^3, then, as we see from (6.3), the Killing spinors will be included once the non-zero modes in the Kaluza-Klein Fourier expansions are taken into account. In this case, the full supersymmetry preservation under Hopf T-duality is reinstated by including all the Kaluza-Klein modes. On the other hand, if the Hopf reduction is on the fibre coordinate of AdS_3, then, as we can see from (6.6) or (6.9), the Killing spinors will still be excluded from the spectrum even after the Kaluza-Klein non-zero modes are included. In this case, therefore, the supersymmetry remains broken even in the full string theory.

7 Non-dilatonic black holes, and Hopf T-duality

Non-dilatonic black holes (for which the dilatons are finite on the horizon) in maximal supergravity arise in $D = 5$ and $D = 4$. In $D = 5$, they are supported by three field strengths, and associated with each is an independent harmonic function. There are a total of 45 possible field configurations that can give rise to five-dimensional non-dilatonic black holes [58], *viz.*

$$D=5: \quad \{F_{(2)ij}, F_{(2)k\ell}, F_{(2)mn}\}_{15}\,, \quad \{*F_{(3)i}, \mathcal{F}^j_{(2)}, F_{(2)ij}\}_{30}\,. \tag{7.1}$$

(We are using the notation of [48,49,58] here. The subscripts on each set of field strengths denotes the multiplicities of the solutions. The Hodge duals indicate that the associated fields carry electric charges if the fields without duals carry magnetic charges, and *vice versa.*) In $D=4$, non-dilatonic black holes are supported by four field strengths, and can arise with the following possible field-strength configurations [58]:

$$\begin{aligned} N=4: \; & \{F_{(2)ij}, F_{(2)k\ell}, F_{(2)mn}, *\mathcal{F}^p_{(2)}\}_{105+105}\,, \quad \{F_{(2)ij}, *F_{(2)ik}, \mathcal{F}^j_{(2)}, *\mathcal{F}^k_{(2)}\}_{210}\,, \\ & \{F_{(2)ij}, F_{(2)k\ell}, *F_{(2)ik}, *F_{(2)j\ell}\}_{210}\,. \end{aligned} \tag{7.2}$$

The near-horizon regions of these $D=5$ and $D=4$ black holes are $AdS_2 \times S^3$ and $AdS_2 \times S^2$ respectively. If we dimensionally oxidise these solutions to $D=6$, they will describe the intersections of p-branes, waves and NUTS. There are four possible near-horizon limits that can arise for these intersections, namely

$$AdS_3 \times S^3,\; AdS_3 \times (S^2 \times S^1),\; (AdS_2 \times S^1) \times S^3,\; (AdS_2 \times S^1) \times (S^2 \times S^1)\,. \tag{7.3}$$

(To be precise, the AdS_3 and the S^3 can in general be factored by cyclic groups, in the manner discussed previously.) If we oxidise these near-horizon solutions further, to $D=10$ or $D=11$, then the additional dimensions provide additional factors of T^4 or T^5 respectively.

For example, the four-charge black hole solution using the field strengths of the last entry in the list (7.2) becomes an intersection of p-branes in $D=6$, and its near-horizon region is $(AdS_2 \times S^1) \times (S^2 \times S^1)$. This is because the Kaluza-Klein vector is not involved in the solution, and so the six-dimensional metric is diagonal. If the solution is built using the set of field strengths in the first entry of the list (7.2), two possibilities can arise. If the Kaluza-Klein field $\mathcal{F}^p_{(2)}$ carries a magnetic charge (implying that the other three field strengths carry electric charges), then the solution becomes $(AdS_2 \times S^1) \times S^3$. In other words, the Kaluza-Klein vector describes a magnetic monopole which corresponds, from the six-dimensional point of view, to a NUT charge that twists the $S^2 \times S^1$ product to give S^3. On other hand, if the Kaluza-Klein field carries an electric charge (and so the other three field strengths carry magnetic charges), then the solution describes $AdS_3 \times (S^2 \times S^1)$. This is because in this case the Kaluza-Klein vector has a configuration which, from the six-dimensional point of view, corresponds to a wave which twists the $AdS_2 \times S^1$ product to give AdS_3. The oxidation of the solutions for the configurations of field strengths listed in the second entry in (7.2) describe $AdS_3 \times S^3$ in $D = 6$. This is because there are two Kaluza-Klein fields in this case, *viz.* $\mathcal{F}^j_{(2)}$ and $\mathcal{F}^k_{(2)}$. One of them carries a magnetic charge and hence twists the $S^2 \times S^1$ product, while the other carries an electric charge, and hence twists the $AdS_2 \times S^1$ product.

We showed in sections 3, 4 and 5 that the near-horizon structures (7.3) of the six-dimensional intersections that come from the oxidations of the non-dilatonic black holes in $D = 5$ and $D = 4$ are related to each other through Hopf T-duality. Furthermore, even the near-horizon limits of the 4-charge solutions which are, owing to sign choices, non-supersymmetric, are related by Hopf T-duality to the supersymmetric ones. As we show in appendix D, the Hopf T-duality not only relates the near-horizon limits listed in and below (7.3), but also relates the associated full solutions. Thus in particular, a Hopf reduction and T-duality has the effect of mapping the solutions (7.1) and (7.2) among each other. The harmonic function associated with the Kaluza-Klein vector coming from the Hopf reduction lacks a constant term; however, it can be introduced by performing an appropriate U-duality transformation[59,8,60,61].

Appendices

A T-duality of the truncated six-dimensional theories

In section 2, we obtained two different consistent truncations of six-dimensional maximal supergravity. One of them, given by (2.2), naturally arose as a consistent truncation of six-dimensional maximal supergravity in the type IIB picture. This truncated theory has an $O(2,2)$ global symmetry. The other theory, given by (2.3), naturally arose as a consistent truncation of six-dimensional maximal supergravity in the type IIA picture. This theory has only an $\mathbb{R} \times \mathbb{R}$ global symmetry. Interestingly, it can be obtained as the dimensional reduction of the seven-dimensional Lagrangian (2.4), which itself can easily be shown to be a consistent truncation of seven-dimensional maximal supergravity. In this appendix, we shall show that the two different six-dimensional Lagrangians (2.2) and (2.3) are T-dual to each other, in the sense that upon dimensional reduction they give rise to the same five-dimensional theory, up to field redefinitions.

We begin by obtaining the dimensional reduction of the Lagrangian (2.2), coming from the consistent truncation in the type IIB picture. We find

$$e^{-1}\,\mathcal{L}_{5B} = R - \tfrac{1}{2}(\partial\phi_1)^2 - \tfrac{1}{2}(\partial\phi_2)^2 - \tfrac{1}{2}(\partial\varphi)^2 - \tfrac{1}{2}e^{2\phi_1}\,(\partial\chi_1)^2 - \tfrac{1}{2}e^{2\phi_2}\,(\partial\chi_2)^2$$

$$
\begin{aligned}
&-\tfrac{1}{12}e^{-\phi_1-\phi_2+\sqrt{\frac{2}{3}}\varphi}(F^{\rm NS}_{(3)})^2-\tfrac{1}{12}e^{\phi_1-\phi_2+\sqrt{\frac{2}{3}}\varphi}(F^{\rm RR}_{(3)})^2\\
&-\tfrac{1}{4}e^{-\phi_1-\phi_2-\sqrt{\frac{2}{3}}\varphi}(F^{\rm NS}_{(2)1})^2-\tfrac{1}{4}e^{\phi_1-\phi_2-\sqrt{\frac{2}{3}}\varphi}(F^{\rm RR}_{(2)1})^2\\
&-\tfrac{1}{4}e^{\sqrt{\frac{8}{3}}\varphi}(\mathcal{F}_{(2)})^2+\chi_2\, dA^{\rm NS}_{(2)}\wedge dA^{\rm RR}_{(1)1}-\chi_2\, dA^{\rm RR}_{(2)}\wedge dA^{\rm NS}_{(1)1}\,. \qquad \text{(A.1)}
\end{aligned}
$$

Here, the field strengths are given by

$$
\begin{aligned}
&F^{\rm NS}_{(3)}=dA^{\rm NS}_{(2)}-dA^{\rm NS}_{(1)1}\wedge\mathcal{A}_{(1)}\,,\qquad F^{\rm NS}_{(2)1}=dA^{\rm NS}_{(1)1}\,,\qquad \mathcal{F}_{(2)}=d\mathcal{A}_{(1)}\,,\\
&F^{\rm RR}_{(3)}=dA^{\rm RR}_{(2)}-dA^{\rm RR}_{(1)1}\mathcal{A}_1+\chi_1(dA^{\rm NS}_{(2)}-dA^{\rm NS}_{(1)1}\wedge\mathcal{A}_{(1)})\,,\\
&F^{\rm RR}_{(2)1}=dA^{\rm RR}_{(1)1}+\chi_1 dA^{\rm NS}_{(1)1}\,. \qquad \text{(A.2)}
\end{aligned}
$$

The five-dimensional Lagrangian coming from the dimensional reduction of the truncated theory (2.3) is given by

$$
\begin{aligned}
e^{-1}\mathcal{L}_{5A}=R&-\tfrac{1}{2}(\partial\phi_1)^2-\tfrac{1}{2}(\partial\phi_2)^2-\tfrac{1}{2}(\partial\varphi)^2\\
&-\tfrac{1}{2}e^{-\frac{1}{2}\phi_1+\frac{3}{2}\phi_2-\sqrt{\frac{3}{2}}\varphi}(\partial\chi')^2-\tfrac{1}{2}e^{\frac{3}{2}\phi_1-\frac{1}{2}\phi_2-\sqrt{\frac{3}{2}}\varphi}(\partial A_{(0)1})^2\\
&-\tfrac{1}{12}e^{-\phi_1-\phi_2+\sqrt{\frac{2}{3}}\varphi}(F_{(3)})^2-\tfrac{1}{12}e^{\frac{1}{2}\phi_1-\frac{3}{2}\phi_2-\frac{1}{\sqrt{6}}\varphi}(F_{(3)1})^2\\
&-\tfrac{1}{4}e^{-\phi_1-\phi_2-\sqrt{\frac{2}{3}}\varphi}(F_{(2)1})^2-\tfrac{1}{4}e^{\frac{3}{2}\phi_1-\frac{1}{2}\phi_2+\frac{1}{\sqrt{6}}\varphi}(F_{(2)})^2\\
&-\tfrac{1}{4}e^{\sqrt{\frac{8}{3}}\varphi}(\mathcal{F}_{(2)})^2+\chi' F_{(3)1}\wedge\mathcal{F}_2+\chi' F_{(3)}\wedge F_{(2)}\,, \qquad \text{(A.3)}
\end{aligned}
$$

where we have dualised the 3-form potential $A_{(3)}$ to an axion χ'. The field strengths in (A.4) are given given by

$$
\begin{aligned}
&F_{(3)1}=dA_{(2)1}+dA_{(1)1}\wedge dA_{(1)}-A_{(0)1}\,dA_{(2)}\,,\qquad F_{(2)1}=dA_{(1)1}\,,\\
&F_{(1)1}=dA_{(0)1}\,,\qquad F_{(3)}=dA_{(2)}-dA_{(1)1}\wedge\mathcal{A}_{(1)}\,,\\
&F_{(2)}=dA_{(1)}-dA_{(0)1}\wedge\mathcal{A}_{(1)}\,. \qquad \text{(A.4)}
\end{aligned}
$$

It can be verified that the two 5-dimensional Lagrangians (A.1) and (A.3) are related to each other by the field redefinition described in Table 1 and (2.6). To be precise, we first make the following field redefinitions:

$$
A'_{(1)}=A_{(1)}-A_{(0)1}\mathcal{A}_{(1)}\,,\quad A'_{(2)}=A_{(2)}-A_{(1)1}\wedge\mathcal{A}_{(1)}\,,\quad A'_{(2)1}=A_{(2)1}+A_{(1)1}\wedge A'_{(1)}\,. \qquad \text{(A.5)}
$$

After doing this we find that the Lagrangian (A.3) can be mapped to the Lagrangian (A.1) by the following transformations

$$A'_{(2)} \longrightarrow A^{\rm NS}_{(2)}\ , \quad A'_{(2)1} \longrightarrow -A^{\rm RR}_{(2)}\ , \quad A_1 \longrightarrow A^{\rm NS}_{(1)1}\ , \quad A'_{(1)} \longrightarrow A^{\rm RR}_{(1)1}\ ,$$
$$A_{(1)1} \longrightarrow \mathcal{A}_{(1)}\ , \quad A_{(0)1} \longrightarrow \chi_1\ , \quad \chi' \longrightarrow \chi_2\ , \qquad \text{(A.6)}$$

together with the transformation of the dilatons given by (2.6).

B $O(2,2)$ symmetry of the truncated six-dimensional theory

Here, we give the explicit $O(2,2) \sim SL(2,\mathbb{R})_1 \times SL(2,\mathbb{R})_2$ global symmetry transformations for the truncated six-dimensional theory (2.2). The factor $SL(2,\mathbb{R})_1$ is an S-duality symmetry that maps between the NS-NS and R-R 2-form potentials, and is realised at the level of the Lagrangian. The factor $SL(2,\mathbb{R})_2$ is an electric/magnetic duality symmetry between the NS-NS and R-R 3-form field strengths, which is realised only at the level of the equations of motion.

To present the global transformation rules, it is useful first to define the two complex scalar fields

$$\tau_1 \equiv \chi_1 + {\rm i}\, e^{-\phi_1}\ , \qquad \tau_2 \equiv \chi_2 + {\rm i}\, e^{-\phi_2}\ . \qquad \text{(B.1)}$$

The two $SL(2,\mathbb{R})$ transformations act non-linearly on the scalar manifold as follows:

$$SL(2,\mathbb{R})_1 :\ \tau_1 \longrightarrow \Lambda_1 \cdot \tau_1 \equiv \frac{a_1\,\tau_1 + b_1}{c_1\,\tau_1 + d_1}\ , \qquad \tau_2 \longrightarrow \tau_2\ ,$$
$$SL(2,\mathbb{R})_2 :\ \tau_2 \longrightarrow \Lambda_2 \cdot \tau_2 \equiv \frac{a_2\,\tau_2 + b_2}{c_2\,\tau_2 + d_2}\ , \qquad \tau_1 \longrightarrow \tau_1\ , \qquad \text{(B.2)}$$

where $a_1\, d_1 - b_1\, c_1 = 1 = a_2\, d_2 - b_2\, c_2$, and we may define the $SL(2,\mathbb{R})$ matrices Λ_1 and Λ_2 in the standard way:

$$\Lambda_1 = \begin{pmatrix} a_1 & b_1 \\ c_1 & d_1 \end{pmatrix}\ , \qquad \Lambda_2 = \begin{pmatrix} a_2 & b_2 \\ c_2 & d_2 \end{pmatrix}\ . \qquad \text{(B.3)}$$

$SL(2,\mathbb{R})_1$ is a symmetry of the Lagrangian, and it acts linearly on the 2-form potentials:

$$A_{(2)} \equiv \begin{pmatrix} A^{\rm NS}_{(2)} \\ A^{\rm RR}_{(2)} \end{pmatrix} \longrightarrow (\Lambda_1^T)^{-1}\, A_{(2)}\ . \qquad \text{(B.4)}$$

$SL(2,\mathbb{R})_2$ is a symmetry only at the level of the equations of motion, and it acts locally only on the field strengths. We shall also, for convenience, present the action of $SL(2,\mathbb{R})_1$ on the field strengths. We begin by defining two field-strength doublets, $H^1_{(3)}$ and $H^2_{(3)}$:

$$H^1_{(3)} = \begin{pmatrix} e^{-\phi_1} F^{\rm NS}_{(3)} + \chi_1 e^{\phi_1} F^{\rm RR}_{(3)} \\ e^{\phi_1} F^{\rm RR}_{(3)} \end{pmatrix}, \qquad H^2_{(3)} = \begin{pmatrix} e^{-\phi_2} *F^{\rm NS}_{(3)} + \chi_2 F^{\rm RR}_{(3)} \\ F^{\rm RR}_{(3)} \end{pmatrix}. \tag{B.5}$$

Their $SL(2,\mathbb{R})$ transformations are:

$$\begin{aligned} SL(2,\mathbb{R})_1 &: \ H^1_{(3)} \longrightarrow \Lambda_1 H^1_{(3)}, \\ SL(2,\mathbb{R})_2 &: \ H^2_{(3)} \longrightarrow \Lambda_2 H^2_{(3)}. \end{aligned} \tag{B.6}$$

Note that the two fields $H^1_{(3)}$ and $H^2_{(3)}$ are not independent, and so given the transformation on one, the transformation on the other is in principle determined. Each was introduced for the specific purpose of encoding one or other of the two $SL(2,\mathbb{R})$ transformations in a simple way, as indicated in (B.6), and we do not need to give the associated transformations on the other $H_{(3)}$ field.

The theory can describe strings that carry four independent charges, namely the electric and the magnetic charges for both the NS-NS and the R-R 3-forms. It is easily seen from the equations of motion following from (2.2), and from the Bianchi identities for $F^{\rm NS}_{(3)}$ and $F^{\rm RR}_{(3)}$, that they are given by

$$\begin{aligned} Q^{\rm NS}_e &= \frac{1}{16\pi^2}\int \Big\{ e^{-\phi_1-\phi_2} *F^{\rm NS}_{(3)} + \chi_2 F^{\rm RR}_{(3)} + \chi_1 \Big(e^{\phi_1-\phi_2} *F^{\rm RR}_{(3)} - \chi_2 F^{\rm NS}_{(3)} \Big) \Big\}, \\ Q^{\rm NS}_m &= \frac{1}{16\pi^2}\int F^{\rm NS}_{(3)}, \\ Q^{\rm RR}_e &= \frac{1}{16\pi^2}\int \Big\{ e^{\phi_1-\phi_2} *F^{\rm RR}_{(3)} - \chi_2 F^{\rm NS}_{(3)} \Big\}, \\ Q^{\rm RR}_m &= \frac{1}{16\pi^2}\int \Big\{ F^{\rm RR}_{(3)} - \chi_1 F^{\rm NS}_{(3)} \Big\}. \end{aligned} \tag{B.7}$$

From the $SL(2,\mathbb{R})_1$ and $SL(2,\mathbb{R})_2$ transformations rules (B.2) and (B.6), we find that these charges transform as:

$$\begin{aligned} SL(2,\mathbb{R})_1 &: \vec{Q}_{ee} \equiv \begin{pmatrix} Q^{\rm NS}_e \\ Q^{\rm RR}_e \end{pmatrix} \longrightarrow \Lambda_1 \vec{Q}_{ee}, \ \vec{Q}_{mm} \equiv \begin{pmatrix} Q^{\rm NS}_m \\ Q^{\rm RR}_m \end{pmatrix} \longrightarrow (\Lambda_1^T)^{-1} \vec{Q}_{mm}, \\ SL(2,\mathbb{R})_2 &: \vec{Q}_{em} \equiv \begin{pmatrix} Q^{\rm NS}_e \\ Q^{\rm RR}_m \end{pmatrix} \longrightarrow \Lambda_2 \vec{Q}_{em}, \ \vec{Q}_{me} \equiv \begin{pmatrix} Q^{\rm NS}_m \\ Q^{\rm RR}_e \end{pmatrix} \longrightarrow (\Lambda_2^T)^{-1} \vec{Q}_{me}. \end{aligned} \tag{B.8}$$

Here we are introducing the notation that $\vec{Q}_{xy}$ is a two-component charge vector, whose upper component is the electric ($x = e$) or magnetic ($x = m$)

NS-NS charge, and whose lower component is the electric ($y = e$) or magnetic ($y = m$) R-R charge.

It is worthwhile pausing at this point, to understand why the charge vectors $\vec{Q}_{mm}$ and $\vec{Q}_{me}$ transform contragrediently in comparison to the transformations of $\vec{Q}_{ee}$ and $\vec{Q}_{em}$. If we introduce an index notation for the two fields, so that $F^i_{(3)}$ denotes the NS-NS field when $i = 1$, and the R-R field when $i = 2$, then the dual fields $\widetilde{F}_{(3)1} = e^{-\phi_1-\phi_2} *F^1_{(3)}$ and $\widetilde{F}_{(3)2} = e^{\phi_1-\phi_2} *F^2_{(3)}$ actually correspond to the first and second components of a doublet $\widetilde{F}_{(3)i}$ with a *downstairs* index i. (This can be seen from the fact that the relevant $SL(2,\mathbb{R})_1$-invariant kinetic terms in the Lagrangian have the form $-\frac{1}{2}\widetilde{F}_{(3)i} \wedge F^i_{(3)}$ [62].) The electric charges therefore can be labelled as $Q_{e\,i} = \{Q_e^{\rm NS}, Q_e^{\rm RR}\}$, while the magnetic charges can be labelled as $Q^i_m = \{Q_m^{\rm NS}, Q_m^{\rm RR}\}$. Since the electric charges transform directly with Λ, we can see that the suppressed matrix indices on Λ_1, if made explicit, are located as follows: $(\Lambda_1)_i{}^j$. It is now clear why it is the contragedient representation that acts on the magnetic charges Q^i_m. An analogous comment applies to the electric and magnetic indices associated with the $SL(2,\mathbb{R})_2$ transformations.

It is convenient to give a 4×4 matrix representation for the $SL(2,\mathbb{R})_1 \times SL(2,\mathbb{R})_2$ transformations, by defining the tensor product of 2×2 matrices, and 2-component vectors, as follows:

$$M \otimes N \equiv \begin{pmatrix} m_{11}\,N & m_{12}\,N \\ m_{21}\,N & m_{22}\,N \end{pmatrix} ,$$

$$\begin{pmatrix} x \\ y \end{pmatrix} \otimes \begin{pmatrix} u \\ v \end{pmatrix} \equiv \begin{pmatrix} x\,u \\ x\,v \\ y\,u \\ y\,v \end{pmatrix} , \qquad \text{(B.9)}$$

where m_{ij} denotes the components of M. In view of the remarks in the previous paragraph, it follows that we can represent the action of the two $SL(2,\mathbb{R})$'s on the charges as $\vec{Q} \longrightarrow \Lambda\,\vec{Q}$, where $\Lambda = \Lambda_2 \otimes \Lambda_1$, and

$$\vec{Q} = \begin{pmatrix} Q_e^{\rm NS} \\ Q_e^{\rm RR} \\ Q_m^{\rm RR} \\ -Q_m^{\rm NS} \end{pmatrix} . \qquad \text{(B.10)}$$

For later convenience, we may also now introduce an alternative parameterisation of the scalar coset manifold. It is sufficient for this purpose to consider a generic $SL(2,\mathbb{R})/O(2)$ coset, and the formalism can then be applied to both $SL(2,\mathbb{R})_1$ and $SL(2,\mathbb{R})_2$. We define the upper-triangular matrix

$$\mathcal{V} = \begin{pmatrix} e^{-\frac{1}{2}\phi} & \chi\, e^{\frac{1}{2}\phi} \\ 0 & e^{\frac{1}{2}\phi} \end{pmatrix} . \tag{B.11}$$

This gives a Borel parameterisation of the coset, whose Lagrangian can now be written as $\frac{1}{4}\mathrm{tr}(\partial_\mu(\mathcal{M}^{-1})\,\partial^\mu \mathcal{M})$, where

$$\mathcal{M} = \mathcal{V}\,\mathcal{V}^T = \begin{pmatrix} e^{-\phi} + \chi^2\, e^{\phi} & \chi\, e^{\phi} \\ \chi\, e^{\phi} & e^{\phi} \end{pmatrix} . \tag{B.12}$$

The $SL(2,\mathbb{R})$ transformation (B.2) of the scalar fields can then be expressed as $\mathcal{V} \longrightarrow \Lambda\,\mathcal{V}\,\mathcal{O}$, where $\mathcal{O}$ is a field-dependent compensating $O(2)$ transformation that restores $\mathcal{V}$ to the Borel gauge. On the matrix $\mathcal{M}$, the $SL(2,\mathbb{R})$ transformation is simply $\mathcal{M} \longrightarrow \Lambda\,\mathcal{M}\,\Lambda^T$. Note that $H^1_{(3)}$ in (B.5) can now be written as $H^1_{(3)} = \mathcal{M}_1\, dA_{(2)}$.

With these preliminaries, we are now in a position to construct, in the next section, an $O(2,2)$ multiplet of dyonic strings

C An $O(2,2;\mathbf{Z})$ multiplet of dyonic strings

We may construct general dyonic string solutions with arbitrary electric and magnetic NS-NS and R-R charges $(Q_e^{\rm NS}, Q_m^{\rm NS}, Q_e^{\rm RR}, Q_m^{\rm RR})$ at an arbitrary modulus point specified by τ_1^0 and τ_2^0, by starting with the simple case of an NS-NS dyonic string in the $\tau_1^0 = \tau_2^0 = \mathrm{i}$ vacuum, with electric and magnetic NS-NS charges q and p. We then act with $O(2)_1$ and $O(2)_2$ transformations

$$\Lambda_1(\theta_1) = \begin{pmatrix} \cos\theta_1 & \sin\theta_1 \\ -\sin\theta_1 & \cos\theta_1 \end{pmatrix} , \qquad \Lambda_2(\theta_2) = \begin{pmatrix} \cos\theta_2 & \sin\theta_2 \\ -\sin\theta_2 & \cos\theta_2 \end{pmatrix} , \tag{C.1}$$

which lie in the stability subgroups of the $SL(2,\mathbb{R})$'s that rotate the charges while leaving the $\tau^0 = \mathrm{i}$ modulus points fixed. Next, we act with the Borel transformations

$$\Lambda_1(\tau_1^0) = \begin{pmatrix} e^{-\frac{1}{2}\phi_1^0} & \chi_1^0\, e^{\frac{1}{2}\phi_1^0} \\ 0 & e^{\frac{1}{2}\phi_1^0} \end{pmatrix} , \qquad \Lambda_2(\tau_2^0) = \begin{pmatrix} e^{-\frac{1}{2}\phi_2^0} & \chi_2^0\, e^{\frac{1}{2}\phi_2^0} \\ 0 & e^{\frac{1}{2}\phi_2^0} \end{pmatrix} , \tag{C.2}$$

which map the original modulus points $\tau_1^0 = \tau_2^0 = \mathrm{i}$ to the arbitrary points $\tau_1^0 = \chi_1^0 + \mathrm{i}\, e^{-\phi_1^0}$ and $\tau_2^0 = \chi_2^0 + \mathrm{i}\, e^{-\phi_2^0}$. The combined effect of the stability-subgroup and Borel-subgroup transformations is to map the original charges $\vec{Q} = (Q_e^{\rm NS}, Q_e^{\rm RR}, Q_m^{\rm RR}, -Q_m^{\rm NS}) = (q, 0, 0, -p)$ to arbitrary charges, which are related to the parameters θ_1, θ_2, q and p (at the given modulus point $(\phi_1^0, \chi_1^0, \phi_2^0, \chi_2^0)$)

in a manner that we shall determine below. Thus we can obtain strings with their four charges lying at arbitrary points on the charge lattice (that satisfy the Dirac quantisation condition) by appropriately choosing the four parameters. (The spirit of this construction is similar to that used in [63] in the discussion of the charge lattice for the $O(6,22;\mathbb{Z})$ multiplet of black holes in the $D=4$ heterotic string, and in [46] for the construction of the $SL(2,\mathbb{Z})$ multiplet of type IIB strings. A general procedure for generating the U-duality multiplets for all p-brane solitons was given in [64]. A group theoretic approach was also introduced in [64], using the homogeneous scaling symmetries of the equations of motion that arise in theories such as the maximal supergravities, in order to give a construction of genuine spectrum-generating groups for BPS states. An explicit construction of U-duality multiplets for BPS states in eight dimensions was given in [65].)

C.1 Unboosted isotropic dyonic strings

Moving now to the details, let us consider an isotropic unboosted dyonic string, supported by the NS-NS 3-form field [5];

$$\begin{aligned} ds^2 &= (H_e\, H_m)^{-1/2}\, dx^\mu\, dx_\mu + (H_e\, H_m)^{1/2}\, d\vec{y}\cdot d\vec{y}\ ,\\ F^{\rm NS}_{(3)} &= 8\,p\,\Omega_{(3)} + 8\,q\,H_m\,H_e^{-1}\,{*Omega}_{(3)}\ , \qquad F^{\rm RR}_{(3)} = 0\ ,\\ \tau_1 &= \chi_1 + {\rm i}\,e^{-\phi_1} = \tau_2 = \chi_2 + {\rm i}\,e^{-\phi_2} = {\rm i}\,(H_e/H_m)^{1/2}\ , \end{aligned} \tag{C.3}$$

where $H_e = 1 + 4q/r^2$ and $H_m = 1 + 4p/r^2$, q and p are the electric and magnetic charges (following the normalisations given in (B.7)), and $\Omega_{(3)}$ is the volume form on the unit S^3. After some algebra, we find, following the steps outlined above, that the solution after performing the stability-subgroup and Borel-subgroup transformations becomes

$$\begin{aligned} ds^2 &= (H_e\, H_m)^{-1/2}\, dx^\mu\, dx_\mu + (H_e\, H_m)^{1/2}\, d\vec{y}\cdot d\vec{y}\ ,\\ \tau_1 &= \chi_1^0 + {\rm i}\,e^{-\phi_1^0}\, \frac{\sqrt{H_e} - {\rm i}\,\tan\theta_1\,\sqrt{H_m}}{\sqrt{H_m} - {\rm i}\,\tan\theta_1\,\sqrt{H_e}}\ ,\\ \tau_2 &= \chi_2^0 + {\rm i}\,e^{-\phi_2^0}\, \frac{\sqrt{H_e} - {\rm i}\,\tan\theta_2\,\sqrt{H_m}}{\sqrt{H_m} - {\rm i}\,\tan\theta_2\,\sqrt{H_e}}\ ,\\ F^{\rm NS}_{(3)} &= e^{\frac12(\phi_1^0+\phi_2^0)}\Big(\cos\theta_1\,\cos\theta_2\,\Theta - \sin\theta_1\,\sin\theta_2\,\frac{H_e}{H_m}\,{*\Theta}\Big)\ ,\\ F^{\rm RR}_{(3)} &= \frac{-e^{\frac12(\phi_2^0-\phi_1^0)}\,H_e}{\sin^2\theta_1\,H_e + \cos^2\theta_1\,H_m}\Big(\sin\theta_1\,\cos\theta_2\,\Theta + \cos\theta_1\,\sin\theta_2\,{*\Theta}\Big)\ , \end{aligned} \tag{C.4}$$

where $\Theta \equiv 8\,p\,\Omega_{(3)} + 8\,q\,{*\Omega}_{(3)}$.

Under the stability-subgroup and Borel-subgroup transformations (C.1) and (C.2), the initial charge 4-vector $\vec{Q}_0 = \{q, 0, 0, -p\}$ is mapped to the final charge vector $\vec{Q}_f$, given by

$$\vec{Q}_f = \Big(\Lambda_2(\tau_2^0) \otimes \Lambda_1(\tau_1^0)\Big) \Big(\Lambda_2(\theta_2) \otimes \Lambda_1(\theta_1)\Big) \vec{Q}_0 \,, \tag{C.5}$$

which implies

$$\Big(\Lambda_2(\theta_2) \otimes \Lambda_1(\theta_1)\Big) \vec{Q}_0 = \Big(\Lambda_2(\tau_2^0)^{-1} \otimes \Lambda_1(\tau_1^0)^{-1}\Big) \vec{Q}_f \equiv \Lambda(\tau_0)^{-1} \vec{Q}_f \,. \tag{C.6}$$

This equation provides the relation between the four parameters $(\theta_1, \theta_2, q, p)$ and the four final charges $\vec{Q}_f$, for any given values of the scalar moduli $\tau_0 = (\tau_1^0, \tau_2^0)$. To obtain the explicit solution, we first note that the stability-subgroup $O(2)$ rotations (C.1) can be written as $\Lambda(\theta) = e^{i\theta\sigma}$, where $\sigma = \begin{pmatrix} 0 & -i \\ i & 0 \end{pmatrix}$. Equation (C.6) can now be written as

$$e^{i\theta_2\sigma} \otimes e^{i\theta_1\sigma} \vec{Q}_0 = \Lambda(\tau_0)^{-1} \vec{Q}_f \,. \tag{C.7}$$

This implies that $e^{i\theta_2\sigma} \otimes e^{i\theta_1\sigma} U_\pm = V_\pm$, where

$$U_\pm = (\mathbf{1} + \sigma) \otimes (\mathbf{1} \pm \sigma) \vec{Q}_0 \,, \qquad V_\pm = (\mathbf{1} + \sigma) \otimes (\mathbf{1} \pm \sigma) \vec{Q}_f \,. \tag{C.8}$$

After elementary algebra, it now follows that

$$\begin{aligned}(q+p)\, e^{i(\theta_2+\theta_1)} &= e^{\frac{1}{2}(\phi_1^0+\phi_2^0)} \,(Q_e^{\rm NS} - \tau_1^0\, Q_e^{\rm RR} - \tau_2^0\, Q_m^{\rm RR} - \tau_1^0\, \tau_2^0\, Q_e^{\rm NS}) \equiv \Delta_+ \,, \\ (q-p)\, e^{i(\theta_2-\theta_1)} &= e^{\frac{1}{2}(\phi_1^0+\phi_2^0)} \,(Q_e^{\rm NS} - \bar{\tau}_1^0\, Q_e^{\rm RR} - \tau_2^0\, Q_m^{\rm RR} - \bar{\tau}_1^0\, \tau_2^0\, Q_e^{\rm NS}) \equiv \Delta_- \end{aligned} \tag{C.9}$$

From here, the solutions for $(\theta_1, \theta_2, q, p)$ immediately follow. Note that the factor $e^{\frac{1}{2}(\phi_1^0+\phi_2^0)}$ is precisely the six-dimensional effective string coupling constant.

It is now easy to determine the formula for the mass m per unit length for the general 4-charge dyonic string. To do this, we note that in the original NS-NS dyonic string solution (C.3), the mass is simply given by $m = q + p$, which we can write in the $O(2,2)$-invariant form

$$m^2 = q^2 + p^2 + 2q\,p = \vec{Q}_0^T \vec{Q}_0 - \vec{Q}_0^T \hat{\Omega} \vec{Q}_0 \,, \tag{C.10}$$

where $\hat{\Omega} \equiv \Omega \otimes \Omega$, and $\Omega = \begin{pmatrix} 0 & 1 \\ -1 & 0 \end{pmatrix}$. In terms of the final charges $\vec{Q}_f$ of the generic 4-charge dyonic string, the mass is therefore given by

$$m^2 = \vec{Q}_f^T \,(\Lambda(\tau^0)^T)^{-1}\, \Lambda(\tau^0)^{-1} \vec{Q}_0 - \vec{Q}_f^T \,(\Lambda(\tau^0)^T)^{-1}\, \hat{\Omega}\, \Lambda(\tau^0)^{-1} \vec{Q}_0$$

$$= \vec{Q}_f^T \left((\Lambda(\tau^0)\,\Lambda(\tau^0)^T)^{-1} - \hat{\Omega} \right) \vec{Q}_f \; .$$
$$= |\Delta_+|^2 \; . \qquad \text{(C.11)}$$

(The expression for m^2 in the last line follows directly from (C.10) and (C.9).)

Note that the second line of the mass formula (C.11) is composed of two independent $O(2,2)$-invariant quantities. The second term is precisely the quantity that appears in the Dirac quantisation condition [66], namely

$$\vec{Q}_f^T\, \hat{\Omega}\, \vec{Q}_f = \text{integer} \; . \qquad \text{(C.12)}$$

This condition implies that the four charges $\vec{Q}_f = (Q_e^{\rm NS}, Q_e^{\rm RR}, Q_m^{\rm RR}, -Q_m^{\rm NS})$ must lie on a discrete lattice, which, for simplicity, may be taken to be the square integer lattice. In our construction of the multiplet of integer-charge solutions, we of course allowed the initial charge parameters q and p and the rotation angles θ_1 and θ_2 to be unrestricted by any quantisation condition. For a given modulus point $\tau^0 = (\tau_1^0, \tau_2^0)$, after restricting the final charges to lie on the Dirac charge lattice, the initial parameters will themselves fill out only a discrete lattice of values.

The points in a given charge lattice, for example the integer square lattice, can be filled out by a discrete $O(2,2;\mathbb{Z})$ spectrum-generating group that leaves the scalar modulus point τ^0 invariant. It is therefore not itself the discretised form of the of the $O(2,2)$ global symmetry transformations that we originally discussed, since this transforms the scalar moduli at the same time as it moves the charges on their lattice. The difference is highlighted by the fact that the spectrum-generating group must, of course, in particular generate strings of different masses, whilst the original global symmetry group leaves the metric, and hence the mass, invariant. This issue was extensively discussed in [64], where it was shown that to get the true spectrum-generating group it is necessary to make use also of an homogeneous scaling "trombone" symmetry of the theory. In [64], the example of the $SL(2,\mathbb{Z})$ multiplet of type IIB strings was discussed in detail. However, in the $O(2,2)$ case we are considering here there is an added subtlety. This can be seen most easily at the classical level. In the case of the type IIB NS-NS and R-R strings, the charge space is two-dimensional, and can be completely spanned, for any given modulus point, by the action of the $O(2)$ denominator group together with the trombone rescaling symmetry. In our present case, however, the charge space is four-dimensional, while the modulus-preserving denominator group is $O(2) \times O(2)$. Together with the trombone symmetry this gives only three parameters, and hence this is insufficient to span the charge vector space. In our construction, we were nevertheless able to construct the complete charge lattice of dyonic strings. This is because we started with the solution (C.3) that had the two free parameters q and p. The trombone symmetry is responsible for the existence of the 1-parameter sub-family of solutions where q and p are uniformly rescaled by the same factor, which has the effect of also rescaling the mass. It is less clear what is the symmetry that is responsible for allowing solutions with different relative ratios between the electric and magnetic charge parameters q and p.

It was argued in [64] that one resolution to this puzzle might be that since the dyonic string can be viewed as a bound state with *zero* binding energy, it is less fundamental than the individual electric and magnetic building blocks. Thus it would be unnecessary to find a symmetry to relate the bound-state solutions with different ratios for the electric and magnetic charge. (Another example where there is no homogeneous scaling symmetry to account for the existence of arbitrary-charge solutions is in the heterotic string. This is because Yang-Mills fields $F \sim dA + A \wedge A$ do not scale uniformly under $A \to \lambda A$. (One requires a scaling symmetry that rescales the metric, so that the mass can be changed, while leaving the scalars invariant, so that the scalar moduli are unchanged.))

In this section, we constructed an $O(2,2;\mathbf{Z})$ multiplet of dyonic strings in $D = 6$. This subgroup of the full $O(5,5;\mathbf{Z})$ or $O(5,21;\mathbf{Z})$ duality groups captures the essence of the complete groups, in that it describes both the NS-NS/R-R duality and the electric/magnetic duality. The extension to the full duality group is straightfoward in principle, using, for example, the techniques described in [64], but its detailed implementation would be tedious and rather unrewarding. The most interesting aspect of the U-duality multiplets is to see how the charges transform under the full U-duality group. The classification of the U-duality orbits of the charges in all dimensions can be found in [67,68].

C.2 Boosted and twisted dyonic strings

In the discussion above, we gave the construction of the $O(2,2;\mathbf{Z})$ multiplet of unboosted, isotropic dyonic strings. It is straightforward to boost in the worldvolume, or to twist in the transverse space, and thereby obtain $O(2,2;\mathbf{Z})$ multiplets of boosted or twisted dyonic strings. We do this by following a strategy analogous to the one we used previously, namely by starting with a boosted and twisted dyonic string supported purely by NS-NS 3-form charges, and taking the scalar moduli to vanish. The form of the solutions for the fields $(\phi_1, \chi_1, \phi_2, \chi_2, F^{\rm NS}_{(3)}, F^{\rm RR}_{(3)})$ is identical to that for the unboosted, untwisted case given in (C.3), except that now the harmonic functions H_e and H_m are modified to $H_e = 1 + q/r$, $H_m = 1 + p/r$. (In other words, they are now harmonic only in the 3-dimensional overall transverse space, rather than the 4-dimensional transverse space for the dyonic strings.[j])

The metric is given by

$$ds^2 = -(H_e H_m)^{-\frac{1}{2}} K_w^{-1} dt^2 + (H_e H_m)^{-\frac{1}{2}} K_w (dz_1 + (K_w^{-1} - 1)dt)^2$$
$$+(H_e H_m)^{\frac{1}{2}} K_N^{-1} (dz_2 + Q_N \cos\theta\, d\phi)^2$$
$$+(H_e H_m)^{\frac{1}{2}} K_N (dr^2 + r^2 d\theta^2 + r^2 \sin^2\theta\, d\phi^2) , \tag{C.13}$$

where the extra harmonic functions for the wave and the boost charges are $K_w = 1 + Q_w/r$ and $K_N = 1 + Q_N/r$. We may now repeat the steps that we followed previously, to generate the entire $O(2,2;\mathbb{Z})$ multiplet of boosted and twisted dyonic strings. The final expressions for the dilatons, axions and 3-form fields will be identical to those given in (C.4), again with the understanding that the H_e and H_m harmonic functions are modified to those given above. The reason for this is that the only other change in the starting point is in the metric, which is a singlet under $O(2,2)$, and so the entire calculation of the $O(2,2;\mathbb{Z})$ multiplet proceeds identically to the one we described previously. Note that the charges p and q are still given by (C.9), as in the case of unboosted isotropic dyonic strings.

The solution (C.13) can be dimensionally reduced on the fibre coordinates z_1 and z_2, giving rise to a non-dilatonic black hole in $D = 4$. The near-horizon limit in $D = 4$ is $\mathrm{AdS}_2 \times S^2$, and so the area of the event horizon is non-vanishing, implying that there is a non-vanishing entropy, even though the solution is extremal. The entropy is of the form

$$S \sim \sqrt{p\, q\, Q_w\, Q_N} . \tag{C.14}$$

From the six-dimensional point of view, the near-horizon limit is also locally non-singular, and is given, after a rescaling of the time coordinate, by

$$ds^2 = \sqrt{p\,q}\, Q_N \Big\{ - e^{2\rho}\, dt^2 + d\rho^2 + \Big(\sqrt{\frac{Q_w}{p\,q\,Q_N}}\, dz_1 + e^{\rho}\, dt\Big)^2$$
$$+ d\theta^2 + \sin^2\theta\, d\phi^2 + (Q_N^{-1}\, dz_2 + \cos\theta\, d\phi)^2 \Big\} . \tag{C.15}$$

It is straightforward to verify that the 4-volume of the spatial metric at $\rho = -\infty$ is of the form (C.14). Note that Q_w measures the momentum of the wave

---

[j]Recall that we are defining charges arising from integrals over S^2 and over S^3 by $Q = \frac{1}{4\pi}\int F_2$ and $Q = \frac{1}{16\pi^2}\int F_3$ respectively, which ensures that the charge is preserved under dimensional reduction. It follows that the harmonic functions in the two cases will be of the form $H = 1 + Q/r$ and $H = 1 + 4Q/r^2$ respectively.

propagating on the world-sheet of the dyonic string, and it, together with the other charges, has the effect of rescaling the fibre coordinate z_1. Q_N, on the other hand, is the NUT charge, and it has the effect of changing the local structure of the S^3 factor to the lens space S^3/Z_{Q_N}. Thus we see that the phenomenon of S^3 being factored to become a cyclic lens space plays a rôle in the understanding of the entropy of 4-dimensional black holes.

If we consider the case where the dyonic string is only twisted, but not boosted, the solution is given by (C.13) with $Q_w = 0$. In this case a dimensional reduction on the fibre coordinate z_2 gives rise to a non-dilatonic string in $D = 5$, which is dual to a non-dilatonic 3-charge black hole. If, on the other hand, we instead consider a case where there is only boosting, but no twisting, of the dyonic string, the solution is given by

$$ds^2 = -(H_e H_m)^{-\frac{1}{2}} K_w^{-1}\, dt^2 + (H_e H_m)^{-\frac{1}{2}} K_w\, (dz_1 + (K_w^{-1} - 1)dt)^2 \\ +(H_e H_m)^{\frac{1}{2}} (dr^2 + r^2\, d\Omega_3^2)\,, \qquad \text{(C.16)}$$

where now the functions H_e, H_m and K_w are harmonic in a 4-dimensional transverse space, and so $H_e = 1{+}4q/r^2$, $H_m = 1{+}4p/r^2$ and $K_w = 1{+}4Q_w/r^2$. Its dimensional reduction on the fibre coordinate z_1 gives a non-dilatonic 3-charge black hole in $D = 5$. The near-horizon limit of the six-dimensional metric (C.16), after rescaling the time coordinate, is

$$ds^2 = \sqrt{p\,q}\Big\{ - e^{2\rho}\, dt^2 + d\rho^2 + \Big(\sqrt{\frac{Q_w}{p\,q}}\, dz_1 + e^\rho\, dt\Big)^2 \\ +d\theta^2 + \sin^2\theta\, d\phi^2 + (dz_2 + \cos\theta\, d\phi)^2\Big\}\,. \qquad \text{(C.17)}$$

The entropy is of the form $S \sim \sqrt{p\,q\,Q_w}$.

D Hopf reductions and isentropic mappings

We have seen already that the Hopf T-duality transformations on the fibre coordinates of AdS_3 or S^3, which can have the effect of (un)twisting and squashing the metrics, preserve the entropy of the solutions. For example, we observe that the near-horizon limit (C.17) is the same as (C.15) with $Q_N = 1$, in which case the lens space S^3/Z_{Q_N} just becomes the 3-sphere. The entropy is therefore given by setting $Q_N = 1$ in (C.14). This shows, from a six-dimensional point of view, that the near-horizon limit of a $D = 5$ isotropic 3-charge black hole is the same as the near-horizon limit of a 4-charge black hole in $D = 4$, where the magnetic Kaluza-Klein charge Q_N is set to 1. In other words, in the

Hopf reduction on the $U(1)$ fibres of the 3-spheres that foliate the transverse space of the 3-charge black hole in $D=5$, a fourth charge ($Q_N=1$) emerges in $D=4$, corresponding to the magnetic charge of the Kaluza-Klein 2-form that governs the curvature of the fibre bundle. This should be contrasted with the more common kind of Kaluza-Klein reduction of the 3-charge black hole in $D=5$, where the reduction is on one of the coordinates of the transverse space. In this situation, it is is necessary first to make a line (or periodic array) of $D=5$ black holes along the intended reduction axis. The consequence of this is that the number of charges is conserved under the reduction process, and thus one arrives at a 3-charge black hole in $D=4$ which is singular on the horizon, and which has zero entropy. Hopf reduction, however, preserves the area of the horizon, and provides a natural "isentropic mapping" between the $D=5$, 3-charge and $D=4$, 4-charge black holes.

To make this property of the Hopf reductions more explicit, let us consider the entire solution for a 3-charge black hole in $D=3$, rather than just looking at the near-horizon limit. The 3-charge solution, which can be obtained, for example, from the dimensional reduction of (C.16) on the z_1 coordinate, is

$$ds_5^2=-(H_eH_mK_w)^{-\frac{2}{3}}\,dt^2+(H_eH_mK_w)^{\frac{1}{3}}\,(dr^2+r^2\,d\Omega_3^2)\,, \tag{D.1}$$

where

$$H_e=1+\frac{4q}{r^2}\,,\qquad H_m=1+\frac{4p}{r^2}\,,\qquad K_w=1+\frac{4Q_w}{r^2}\,. \tag{D.2}$$

Writing the unit 3-sphere metric $d\Omega_3^2$ in the fibre bundle form (3.5), we now perform the reduction on the fibre coordinate z, using the usual Kaluza-Klein ansatz $ds_5^2=e^{-\varphi/\sqrt{3}}\,ds_4^2+e^{2\varphi/\sqrt{3}}\,(dz+B)^2$, giving

$$ds_4^2=-\tfrac{1}{2}r\,(H_eH_mK_w)^{-\frac{1}{2}}\,dt^2+\tfrac{1}{2}r\,(H_eH_mK_w)^{\frac{1}{2}}\,(dr^2+\tfrac{1}{4}r^2\,d\Omega_2^2)\,. \tag{D.3}$$

The new Kaluza-Klein potential B is such that $\mathcal{F}_{(2)}=dB=\Omega_{(2)}$, and hence the Kaluza-Klein magnetic charge is $Q_N=\frac{1}{4\pi}\int\mathcal{F}_{(2)}=1$.

This is not quite like a normal black hole solution in $D=4$ for two reasons. Firstly, there is a conformal factor of r multiplying the entire metric, and secondly, the three-dimensional transverse space has the non-standard metric $dr^2+\frac{1}{4}r^2\,d\Omega_2^2$. This metric suffers from a diverging curvature as r approaches zero, since the foliating 2-spheres are of the wrong radius to "nest" nicely around the origin. We see that it is natural to perform a coordinate transformation in which we define a new radial coordinate $\rho=\frac{1}{4}r^2$. In terms of this, the metric (D.3) becomes

$$ds_4^2=-\Big(\frac{1}{\rho}\,H_eH_mK_w\Big)^{-\frac{1}{2}}\,dt^2+\Big(\frac{1}{\rho}\,H_eH_mK_w\Big)^{\frac{1}{2}}\,(d\rho^2+\rho^2\,d\Omega_2^2)\,. \tag{D.4}$$

Note that in terms of the new radial coordinate, the original harmonic functions H_e, H_m and K_w given in (D.2) become

$$H_e = 1 + \frac{q}{\rho}\,, \qquad H_m = 1 + \frac{p}{\rho}\,, \qquad K_w = 1 + \frac{Q_w}{\rho}\,, \tag{D.5}$$

which are harmonic in the new 3-dimensional transverse space. Thus we see that the effect of performing a Hopf reduction on a 3-charge black hole in $D = 5$ is to give the solution

$$ds_4^2 = -\Big(K_N H_e H_m K_w\Big)^{-\frac{1}{2}} dt^2 + \Big(K_N H_e H_m K_w\Big)^{\frac{1}{2}} (d\rho^2 + \rho^2\, d\Omega_2^2)\,, \tag{D.6}$$

which is precisely of the form of a standard 4-charge black hole, except that the fourth harmonic function $K_N = 1/\rho$, which has charge equal to 1, is lacking a constant term. Of course if we consider the near-horizon limit where ρ approaches zero, the absence of the constant term becomes immaterial. In fact, it is also possible to introduce a constant term in the harmonic function K_N, by performing an appropriate U-duality transformation [59,8,60,61]. Thus we see that the Hopf dimensional reduction provides a natural isentropic mapping between a 3-charge black hole in $D = 5$ and a 4-charge black hole in $D = 4$. The Hopf reduction preserves the supersymmetry of the solution. Although the four-dimensional black hole involves 4 charges, it still preserves the same fraction 1/8 of the supersymmetry as does the 3-charge black hole in $D = 5$. (With the opposite orientation for the Hopf reduction, all the supersymmetry is broken, which is consistent with the fact that 4-charge black holes in $D = 4$ also occur with no preserved supersymmetry, if the sign of the fourth charge is reversed [18,57].)

The crucial feature in the above discussion is that the foliating spheres in the transverse space are 3-dimensional. One can perform a Hopf reduction on the $U(1)$ fibres of any odd-dimensional sphere S^{2n+1}, but in general the base space will then be CP^n. However, it is only in the special case $n = 1$ that the transverse space after the Hopf reduction can become flat, with no singularity at the origin. This is because a metric of the form $d\rho^2 + \rho^2\, d\Sigma_n^2$, where $d\Sigma_n^2$ is the metric on CP^n, is non-singular at the origin only if $d\Sigma_n^2$ is a metric on the unit sphere. Only for $n = 1$ is CP^n isomorphic to a sphere $(CP^1 \sim S^2)$.

In fact, we can apply the Hopf reduction to any N-charge p-brane solution whose transverse space is 4-dimensional, and thereby obtain an $(N+1)$-charge p-brane solution in one dimension less, with the strength of the extra charge being 1. (As in the previous example of $D = 5$ black holes, a further U-duality transformation is needed in order to introduce a constant term in the associated harmonic function.) Note that the procedure can equally well be applied to the case of non-extremal p-branes.

Let us consider a general isotropic N-charge non-extremal p-brane solution in $D+1$ dimensions, where the transverse space has dimension 4. The metric will have the form [69]

$$ds^2_{D+1} = \prod_{i=1}^{N} H_i^{-2/(D-1)} \Big(-e^{2f}\, dt^2 + d\vec{x}\cdot d\vec{x}\Big) + \prod_{i=1}^{N} H_i^{(D-3)/(D-1)} \Big(e^{-2f}\, dr^2 + r^2\, d\Omega_3^2\Big) \tag{D.7}$$

where the "harmonic" functions take the form $H_i = 1 + 4k \sinh^2 \mu_i\, r^{-2}$, and the "blackening function" f is given by $e^{2f} = 1 - k\, r^{-2}$. (Full details of the solutions, and the relation of the charges and the mass to the parameters μ_i and k can be found in [69].) Following the same steps as we described above for the reduction of the 3-charge $D = 5$ black hole, we find that the Hopf-reduced solution in D dimensions has the metric

$$ds^2_D = \prod_{i=1}^{N+1} H_i^{-1/(D-2)} \Big(-e^{2f}\, dt^2 + d\vec{x}\cdot d\vec{x}\Big) + \prod_{i=1}^{N+1} H_i^{(D-3)/(D-2)} \Big(e^{-2f}\, d\rho^2 + \rho^2\, d\Omega_2^2\Big) \tag{D.8}$$

where $\rho = \frac{1}{4} r^2$, and we now have $e^{2f} = 1 - k/(4\rho)$ and

$$H_i = 1 + \frac{k}{\rho} \sinh^2 \mu_i\ , \quad (1 \le i \le N)\ , \qquad H_{N+1} = \frac{1}{\rho}\ . \tag{D.9}$$

As before, we can then perform a suitable U-duality transformation on this solution in order to introduce a constant term in the new harmonic function H_{N+1} [59,8,60,61]. The metric (D.9) is then precisely of the form of an $(N+1)$-charge p-brane in D dimensions, with unit strength for the extra harmonic function. The discussion extends to any intersection of p-branes, NUTs and waves where the overall transverse space is 4-dimensional.

In general, the near-horizon limit of an N-charge extremal p-brane is singular. It follows in these cases that the Hopf reduction will give rise to a singular dilaton in the lower dimension. Thus the supersymmetry will be further broken in such cases.

An application of our discussion of Hopf reductions for 4-dimensional transverse spaces is to 5-branes in $D = 10$. There are NS-NS 5-branes, and R-R D5-branes. For the case of NS-NS 5-branes, Hopf reduction on the $U(1)$ fibres of S^3 gives rise to 2-charge harmonic 5-branes in $D = 9$. Performing a T-duality transformation, the solution can be oxidised to $D = 10$, when it again becomes an NS-NS single-charge 5-brane, this time with unit charge. At the same time, the foliating 3-spheres become lens spaces S^3/Z_n, where n is the magnetic charge of the original 5-brane. Thus a single (*i.e.* unit charge) 5-brane is invariant under the Hopf T-duality. The picture is more complicated for the case of R-R 5-branes. The Hopf reduction to $D = 9$ gives the same metric form as in the NS-NS case, but after T-duality it oxidises back to the intersection of an NS-NS 5-brane and a D6-brane. In this case, the original 3-sphere is untwisted to $S^2 \times S^1$ (but with a non-direct-product metric, since there will be different harmonic functions multiplying the two factors).

Acknowledgment

H.L. and C.N.P. are grateful to CERN for hospitality. We thank Sudipta Mukherji for useful discussions.

References

1. M.J. Duff, G.W. Gibbons and P.K. Townsend, *Macroscopic superstrings as interpolating solitons*, Phys. Lett. **B332** (1994) 321, hep-th/9405124.
2. G.W. Gibbons, G.T. Horowitz and P.K. Townsend, *Higher-dimensional resolution of dilatonic black hole singularities*, Class. Quant. Grav. **12** (1995) 297, hep-th/9410073.
3. M.J. Duff and J.X. Lu, *Black and super p-branes in diverse dimensions*, Nucl. Phys. **B416** (1994) 301, hep-th/9306052.
4. M.J. Duff, R.R. Khuri and J.X. Lu, *String Solitons*, Phys. Rep. **259** (1995) 213, hep-th/9412184.
5. M.J. Duff, S. Ferrara, R. Khuri and J. Rahmfeld, *Supersymmetry and dual string solitons*, Phys. Lett. **B356** (1995) 479, hep-th/9506057.
6. M.J. Duff, H. Lü, C.N. Pope, *Heterotic phase transitions and singularities of the gauge dyonic string*, hep-th/9603037.
7. M.J. Duff, James T. Liu, H. Lü, C.N. Pope, *Gauge dyonic strings and their global limit*, hep-th/9711089.
8. H.J. Boonstra, B. Peeters and K. Skenderis, *Duality and asymptotic geometries*, Phys. Lett. **B411** (1997) 59, hep-th/9706192.
9. H.J. Boonstra, B. Peeters and K. Skenderis, *Brane intersections, anti-de Sitter spacetimes and dual superconformal theories*, hep-th/9803231.
10. G.T. Horowitz, J. Maldacena and A. Strominger, *Nonextremal black hole microstates and U duality*, hep-th/9603109.
11. G.T. Horowitz and A. Strominger, *Counting states of near extremal black holes*, hep-th/9602051.

12. J. Maldacena and A. Strominger, *AdS_3 black holes and a stringy exclusion principle*, hep-th/9804085.
13. N. Kaloper, *Entropy count for extremal three-dimensional black strings*, hep-th/9804062.
14. C. Callan and J. Maldacena, *D-brane approach to black hole quantum mechanics*, hep-th/9602043.
15. A.A. Tseytlin, *Extreme dyonic black holes in string theory*, Mod. Phys. Lett. **A11** (1996) 689, hep-th/9601177.
16. M. Cvetic and A.A. Tseytlin, *General class of BPS saturated black holes as exact superstring solutions*, Phys. Lett. **B366** (1996) 95, hep-th/9510097.
17. M. Cvetic and A.A. Tseytlin, *Solitonic strings and BPS saturated dyonic black holes*, Phys. Rev. **D53** (1996) 5619, hep-th/9512031.
18. H. Lü and C.N. Pope, *Multi-scalar p-brane solitons*, Int. J. Mod. Phys. **A12** (1997) 437, hep-th/9512153.
19. A. Strominger and C. Vafa, *Microscopic origin of the Beckenstein-Hawking entropy*, hep-th/9601029
20. D. Birmingham, I. Sachs and S. Sen, *Entropy of three-dimensional black holes in string theory*, hep-th/9801019.
21. J. Maldacena, *The large N limit of superconformal field theories and supergravity*, hep-th/9711200.
22. K. Behrndt, I. Brunner and I. Gaide, *Entropy and conformal field theories of AdS_3 models*, hep-th/9804159.
23. J. M. Evans, M. R. Gaberdiel and M. J. Perry, *The no-ghost theorem for AdS_3 and the stringy exclusion principle*, hep-th/9806024.
24. L. Andrianopoli, R. D'Auria, S. Ferrara and M.A. Lledo *Horizon geometry, duality and fixed scalars in six dimensions*, hep-th/9802147.
25. J. Navarro-Salas and P. Navarro, *A note on AdS_3 and boundary conformal field theory*, hep-th/9807019.
26. H.W. Lee, N.J. Kim and Y.S. Myung, *Dilaton test of connection between $AdS_3 \times S^3$ and $5D$ black hole*, hep-th/9805050.
27. B. Andreas, G. Curiol and D. Lust, *The Neveu-Schwarz fivebrane and its dual geometries*, hep-th/9807008.
28. S. Deger, A. Kaya, E. Sezgin and P. Sundell, *Spectrum of $D = 6$ $N = 4b$ supergravity on $AdS_3 \times S^3$*, hep-th/9804166.
29. K. Benakli, *Heterotic string theory on anti-de Sitter spaces*, hep-th/9805099.
30. J. de Boer, *Six-dimensional supergravity on $S^3 \times AdS_3$*, hep-th/9806104.
31. C. Vafa, *Puzzles at large N*, hep-th/9804172.
32. H.J. Boonstra, B. Peters, K. Skenderis, *Branes and anti de Sitter space-*

time, hep-th/9801076.
33. P. Claus, R. Kallosh, J. Kumar, P.K. Townsend and A. Van Proeyen, *Supergravity and the large N limit of theories with sixteen supercharges*, hep-th/9801206.
34. M. Banados and M. Ortiz, *The central charge in three dimensional anti-de Sitter space*, hep-th/9806087.
35. A. Giveon, D. Kutasov and N. Seiberg, *Comments on string theory on* AdS_3, hep-th/9806194.
36. E. Martinec, *Matrix models of AdS gravity*, hep-th/9804111
37. M.J. Duff, H. Lü and C.N. Pope, $AdS_5 \times S^5$ *untwisted*, hep-th/9803061.
38. M.J. Duff, H. Lü and C.N. Pope, *Supersymmetry without supersymmetry*, Phys. Lett. **B409** (1997) 136, hep-th/9704186.
39. M.J. Duff, H. Lü, C.N. Pope and E. Sezgin, *Supermembranes with fewer supersymmetries*, Phys. Lett. **B371** (1996) 206, hep-th/9511162.
40. S. Ferrara, A. Kehagias, H. Partouche and A. Zaffaroni, *Membranes and fivebranes with lower supersymmetry and their AdS duals*, hep-th/9803109
41. L. Castellani, A. Ceresole, R. D'Auria, S. Ferrara, P. Fré and M. Trigiante, *G/H M-branes and* AdS_{p+2} *geometries*, hep-th/9803039.
42. S. Kachru and E. Silverstein, *4d conformal field theories and strings on orbifolds*, hep-th/9802183.
43. Y. Oz and J. Terning, *Orbifolds of* $AdS(5) \times S^5$ *and 4-D conformal field theories*, hep-th/9803167.
44. I. Klebanov and E. Witten, *Superconformal field theory on three-branes at a Calabi-Yau singularity*, hep-th/9807080.
45. I.V. Lavrinenko, H. Lü, C.N. Pope and T.A Tran, *U-duality as general coordinate transformations, and spacetime geometry*, hep-th/9807006.
46. J.H. Schwarz, *An* $SL(2, \mathbb{Z})$ *multiplet of superstrings*, Phys. Lett. **B360** (1995) 13; Erratum-ibid. **B364** (1995) 252, hep-th/9508143.
47. E. Alvarez, L. Alvarez-Gaume, J.L.F. Barbon and Y. Lozano, *Some global aspects of duality in string theory*, Nucl. Phys. **B415** (1994) 71, hep-th/9309039.
48. H. Lü and C.N. Pope, *p-brane solitons in maximal supergravities*, Nucl. Phys. **B465** (1996) 127, hep-th/9512012.
49. E. Cremmer, B. Julia, H. Lü and C.N. Pope, *Dualisation of dualities*, Nucl. Phys. **B523** (1998) 73, hep-th/9710119.
50. T.H. Buscher, *A symmetry of the string background field equations*, Phys. Lett. **B194** (1987) 59.
51. B.E.W. Nilsson and C.N. Pope, *Hopf fibration of* $D = 11$ *supergravity*, Class. Quantum Grav. **1** (1984) 499.

52. G.T. Horowitz and D.L. Welch, *Exact three-dimensional black holes in string theory*, Phys. Rev. Lett. **71** (1993) 328, hep-th/9302126.
53. N. Kaloper, *Miens of the three-dimensional black hole*, Phys. Rev. **D48** (1993) 2598, hep-th/9303007.
54. H. Lü, C.N. Pope and P.K. Townsend, *Domain walls from anti-de Sitter spacetime*, Phys. Lett. **B391** (1997) 39, hep-th/9607164.
55. Y. Fujii and K. Yamagishi, *Killing spinors on spheres and hyperbolic manifolds*, J. Math. Phys. **27** (1986) 979.
56. H. Lü, C.N. Pope and J. Rahmfeld, *A construction of Killing spinors on* S^n, hep-th/9805151.
57. R.R. Khuri and T. Ortin, *A non-supersymmetric dyonic extreme Reissner-Nordstrom black hole*, Phys. Lett. **B373** (1996) 56, hep-th/9512178.
58. H. Lü, C.N. Pope, T.A. Tran and K.-W. Xu, *Classification of p-branes, NUTs, waves and intersections*, Nucl. Phys. **B511** (1998) 98, hep-th/9708055.
59. S. Hyun, *U-duality between three and higher dimensional black holes*, hep-th/9704005.
60. E. Cremmer, I.V. Lavrinenko, H. Lü, C.N. Pope, K.S. Stelle and T.A. Tran, *Euclidean-signature supergravities, dualities and instantons*, to appear in Nucl. Phys. **B**, hep-th/9803259.
61. E. Bergshoeff and K. Behrndt, *D-instantons and asymptotic geometries*, Class. Quant. Grav. **15** (1998) 1801, hep-th/9803090.
62. E. Cremmer, B. Julia, H. Lü and C.N. Pope, *Dualisation of dualities II: Twisted self-duality of doubled fields and superdualities*, hep-th/9806106.
63. M.J. Duff and J. Rahmfeld, *Massive string states as extreme black holes*, Phys. Lett. **B345** (1995) 441, hep-th/9406105.
64. E. Cremmer, H. Lü, C.N. Pope and K.S. Stelle, *Spectrum-generating symmetries for BPS solitons*, Nucl. Phys. **B520** (1998) 132, hep-th/9707207.
65. J.X. Lu and S. Roy, *U-duality p-branes in diverse dimensions*, hep-th/9805180.
66. M.S. Bremer, H. Lü, C.N. Pope and K.S. Stelle, *Dirac quantisation conditions and Kaluza-Klein reduction*, to appear in Nucl. Phys. **B**, hep-th/9710244.
67. H. Lü, C.N. Pope and K.S. Stelle, *Multiplet structures of BPS solitons*, Class. Quant. Grav. **15** (1998) 537, hep-th/9708109.
68. S. Ferrara and M. Gunaydin, *Orbits of exceptional groups, duality and BPS states in string theory*, Int. J. Mod. Phys. **A13** (1998) 2075, hep-th/9708025.

69. M.J. Duff, H. Lü and C.N. Pope, *The black branes of M-theory*, Phys. Lett. **B382** (1996) 73, hep-th/9604052.
70. E. Kiritsis, *Exact duality symmetries in CFT and string theory*, Nucl. Phys. **B 405** (1993) 109, hep-th/9302033.
71. I. Antoniadis, C. Bachas and A. Sagnotti, *Gauged supergravity vacua in string theory*, Phys. Lett. **B 235** (1990) 255.

SUPERMEMBRANES IN CURVED SUPERSPACE AND NEAR-HORIZON GEOMETRIES

B. DE WIT

Institute for Theoretical Physics, Utrecht University
Princetonplein 5, 3508 TA Utrecht, The Netherlands
E-mail: bdewit@phys.uu.nl

We give a description of supermembranes in nontrivial target-space geometries. A special class are the $AdS_4 \times S^7$ and $AdS_7 \times S^4$ spaces that have the maximal number of 32 supersymmetries.

1 Introduction

The original study of supermembranes[1] was motivated by the desire to find a consistent quantum-mechanical extension of 11-dimensional supergravity[2] along the same lines that string theory defines a quantum-mechanical extension of 10-dimensional supergravity theories. In 11 spacetime dimensions the supermembrane can consistently couple to a superspace background that satisfies a number of constraints which are equivalent to the supergravity equations of motion. The supermembrane action can also exist in $4, 5$ and 7 spacetime dimensions, just as the Green-Schwarz superstring[3] is classically consistent in $3, 4, 6$ and 10 dimensions. Guided by string theory it was natural to expect that the massless states of the supermembrane would correspond to those of 11-dimensional supergravity. However, the supermembrane mass spectrum turned out to be continuous[4] and in that situation it was cumbersome to directly prove or disprove the possible existence of massless states[5,6]. The difficulties in making sense of the supermembrane mass spectrum and the fact that 11-dimensional supergravity seemed to have no place in string theory, formed an obstacle for subsequent development of the theory. More recently, however, interest in supermembranes was rekindled by the realization that 11-dimensional supergravity does have its role to play as the long-distance approximation to M-theory[7,8,9,10]. M-theory is the conjectured framework for unifying all five superstring theories and 11-dimensional supergravity. It turns out that supermembranes, M-theory and super matrix theory are all intricately related.

An important observation was that it is possible to regularize the supermembrane in terms of a super matrix model based on some finite group, such as U(N). In the limit of infinite N one then recovers the supermembrane[5]. These supersymmetric matrix models were constructed long ago[11] and can be obtained from supersymmetric Yang-Mills theories in the zero-volume limit. More recently it was realized that these models describe the short-distance behaviour of N Dirichlet particles[12]. The continuity of the spectrum is then understood directly in terms of the spectrum of N-particle states. A bold conjecture was that the super matrix models capture the degrees of freedom of M-theory[13]. In the large-N limit, where one considers the states with an infinite number of particles, the supermembranes should then re-emerge. Furthermore there is new evidence meanwhile that the supermembrane has massless states, which will presumably correspond to the states of 11-dimensional supergravity, although proper asymptotic states do not exist. The evidence is based on the matrix model regularization of the supermembrane[14]. For fixed value of N the existence of such states was foreseen on the basis of identifying the Kaluza-Klein states of M-theory compactified on S^1 with the Dirichlet particles and their bound states in type-IIA string theory.

From this viewpoint it is natural to consider the supermembrane in curved backgrounds associated with 11-dimensional supergravity and it is the purpose of this talk to report on progress in this direction. Such backgrounds consist of a nontrivial metric, a three-index gauge field and a gravitino field. This provides us with an action that transforms as a scalar under the combined (local) supersymmetry transformations of the background fields and the supermembrane embedding coordinates. Here it is important to realize that the supersymmetry transformations of the embedding coordinates will themselves depend on the background. When the background is supersymmetric, then the action will be supersymmetric as well. In the light-cone formulation this action leads to models invariant under area-preserving diffeomorphisms, which in certain situations can be approximated by matrix models in curved backgrounds. The area-preserving diffeomorphisms are then replaced by a finite group, such as U(N), but target-space diffeomorphisms are no longer manifestly realized. Matrix models in curved space have already been the subject of independent studies[15]. Also in view of the relation between near-horizon geometries and conformal field theories[16] interesting classes of backgrounds are the ones where the target space factorizes locally into the product of an anti-de Sitter space and some compact space. Examples of these are the $AdS_4 \times S^7$ and $AdS_7 \times S^4$ backgrounds that we discuss at the end of the talk.

2 Supermembranes

Fundamental supermembranes can be described in terms of actions of the Green-Schwarz type, possibly in a nontrivial but restricted (super)spacetime background[1]. Such actions exist for supersymmetric p-branes, where $p =$

$0, 1, \ldots, d-1$ defines the spatial dimension of the brane. Thus for $p = 0$ we have a superparticle, for $p = 1$ a superstring, for $p = 2$ a supermembrane, and so on. The dimension of spacetime in which the superbrane can live is very restricted. These restrictions arise from the fact that the action contains a Wess-Zumino-Witten term, whose supersymmetry depends sensitively on the spacetime dimension. If the coefficient of this term takes a particular value then the action possesses an additional fermionic gauge symmetry, the so-called κ-symmetry. This symmetry is necessary to ensure the matching of (physical) bosonic and fermionic degrees of freedom. In the following we restrict ourselves to supermembranes (i.e., $p = 2$) in 11 dimensions.

The 11-dimensional supermembrane is written in terms of superspace embedding coordinates $Z^M(\zeta) = (X^\mu(\zeta), \theta^\alpha(\zeta))$, which are functions of the three world-volume coordinates ζ^i ($i = 0, 1, 2$). It couples to the superspace geometry of 11-dimensional supergravity, encoded by the supervielbein $E_M{}^A$ and the antisymmetric tensor gauge superfield B_{MNP}, through the action[a]

$$S[Z(\zeta)] = \int \mathrm{d}^3\zeta \left[- \sqrt{-g(Z(\zeta))} - \tfrac{1}{6}\varepsilon^{ijk}\, \Pi_i^A \Pi_j^B \Pi_k^C\, B_{CBA}(Z(\zeta)) \right], \qquad (1)$$

where $\Pi_i^A = \partial Z^M/\partial\zeta^i\, E_M{}^A$ is the pull-back of the supervielbein to the membrane worldvolume. Here the induced metric equals $g_{ij} = \Pi_i^r \Pi_j^s\, \eta_{rs}$, with η_{rs} being the constant Lorentz-invariant metric. This action is invariant under local fermionic κ-transformations alluded to above, given that certain constraints on the background fields hold, which are equivalent to the equations of motion of 11-dimensional supergravity[1].

Flat superspace is characterized by

$$\begin{aligned}
E_\mu{}^r &= \delta_\mu{}^r\,, & E_\mu{}^a &= 0\,,\\
E_\alpha{}^a &= \delta_\alpha{}^a\,, & E_\alpha{}^r &= -(\bar\theta\Gamma^r)_\alpha\,,\\
B_{\mu\nu\alpha} &= (\bar\theta\Gamma_{\mu\nu})_\alpha\,, & B_{\mu\alpha\beta} &= (\bar\theta\Gamma_{\mu\nu})_{(\alpha}\,(\bar\theta\Gamma^\nu)_{\beta)}\,,\\
B_{\alpha\beta\gamma} &= (\bar\theta\Gamma_{\mu\nu})_{(\alpha}\,(\bar\theta\Gamma^\mu)_\beta\,(\bar\theta\Gamma^\nu)_{\gamma)}\,, & B_{\mu\nu\rho} &= 0\,.
\end{aligned} \qquad (2)$$

The gamma matrices are denoted by Γ^r; gamma matrices with more than one index denote antisymmetrized products of gamma matrices with unit weight. In flat superspace the supermembrane Lagrangian, written in components, reads (in the notation and conventions of [5]),

$$\begin{aligned}
\mathcal{L} = &- \sqrt{-g(X,\theta)}\\
&-\varepsilon^{ijk}\, \bar\theta\Gamma_{\mu\nu}\partial_k\theta \left[\tfrac{1}{2}\, \partial_i X^\mu (\partial_j X^\nu + \bar\theta\Gamma^\nu\partial_j\theta) + \tfrac{1}{6}\, \bar\theta\Gamma^\mu\partial_i\theta\, \bar\theta\Gamma^\nu\partial_j\theta \right],
\end{aligned} \qquad (3)$$

[a] Our notation and conventions are as follows. Tangent-space indices are $A = (r, a)$, whereas curved indices are denoted by $M = (\mu, \alpha)$. Here r, μ refer to commuting and a, α to anticommuting coordinates. Moreover we take $\epsilon_{012} = -\epsilon^{012} = 1$.

The target space can have compact dimensions which permit winding membrane states. In flat superspace the induced metric,

$$g_{ij} = (\partial_i X^\mu + \bar\theta\Gamma^\mu\partial_i\theta)(\partial_j X^\nu + \bar\theta\Gamma^\nu\partial_j\theta)\,\eta_{\mu\nu}\,, \tag{4}$$

is supersymmetric. Therefore the first term in (3) is trivially invariant under spacetime supersymmetry,

$$\delta X^\mu = -\bar\epsilon\Gamma^\mu\theta\,, \qquad \delta\theta = \epsilon\,. \tag{5}$$

In 4, 5, 7, or 11 spacetime dimensions the second term in the action proportional to ε^{ijk} is also supersymmetric (up to a total divergence) and the full action is invariant under κ-symmetry.

Let us now consider the supermembrane action for nontrivial backgounds, such as those induced by a nontrivial target-space metric, a target-space tensor field and a target-space gravitino field, corresponding to the fields of (on-shell) 11-dimensional supergravity. This background can in principle be cast into superspace form by a procedure known as 'gauge completion'. For 11-dimensional supergravity, the first steps of this procedure were carried out long ago[17] and recently[18] the results were determined to second order in the fermionic coordinates θ.

To elucidate the generic effects of nontrivial backgrounds for membrane theories, let us confine ourselves for the moment to the purely bosonic theory and present the light-cone formulation of the membrane in a background consisting of the metric $G_{\mu\nu}$ and the tensor gauge field $C_{\mu\nu\rho}$. In the subsequent sections we will include the fermionic coordinates. The Lagrangian density for the bosonic membrane follows directly from (1),

$$\mathcal{L} = -\sqrt{-g} - \tfrac{1}{6}\varepsilon^{ijk}\partial_i X^\mu\,\partial_j X^\nu\,\partial_k X^\rho\,C_{\rho\nu\mu}\,, \tag{6}$$

where $g_{ij} = \partial_i X^\mu\,\partial_j X^\nu\,\eta_{\mu\nu}$. In the light-cone formulation, the coordinates are decomposed in the usual fashion as (X^+, X^-, X^a) with $a = 1\ldots 9$. Furthermore we use the diffeomorphisms in the target space to bring the metric in a convenient form[19],

$$G_{--} = G_{a-} = 0\,. \tag{7}$$

Just as for a flat target space, we identify the time coordinate of the target space with the world-volume time, by imposing the condition $X^+ = \tau$. Moreover we denote the spacesheet coordinates of the membrane by σ^r, $r = 1, 2$. Following the same steps as for the membrane in flat space, one arrives at a Hamiltonian formulation of the theory in terms of coordinates and momenta. These phase-space variables are subject to a constraint, which takes the same form as for the membrane theory in flat space, namely,

$$\phi_r = P_a\,\partial_r X^a + P_-\,\partial_r X^- \approx 0\,. \tag{8}$$

Of course, the definition of the momenta in terms of the coordinates and their derivatives does involve the background fields, but at the end all explicit dependence on the background cancels out in the phase-space constraints.

The gauge choice $X^+ = \tau$ still allows for τ-dependent reparametrizations of the world-space coordinates σ^r In addition there remains the freedom of performing tensor gauge transformations of the target-space three-form $C_{\mu\nu\rho}$. In order to write the theory as a gauge theory of area-preserving diffeomorphisms it is desirable to obtain a Hamiltonian which is polynomial in momenta and coordinates. Obviously one wants the background fields to be independent of $X^\pm$. With suitable choices for the gauge condition one thus derives[18],

$$\begin{aligned} H = \int \mathrm{d}^2\sigma \Big\{ & \frac{G_{+-}}{P_- - C_-}\Big[\tfrac{1}{2}\Big(P_a - C_a - \frac{P_- - C_-}{G_{+-}}\,G_{a+}\Big)^2 + \tfrac{1}{4}(\varepsilon^{rs}\,\partial_r X^a\,\partial_s X^b)^2\Big] \\ & - \frac{P_- - C_-}{2\,G_{+-}}\,G_{++} - C_+ \\ & + \frac{1}{P_- - C_-}\Big[\varepsilon^{rs}\partial_r X^a \partial_s X^b\,P_a\,C_{+-b} + C_-\,C_{+-}\Big]\Big\}, \end{aligned} \tag{9}$$

where $P_- - C_-$ equals a constant times $\sqrt{w}$ and C_{-ab} has been set to a constant by a choice of gauge. Furthermore we made use of the definitions,

$$\begin{aligned} C_a &= -\varepsilon^{rs}\,\partial_r X^-\,\partial_s X^b\,C_{-ab} + \tfrac{1}{2}\varepsilon^{rs}\,\partial_r X^b \partial_s X^c\,C_{abc}\,, \\ C_\pm &= \tfrac{1}{2}\varepsilon^{rs}\,\partial_r X^a \partial_s X^b\,C_{\pm ab}\,, \qquad C_{+-} = \varepsilon^{rs}\,\partial_r X^-\,\partial_s X^a\,C_{+-a}\,. \end{aligned} \tag{10}$$

At this point one can impose further gauge choices and set $G_{+-} = 1$ and $C_{+-a} = 0$. Taking also $C_{-ab} = 0$ the corresponding Hamiltonian can be cast in Lagrangian form in terms of a gauge theory of area-preserving diffeomorphisms[20],

$$\begin{aligned} w^{-1/2}\,\mathcal{L} = {} & \tfrac{1}{2}(D_0 X^a)^2 + D_0 X^a\left(\tfrac{1}{2}C_{abc}\{X^b, X^c\} + G_{a+}\right) \\ & -\tfrac{1}{4}\{X^a, X^b\}^2 + \tfrac{1}{2}G_{++} + \tfrac{1}{2}C_{+ab}\{X^a, X^b\}\,, \end{aligned} \tag{11}$$

where the derivative is covariant with respect to area-preserving diffeomorphisms. For convenience we have set $(P_-)_0 = 1$. In the case of compact dimensions, it may not always be possible to set C_{+-a} and C_{-ab} to zero, although they can be restricted to constants. One can still follow the same procedure as above, but the Lagrangian then depends explicitly on X^-, a feature that was already exhibited earlier for the winding membrane. However, in the case at hand, the constraint makes the resulting expression for X^- extremely nontrivial. The antisymmetric constant matrix C_{-ab} was conjectured to play a role for the matrix model compactification on a noncommutative torus[21]. It should be interesting to see what the role is of (9) in this context.

With a reformulation of the membrane in background fields as a gauge theory of area-preserving diffeomorphisms at one's disposal, one may consider its regularization through a matrix model by truncating the mode expansion for coordinates and momenta. This leads to a replacement of Poisson brackets by commutators, integrals by traces and products of commuting fields by symmetrized products of the corresponding matrices. At that point the original target-space covariance is affected, as the matrix reparametrizations in terms of symmetrized products of matrices do not possess a consistent multiplication structure; this is just one of the underlying difficulties in the construction of matrix models in curved space[15]. With the above results at hand one may study interactions between membranes by considering the behaviour of a test membrane in a background field induced by another membrane[22].

3 Superspace backgronds

It is possible to express superspace backgrounds in terms of the component fields of 11-dimensional supergravity, by means of a technique called gauge completion. We refer to ref. 18 for the details. The superspace geometry with coordinates $Z^M = (x^\mu, \theta^\alpha)$ is encoded in the supervielbein $E_M{}^A$ and a spin-connection field $\Omega_M{}^{AB}$. In what follows we will not pay much attention to the spin-connection, which is not an independent field. Furthermore we have an antisymmetric tensor gauge field B_{MNP}, subject to tensor gauge transformations,

$$\delta B_{MNP} = 3\,\partial_{[M}\Xi_{NP]}\,. \tag{12}$$

Unless stated otherwise the derivatives with respect to θ are always left derivatives.

Under superspace diffeomorphisms corresponding to $Z^M \to Z^M + \Xi^M(Z)$, the super-vielbein and tensor gauge field transform as

$$\begin{aligned}\delta E_M{}^A &= \Xi^N \partial_N E_M{}^A + \partial_M \Xi^N E_N{}^A\,,\\ \delta B_{MNP} &= \Xi^Q \partial_Q B_{MNP} + 3\,\partial_{[M}\Xi^Q B_{|Q|NP]}\,.\end{aligned} \tag{13}$$

Tangent-frame rotations are Z-dependent Lorentz transformations that act on the vielbein according to

$$\delta E_M{}^A = \tfrac{1}{2}(\Lambda^{rs} L_{rs})^A{}_B\, E_M{}^B\,, \tag{14}$$

where the Lorentz generators L_{rs} are defined by

$$\tfrac{1}{2}(\Lambda^{rs} L_{rs})^t{}_u = \Lambda^t{}_u\,, \qquad \tfrac{1}{2}(\Lambda^{rs} L_{rs})^a{}_b = \tfrac{1}{4}\Lambda^{rs}(\Gamma_{rs})^a{}_b\,. \tag{15}$$

The superspace that we are dealing with is not unrestricted but is subject to certain constraints and gauge conditions. Furthermore, we will not describe an off-shell situation as all superfields will be expressed entirely in terms of the three component fields of on-shell 11-dimensional supergravity, the elfbein $e_\mu{}^r$, the antisymmetric tensor gauge field $C_{\mu\nu\rho}$ and the gravitino field ψ_μ. As a result of these restrictions the residual symmetry transformations are confined to 11-dimensional diffeomorphisms with parameters $\xi^\mu(x)$, local Lorentz transformations with parameters $\lambda^{rs}(x)$, tensor-gauge transformations with parameters $\xi_{\mu\nu}(x)$ and local supersymmetry transformations with parameters $\epsilon(x)$. To derive how the superfields are parametrized in terms of the component fields it is necessary to also determine the form of the superspace transformation parameters, Ξ^M, Λ^{rs} and Ξ_{MN}, that generate the supersymmetry transformations. Here it is important to realize that we are dealing with a gauge-fixed situation. For that reason the superspace parameters depend on both the x-dependent component parameters defined above as well as on the component fields. This has the consequence that local supersymmetry transformations reside in the superspace diffeomorphisms, the Lorentz transformations and the tensor gauge transformations. Thus, when considering supersymmetry variations of the various fields, one must in principle include each of the three possible superspace transformations.

For further details we refer to ref. 18 and we proceed directly to the results. For the supervielbein $E_M{}^A$ the following expressions were found,

$$\begin{aligned}
E_\mu^{\;r} &= e_\mu{}^r + 2\,\bar\theta\,\Gamma^r\psi_\mu \\
&\quad + \bar\theta\,\Gamma^r\Big[-\tfrac14\,\hat\omega_\mu{}^{st}\,\Gamma_{st} + T_\mu{}^{\nu\rho\sigma\lambda}\,\hat F_{\nu\rho\sigma\lambda}\Big]\theta + \mathcal{O}(\theta^3)\,, \\
E_\mu{}^a &= \psi_\mu{}^a - \tfrac14\,\hat\omega_\mu^{rs}\,(\Gamma_{rs}\theta)^a + (T_\mu{}^{\nu\rho\sigma\lambda}\theta)^a\,\hat F_{\nu\rho\sigma\lambda} + \mathcal{O}(\theta^2)\,, \\
E_\alpha^{\;r} &= -(\bar\theta\,\Gamma^r)_\alpha + \mathcal{O}(\theta^3)\,, \\
E_\alpha{}^a &= \delta_\alpha^a + M_\alpha{}^a + \mathcal{O}(\theta^3)\,,
\end{aligned} \tag{16}$$

where $M_\alpha{}^a$ characterizes the $\hat F\theta^2$-contributions, which have not been evaluated. Observe that $E_\mu{}^a$ was determined only up to terms of order θ^2. The result for the tensor field B_{MNP} reads as follows,

$$\begin{aligned}
B_{\mu\nu\rho} &= C_{\mu\nu\rho} - 6\,\bar\theta\Gamma_{[\mu\nu}\psi_{\rho]} - 3\,\bar\theta\,\Gamma_{[\mu\nu}\Big[-\tfrac14\hat\omega_{\rho]}{}^{rs}\,\Gamma_{rs} + T_{\rho]}{}^{\sigma\lambda\kappa\tau}\,\hat F_{\sigma\lambda\kappa\tau}\Big]\theta \\
&\quad - 12\,\bar\theta\,\Gamma_{\sigma[\mu}\psi_\nu\;\bar\theta\,\Gamma^\sigma\psi_{\rho]} + \mathcal{O}(\theta^3)\,, \\
B_{\mu\nu\alpha} &= (\bar\theta\,\Gamma_{\mu\nu})_\alpha - \tfrac83\bar\theta\,\Gamma^\rho\psi_{[\mu}\;(\bar\theta\,\Gamma_{\nu]\rho})_\alpha + \tfrac43(\bar\theta\,\Gamma^\rho)_\alpha\;\bar\theta\,\Gamma_{\rho[\mu}\psi_{\nu]} + \mathcal{O}(\theta^3)\,, \\
B_{\mu\alpha\beta} &= (\bar\theta\,\Gamma_{\mu\nu})_{(\alpha}\,(\bar\theta\,\Gamma^\nu)_{\beta)} + \mathcal{O}(\theta^3)\,, \\
B_{\alpha\beta\gamma} &= (\bar\theta\Gamma_{\mu\nu})_{(\alpha}\,(\bar\theta\Gamma^\mu)_\beta\,(\bar\theta\Gamma^\nu)_{\gamma)} + \mathcal{O}(\theta^3)\,.
\end{aligned} \tag{17}$$

For completeness we included the θ^3-term in $B_{\alpha\beta\gamma}$ which were already known from the flat-superspace results (2).

Then we turn to some of the transformation parameters. The supersymmetry transformations consistent with the fields specified above, are generated by superspace diffeomorphisms, local Lorentz transformations and tensor gauge transformations. We only present the parameters for the superspace diffeomorphisms here and refer to the literature for more complete results[18],

$$\begin{aligned}\Xi^\mu(\epsilon) &= \bar\theta\,\Gamma^\mu\epsilon - \bar\theta\,\Gamma^\nu\epsilon\,\bar\theta\,\Gamma^\mu\psi_\nu + \mathcal{O}(\theta^3)\,,\\ \Xi^\alpha(\epsilon) &= \epsilon^\alpha - \bar\theta\,\Gamma^\mu\epsilon\,\psi_\mu^\alpha\\ &\quad+\bar\theta\,\Gamma^\nu\epsilon\,\bar\theta\,\Gamma^\mu\psi_\nu\,\psi_\mu{}^\alpha + \tfrac14\bar\theta\,\Gamma^\nu\epsilon\,\hat\omega_\nu^{rs}\,(\Gamma_{rs}\theta)^\alpha + \epsilon^\beta\,N_\beta{}^\alpha + \mathcal{O}(\theta^3)\,,\end{aligned} \tag{18}$$

where $N_\beta{}^\alpha$ encodes unknown terms proportional to $\hat F\theta^2$.

Substituting the above results into the initial supermembrane action (1) we obtain the action for a supermembrane in a supergravity background to order θ^2. While the original action was manifestly covariant under independent superspace diffeomorphisms, tangent-space Lorentz transformations and tensor gauge transformations, this is no longer the case and one has to restrict oneself to the superspace transformations corresponding to the component supersymmetry, general-coordinate, local Lorentz and tensor gauge transformations. When writing (1) in components, utilizing the expressions found above, one thus obtains an action that is covariant under the restricted superspace diffeomorphisms (18) acting on the superspace coordinates $Z^M = (X^\mu,\theta^\alpha)$ (including the spacetime arguments of the background fields) combined with usual transformations on the component fields. Note that the result does not constitute an invariance. Rather it implies that actions corresponding to two different sets of background fields that are equivalent by a component gauge transformation, are the same modulo a reparametrization of the supermembrane embedding coordinates. In order to be precise let us briefly turn to an example and consider the action of a particle moving in a curved spacetime background with metric $g_{\mu\nu}$,

$$S[X^\mu, g_{\mu\nu}(X)] = -m\int \mathrm{d}t\,\sqrt{-g_{\mu\nu}(X(t))\,\dot X^\mu(t)\,\dot X^\nu(t)}\,. \tag{19}$$

This action, which is obviously invariant under world-line diffeomorphisms, satisfies $S[X'^\mu, g'_{\mu\nu}(X')] = S[X^\mu, g_{\mu\nu}(X)]$, where X'^μ and X^μ are related by a target-space general coordinate transformation which also governs the relation between $g'_{\mu\nu}$ and $g_{\mu\nu}$. Of course, when considering a background that is invariant under (a subset of the) general coordinate transformations (so that $g = g'$), then the action will be invariant under the corresponding change of the coordinates. This is the situation that we will address in the next section, where we take a specific background metric with certain isometries. In that context the relevant target space for (19) is an anti-de Sitter (AdS_d) space, which has isometries that constitute the group SO($d-1,2$), where d is the spacetime dimension. Then (19) describes a one-dimensional field theory with an SO($d-1,2$) invariance group. In the particular case of $d=2$ this invariance can be re-interpreted as a conformal invariance for a supersymmetric quantum mechanical system.[b]

Using the previous results one may now write down the complete action of the supermembrane coupled to background fields up to order θ^2. Direct substitution leads to the following result for the supervielbein pull-back,

$$\begin{aligned}\Pi_i^r &= \partial_i X^\mu \left(e_\mu{}^r + 2\,\bar\theta\,\Gamma^r\psi_\mu - \tfrac{1}{4}\bar\theta\,\Gamma^{rst}\theta\,\hat\omega_{\mu st} + \bar\theta\,\Gamma^r T_\mu{}^{\nu\rho\sigma\lambda}\theta\,\hat F_{\nu\rho\sigma\lambda}\right)\\ &\quad + \bar\theta\Gamma^r\,\partial_i\theta + \mathcal{O}(\theta^3)\,,\\ \Pi_i^a &= \partial_i X^\mu \left(\psi_\mu{}^a - \tfrac{1}{4}\,\hat\omega_\mu^{rs}\,(\Gamma_{rs}\theta)^a + (T_\mu{}^{\nu\rho\sigma\lambda}\theta)^a\,\hat F_{\nu\rho\sigma\lambda}\right)\\ &\quad + \partial_i\theta^a + \mathcal{O}(\theta^2)\,.\end{aligned} \tag{20}$$

Consequently the induced metric is known up to terms of order θ^3. Furthermore the pull-back of the tensor field equals

$$\begin{aligned}&-\tfrac{1}{6}\varepsilon^{ijk}\,\Pi_i^A\,\Pi_j^B\,\Pi_k^C\,B_{CBA} = -\tfrac{1}{6}\varepsilon^{ijk}\partial_i Z^M\,\partial_j Z^N\,\partial_k Z^P\,B_{PNM} =\\ &\tfrac{1}{6}\,dX^{\mu\nu\rho}\Big[C_{\mu\nu\rho} - 6\,\bar\theta\Gamma_{\mu\nu}\psi_\rho + \tfrac{3}{4}\bar\theta\,\Gamma_{rs}\Gamma_{\mu\nu}\theta\,\hat\omega_\rho{}^{rs}\\ &\qquad -3\,\bar\theta\,\Gamma_{\mu\nu}T_\rho{}^{\sigma\lambda\kappa\tau}\theta\,\hat F_{\sigma\lambda\kappa\tau} - 12\,\bar\theta\,\Gamma_{\sigma\mu}\psi_\nu\,\bar\theta\,\Gamma^\sigma\psi_\rho\Big]\\ &-\varepsilon^{ijk}\,\bar\theta\,\Gamma_{\mu\nu}\partial_k\theta\Big[\tfrac{1}{2}\,\partial_i X^\mu(\partial_j X^\nu + \bar\theta\,\Gamma^\nu\partial_j\theta) + \tfrac{1}{6}\,\bar\theta\,\Gamma^\mu\partial_i\theta\,\bar\theta\,\Gamma^\nu\partial_j\theta\Big]\\ &+\tfrac{1}{3}\varepsilon^{ijk}\partial_i X^\mu\,\partial_j X^\nu\Big[4\,\bar\theta\,\Gamma_{\rho\mu}\partial_k\theta\,\bar\theta\,\Gamma^\rho\psi_\nu - 2\,\bar\theta\,\Gamma^\rho\partial_k\theta\,\bar\theta\,\Gamma_{\rho\mu}\psi_\nu\Big] + \mathcal{O}(\theta^3)\,,\end{aligned} \tag{21}$$

where we have introduced the abbreviation $dX^{\mu\nu\rho} = \varepsilon^{ijk}\,\partial_i X^\mu\,\partial_j X^\nu\,\partial_k X^\rho$ for the world-volume form. Observe that we included also the terms of higher-order θ-terms that were determined in previous sections and listed in (17). The first formula of (20) and (21) now determine the supermembrane action (1) up to order θ^3. It has been shown that the resulting expressions are consistent with supersymmetry and κ-invariance[18].

[b]This situation arises generically for any p-brane moving in a target space that is locally the product of AdS_{p+2} and some compact space. The conformal interpretation was emphasized in ref. 23.

4 Near-horizon geometries

In the previous section we discussed the determination of superspace quantities, i.e. the superspace vielbein and the tensor gauge field, in terms of the fields of 11-dimensional on-shell supergravity. The corresponding expressions are obtained by iteration order-by-order in θ coordinates, but except for the leading terms it is hard to proceed with this program. Nevertheless these results enable one to write down the 11-dimensional supermembrane action coupled to a nontrivial supergravity component-field background to second order in θ, so that one can start a study of the supermembrane degrees of freedom in the corresponding background geometries.

However, in specific backgrounds with a certain amount of symmetry, it is possible to obtain results to all orders in θ. Interesting candidates for such backgrounds are the membrane[24] and the five-brane solution[25] of 11-dimensional supergravity, as well as solutions corresponding to the product of anti-de Sitter spacetimes with compact manifolds. Coupling to the latter solutions, which appear near the horizon of black D-branes[26], seem especially appealing in view of the recent results on a connection between large-N superconformal field theories and supergravity on a product of an AdS space with a compact manifold[16]. The target-space geometry induced by the p-branes interpolates between $AdS_{p+2} \times B$ near the horizon, where B denotes some compact manifold (usually a sphere), and flat $(p+1)$-dimensional Minkowski space times a cone with base B.

This program has been carried out recently for the type-IIB superstring and the D3-brane in a IIB-supergravity background of this type[27,28,29]. In the context of 11-dimensional supergravity the $AdS_4 \times S^7$ and $AdS_7 \times S^4$ backgrounds stand out as they leave all 32 supersymmetries invariant[30,31]. These backgrounds are associated with the near-horizon geometries corresponding to two- and five-brane configurations and thus to possible conformal field theories in 3 and 6 spacetime dimensions with 16 supersymmetries, whose exact nature is not yet completely known. In this section we consider the supermembrane in these two backgrounds[32]. As the corresponding spaces are local products of homogeneous spaces, their geometric information can be extracted from appropriate coset representatives leading to standard invariant one-forms corresponding to the vielbeine and spin-connections. The approach of ref. 32 differs from that of ref. 33; in the latter one constructs the geometric information exploiting simultaneously the κ-symmetry of the supermembrane action, while in ref. 32 the geometric information is determined independently from the supermembrane action. The results for the geometry coincide with those of ref. 34.

As is well known, the compactifications of the theory to $AdS_4 \times S^7$ and $AdS_7 \times S^4$ are induced by the antisymmetric 4-rank field strength of M-theory. These two compactifications are thus governed by the Freund-Rubin field f, defined by (in Pauli-Källén convention, so that we can leave the precise signature of the spacetime open),

$$F_{\mu\nu\rho\sigma} = 6f\, e\, \varepsilon_{\mu\nu\rho\sigma}\,, \tag{22}$$

with e the vierbein determinant. When f is purely imaginary we are dealing with an $AdS_4 \times S^7$ background while for real f we have an $AdS_7 \times S^4$ background. The nonvanishing curvature components corresponding to the 4- and 7-dimensional subspaces are equal to

$$\begin{aligned} R_{\mu\nu\rho\sigma} &= -4f^2(g_{\mu\rho}\, g_{\nu\sigma} - g_{\mu\sigma}\, g_{\nu\rho})\,, \\ R_{\mu'\nu'\rho'\sigma'} &= f^2(g_{\mu'\rho'}\, g_{\nu'\sigma'} - g_{\mu'\sigma'}\, g_{\nu'\rho'})\,. \end{aligned} \tag{23}$$

Here μ, ν, ρ, σ and $\mu', \nu', \rho', \sigma'$ are 4- and 7-dimensional world indices, respectively. We also use $m_{4,7}$ for the inverse radii of the two subspaces, defined by $|f|^2 = {m_7}^2 = \frac{1}{4}{m_4}^2$. The Killing-spinor equations associated with the 32 supersymmetries in this background take the form

$$\Big(D_\mu - f\gamma_\mu\gamma_5 \otimes \mathbf{1}\Big)\epsilon = \Big(D_{\mu'} + \tfrac{1}{2}f\mathbf{1} \otimes \gamma'_{\mu'}\Big)\epsilon = 0\,, \tag{24}$$

where we make use of the familiar decomposition of the (hermitean) gamma matrices γ_μ and $\gamma'_{\mu'}$, appropriate to the product space of a 4- and a 7-dimensional subspace. Here D_μ and $D_{\mu'}$ denote the covariant derivatives containing the spin-connection fields corresponding to SO(3,1) or SO(4) and SO(7) or SO(6,1), respectively.

The algebra of isometries of the $AdS_4 \times S^7$ and $AdS_7 \times S^4$ backgrounds is given by $osp(8|4;\mathbf{R})$ and $osp(6,2|4)$. Their bosonic subalgebra consists of $so(8) \oplus sp(4) \simeq so(8) \oplus so(3,2)$ and $so(6,2) \oplus usp(4) \simeq so(6,2) \oplus so(5)$, respectively. The spinors transform in the $(8,4)$ of this algebra.

One may decompose the generators of $osp(8|4)$ or $osp(6,2|4)$ in terms of irreducible representations of the bosonic $so(7) \oplus so(3,1)$ and $so(6,1) \oplus so(4)$ subalgebras. In that way one obtains the bosonic (even) generators P_r, M_{rs}, which generate $so(3,2)$ or $so(5)$, and $P_{r'}$, $M_{r's'}$, which generate $so(8)$ or $so(6,2)$. All the bosonic generators are taken antihermitean (in the Pauli-Källén sense). The fermionic (odd) generators $Q_{aa'}$ are Majorana spinors, where we denote the spinorial tangent-space indices by $a, b, \ldots$ and $a', b', \ldots$ for 4- or 7-dimensional indices. The commutation relations between even generators are

$$\begin{aligned}
[P_r, P_s] &= -4f^2 M_{rs}\,, & [P_{r'}, P_{s'}] &= f^2 M_{r's'}\,,\\
[P_r, M_{st}] &= \delta_{rs}\, P_t - \delta_{rt}\, P_s\,, & [P_{r'}, M_{s't'}] &= \delta_{r's'}\, P_{t'} - \delta_{r't'}\, P_{s'}\,,\\
[M_{rs}, M_{tu}] &= \delta_{ru}\, M_{st} + \delta_{st}\, M_{ru} & [M_{r's'}, M_{t'u'}] &= \delta_{r'u'}\, M_{s't'} + \delta_{s't'}\, M_{r'u'}\\
&\quad -\delta_{rt}\, M_{su} - \delta_{su}\, M_{rt}\,, & &\quad -\delta_{r't'}\, M_{s'u'} - \delta_{s'u'}\, M_{r't'}\,.
\end{aligned} \tag{25}$$

The odd-even commutators are given by

$$\begin{aligned}
[P_r, Q_{aa'}] &= -f(\gamma_r\gamma_5)_a{}^b\, Q_{ba'}\,, & [P_{r'}, Q_{aa'}] &= -\tfrac{1}{2} f(\gamma'_{r'})_{a'}{}^{b'}\, Q_{ab'}\,,\\
[M_{rs}, Q_{aa'}] &= -\tfrac{1}{2}(\gamma_{rs})_a{}^b\, Q_{ba'}\,, & [M_{r's'}, Q_{aa'}] &= -\tfrac{1}{2}(\gamma'_{r's'})_{a'}{}^{b'}\, Q_{ab'}\,.
\end{aligned} \tag{26}$$

Finally, we have the odd-odd anti-commutators,

$$\begin{aligned}
\{Q_{aa'}, Q_{bb'}\} = &-(\gamma_5 C)_{ab}\Big(2(\gamma'_{r'}C')_{a'b'}\, P^{r'} - f(\gamma'_{r's'}C')_{a'b'}\, M^{r's'}\Big)\\
&-C'_{a'b'}\Big(2(\gamma_r C)_{ab}\, P^r + 2f(\gamma_{rs}\gamma_5 C)_{ab}\, M^{rs}\Big).
\end{aligned} \tag{27}$$

All other (anti)commutators vanish. The normalizations of the above algebra were determined by comparison with the supersymmetry algebra in the conventions of [18] in the appropriate backgrounds.

However, one can return to 11-dimensional notation and drop the distinction between 4- and 7-dimensional indices so that the equations obtain a more compact form. In that case the above (anti)commutation relations that involve the supercharges can be concisely written as,

$$\begin{aligned}
&[P_r, \bar Q] = \bar Q\, T_r{}^{stuv} F_{stuv}\,, \qquad [M_{rs}, \bar Q] = \tfrac{1}{2}\bar Q\, \Gamma_{rs}\,,\\
&\{Q, \bar Q\} = -2\Gamma_r\, P^r + \tfrac{1}{144}\Big[\Gamma^{rstuvw} F_{tuvw} + 24\,\Gamma_{tu} F^{rstu}\Big] M_{rs}\,,
\end{aligned} \tag{28}$$

where the tensor T equals

$$T_r{}^{stuv} = \tfrac{1}{288}\Big(\Gamma_r{}^{stuv} - 8\,\delta_r^{[s}\,\Gamma^{tuv]}\Big)\,. \tag{29}$$

Note, however, that the above formulae are only applicable in the background where the field strength takes the form given in (22). In what follows, we will only make use of (28).

5 Coset-space representatives of $AdS_4 \times S^7$ and $AdS_7 \times S^4$

Both backgrounds that we consider correspond to homogenous spaces and can thus be formulated as coset spaces. In the case at hand these (reductive) coset spaces G/H are $OSp(8|4;\mathbf{R})/SO(7)\times SO(3,1)$ and $OSp(6,2|4)/SO(6,1)\times SO(4)$. To each element of the coset G/H one associates an element of G, which we denote by $L(Z)$. Here Z^A stands for the coset-space coordinates x^r, θ^a (or, alternatively, x^r, $y^{r'}$ and $\theta^{aa'}$). The coset representative L transforms from the left under constant G-transformations corresponding to the isometry group of the coset space and from the right under local H-transformations: $L\to L'=g\,L\,h^{-1}$.

The vielbein and the torsion-free H-connection one-forms, E and Ω, are defined through[c]

$$\mathrm{d}L+L\,\Omega=L\,E\,, \tag{30}$$

where

$$E=E^r P_r+\bar{E}Q\,,\qquad \Omega=\tfrac{1}{2}\Omega^{rs}M_{rs}. \tag{31}$$

The integrability of (30) leads to the Maurer-Cartan equations,

$$\begin{aligned}
&\mathrm{d}\Omega-\Omega\wedge\Omega-\tfrac{1}{2}E^r\wedge E^s\,[P_r,P_s]\\
&\qquad-\tfrac{1}{288}\bar{E}\Big[\Gamma^{rstuvw}F_{tuvw}+24\,\Gamma_{tu}F^{rstu}\Big]E\,M_{rs}=0\,,\\
&\mathrm{d}E^r-\Omega^{rs}\wedge E_s-\bar{E}\,\Gamma^r\wedge E=0\,,\\
&\mathrm{d}E+E^r\wedge T_r{}^{tuvw}E\,F_{tuvw}-\tfrac{1}{4}\Omega^{rs}\wedge\Gamma_{rs}E=0\,,
\end{aligned} \tag{32}$$

where we suppressed the spinor indices on the anticommuting component E^a. The first equation in a fermion-free background reproduces (23) upon using the commutation relations (25).

Now the question is how to determine the vielbeine and connections to all orders in θ for the spaces of interest. First, observe that the choice of the coset representative amounts to a gauge choice that fixes the parametrization of the coset space. We will not insist on an explicit parametrization of the bosonic part of the space. It turns out to be advantageous to factorize $L(Z)$ into a group element of the bosonic part of G corresponding to the bosonic coset space, whose parametrization we leave unspecified, and a fermion factor. Hence one may write

$$L(Z)=\ell(x)\,\hat{L}(\theta)\,,\qquad \text{with }\hat{L}(\theta)=\exp[\bar{\theta}Q]\,. \tag{33}$$

There exists a convenient trick[27,28,35] according to which one first rescales the odd coordinates according to $\theta\to t\,\theta$, where t is an auxiliary parameter that we will put to unity at the end. Taking the derivative with respect to t of (30)

---

[c]A one-form V stands for $V\equiv\mathrm{d}Z^A V_A$ and an exterior derivative acts according to $\mathrm{d}V\equiv-\mathrm{d}Z^B\wedge\mathrm{d}Z^A\,\partial_A V_B$. Fermionic derivatives are thus always left-derivatives.

then leads to a first-order differential equation for E and Ω (in 11-dimensional notation),

$$\dot{E} - \dot{\Omega} = \mathrm{d}\bar{\theta}\, Q + (E-\Omega)\,\bar{\theta}Q - \bar{\theta}Q\,(E-\Omega) \tag{34}$$

After expanding E and Ω on the right-hand side in terms of the generators and using the (anti)commutation relations (28) we find a set of coupled first-order linear differential equations,

$$\begin{aligned} \dot{E}^a &= \Big(\mathrm{d}\theta + E^r\, T_r{}^{stuv}\theta\, F_{stuv} - \tfrac{1}{4}\Omega^{rs}\,\Gamma_{rs}\theta\Big)^a \,, \\ \dot{E}^r &= 2\,\bar{\theta}\,\Gamma^r E\,, \\ \dot{\Omega}^{rs} &= \tfrac{1}{72}\bar{\theta}\Big[\Gamma^{rstuvw}F_{tuvw} + 24\,\Gamma_{tu}F^{rstu}\Big]E\,. \end{aligned} \tag{35}$$

These equations can be solved straightforwardly[28] and one finds

$$\begin{aligned} E(x,\theta) &= \sum_{n=0}^{16} \frac{1}{(2n+1)!}\,\mathcal{M}^{2n}\,D\theta\,, \\ E^r(x,\theta) &= \mathrm{d}x^\mu\, e_\mu{}^r + 2\sum_{n=0}^{15}\frac{1}{(2n+2)!}\,\bar{\theta}\,\Gamma^r\,\mathcal{M}^{2n}\,D\theta \\ \Omega^{rs}(x,\theta) &= \mathrm{d}x^\mu\,\omega_\mu{}^{rs} \\ &\quad + \tfrac{1}{72}\sum_{n=0}^{15}\frac{1}{(2n+2)!}\,\bar{\theta}\,[\Gamma^{rstuvw}F_{tuvw} + 24\,\Gamma_{tu}F^{rstu}]\,\mathcal{M}^{2n}\,D\theta\,, \end{aligned} \tag{36}$$

where the matrix $\mathcal{M}^2$ equals,

$$\begin{aligned} (\mathcal{M}^2)^a{}_b = {}& 2\,(T_r{}^{stuv}\,\theta)^a\,F_{stuv}\,(\bar{\theta}\,\Gamma^r)_b \\ & -\tfrac{1}{288}(\Gamma_{rs}\,\theta)^a\,(\bar{\theta}\,[\Gamma^{rstuvw}F_{tuvw} + 24\,\Gamma_{tu}F^{rstu}])_b\,. \end{aligned} \tag{37}$$

and

$$D\theta^a \equiv \dot{E}^a\Big|_{t=0} = \Big(\mathrm{d}\theta + e^r\,T_r{}^{stuv}\theta\,F_{stuv} - \tfrac{1}{4}\omega^{rs}\,\Gamma_{rs}\theta\Big)^a\,. \tag{38}$$

It is straightforward to write down the lowest-order terms in these expansions,

$$\begin{aligned} E^r &= e^r + \bar{\theta}\Gamma^r\mathrm{d}\theta + \bar{\theta}\Gamma^r(e^m\,T_m{}^{stuv}F_{stuv} - \tfrac{1}{4}\omega^{st}\,\Gamma_{st})\theta + \mathcal{O}(\theta^4)\,, \\ E &= \mathrm{d}\theta + (e^r\,T_r{}^{stuv}F_{stuv} - \tfrac{1}{4}\omega^{rs}\,\Gamma_{rs})\theta + \mathcal{O}(\theta^3)\,, \\ \Omega^{rs} &= \omega^{rs} + \tfrac{1}{144}\bar{\theta}\,[\Gamma^{rstuvw}F_{tuvw} + 24\,\Gamma_{tu}F^{rstu}]\,\mathrm{d}\theta + \mathcal{O}(\theta^4)\,, \end{aligned} \tag{39}$$

which agree completely with those given in section 8 (and, for the spin-connection field, in ref. 17).

This information can now be substituted into the first part of the supermembrane action (1). By similar techniques one can also determine the Wess-Zumino-Witten part of the action by first considering the most general ansatz for a four-form invariant under tangent-space transformations. Using the lowest-order expansions of the vielbeine (39) and comparing with ref. 18 shows that only two terms can be present. Their relative coefficient is fixed by requiring that the four-form is closed, something that can be verified by making use of the Maurer-Cartan equations (32). The result takes the form

$$F_{(4)} = \frac{1}{4!}\Big[E^r \wedge E^s \wedge E^t \wedge E^u F_{rstu} - 12\,\bar{E} \wedge \Gamma_{rs} E \wedge E^r \wedge E^s\Big]. \qquad (40)$$

To establish this result we also needed the well-known quartic-spinor identity in 11 dimensions. The overall factor in (40) is fixed by comparing to the normalization of the results given in ref. 18.

Because $F_{(4)}$ is closed, it can be written locally as $F_{(4)} = \mathrm{d}B$. The general solution for B can be found by again exploiting the one-forms with rescaled θ coordinates according to $\theta \to t\,\theta$ and deriving a differential equation. Using the lowest order result for B this equation can be solved and yields

$$B = \tfrac{1}{6}\, e^r \wedge e^s \wedge e^t\, C_{rst} - \int_0^1 \mathrm{d}t\; \bar{\theta}\,\Gamma_{rs} E \wedge E^r \wedge E^s\,, \qquad (41)$$

where the vielbein components contain the rescaled θ's. This answer immediately reproduces the flat-space result upon substitution of $F_{rstu} = \omega^{rs} = 0$.

In order to obtain the supermembrane action one substitutes the above expressions in the action (1). The resulting action is then invariant under local fermionic κ-transformations as well as under the superspace isometries corresponding to $osp(8|4)$ or $osp(6,2|4)$.

We have already emphasized that the choice of the coset representative amounts to adopting a certain gauge choice in superspace. The choice that was made in ref. 32 connects directly to the generic 11-dimensional superspace results, written in a Wess-Zumino-type gauge, in which there is no distinction between spinorial world and tangent-space indices. In specific backgrounds, such as the ones discussed here, gauge choices are possible which allow further simplifications. For this we refer to refs. 28 and 36.

The results of this section provide a convincing check of the low-order θ results obtained by gauge completion for general backgrounds[18,17]. A great amount of clarity was gained by expressing our results in 11-dimensional language, so that both the $AdS_4 \times S^7$ and the $AdS_7 \times S^4$ solution could be covered in one go. Note that in both these backgrounds the gravitino vanishes.

We have no reasons to expect that the 11-dimensional form of our results will coincide with the expressions for a generic 11-dimensional superspace (with the gravitino set to zero) at arbitrary orders in θ.

From a more technical viewpoint it is gratifying that explicit constructions of supermembranes in certain nontrivial backgrounds are now possible. The complete M-theory two-brane action in $AdS_4 \times S^7$ and $AdS_7 \times S^4$ to all orders in θ represents a further step in the program of finding the complete anti-de-Sitter background actions for the superstring[27,28] and the M2-, D3-[29] and M5-branes initiated for the bosonic part in ref. 23. Furthermore, by studying the interaction between a test membrane in the background of an M2- or an M5-brane, one may hope to learn more about the interactions between branes. Some of these issues have already been considered recently[22].

Acknowledgments

The results described in this talk were obtained in collaboration with Kasper Peeters, Jan Plefka and Alexander Sevrin.

References

1. E. Bergshoeff, E. Sezgin and P.K. Townsend, Phys. Lett. **189B** (1987) 75; Ann. Phys. **185** (1988) 330.
2. E. Cremmer, B. Julia and J. Scherk, Phys. Lett. **76B** (1978) 409.
3. M.B. Green and J.H. Schwarz, *Phys. Lett.* B1361984367; *Nucl. Phys.* B2431984285.
4. B. de Wit, M. Lüscher and H. Nicolai, Nucl. Phys. **B320** (1989) 135.
5. B. de Wit, J. Hoppe, H. Nicolai, Nucl. Phys. **B305** [FS23] (1988) 545.
6. B. de Wit and H. Nicolai, in proc. Trieste Conference on Supermembranes and Physics in 2 + 1 dimensions, p. 196, eds. M.J. Duff, C.N. Pope and E. Sezgin (World Scient., 1990);
   B. de Wit, Nucl. Phys. B (Proc. Suppl.) **56B** (1997) 76, `hep-th/9701169`.
7. C.M. Hull and P.K. Townsend, Nucl. Phys **B438** (1995) 109, `hep-th/9410167`.
8. P.K. Townsend, Phys. Lett. **B350** (1995) 184, `hep-th/9501068`, Phys. Lett. **B373** (1996) 68, `hep-th/9512062`.
9. E. Witten, Nucl. Phys. **B443** (1995) 85, `hep-th/9503124`.
10. P. Hořava and E. Witten, Nucl. Phys. **B460** (1996) 506, `hep-th/9510209`, **B475** (1996) 94, `hep-th/9603142`.
11. M. Claudson and M.B. Halpern, Nucl. Phys. **B250** (1985) 689;
    R. Flume, Ann. Phys. **164** (1985) 189;
    M. Baake, P. Reinicke, and V. Rittenberg, J. Math. Phys. **26.** (1985)

1070.
12. E. Witten, Nucl. Phys. **B460** (1996) 335, hep-th/9510135.
13. T. Banks, W. Fischler, S.H. Shenker and L. Susskind, *Phys. Rev.* D5519975112, hep-th/9610043.
14. P. Yi, *Nucl. Phys.* B5051997307, hep-th/9704098;
S. Sethi and M. Stern, Commun. Math. Phys. **194** (1998) 675, hep-th/9705046;
M. Porrati and A. Rosenberg, Nucl. Phys. **B515** (1998) 184, hep-th/9708119.
15. M.R. Douglas, H. Ooguri and S.H. Shenker, Phys. Lett. **B402** (1997) 36, hep-th/9702203;
M.R. Douglas, *D-branes in curved space*, hep-th/9703056;
M.R. Douglas, A. Kato and H. Ooguri, *D-brane actions on Kähler manifolds*, hep-th/9708012;
M.R. Douglas and H. Ooguri, Phys. Lett. **B425** (1998) 71, hep-th/9710178.
16. J. Maldacena, Adv. Theor. Math. Phys. **2** (1998) 231, hep-th/9711200;
S.S. Gubser, I.R. Klebanov and A.M. Polyakov, *Phys. Lett.* B4281998105, hept-th/9802109;
E. Witten, Adv. Theor. Math. Phys. **2** (1998) 253, hep-th/9802150.
17. E. Cremmer and S. Ferrara, Phys. Lett. **91B** (1980) 61.
18. B. de Wit, K. Peeters and J.C. Plefka, Nucl. Phys. **B532** (1998) 99, hep-th/9803209.
19. M. Goroff and J.H. Schwarz, Phys. Lett. **127B** (1983) 61.
20. B. de Wit, K. Peeters and J.C. Plefka, in proc. of the València 97 workshop *Beyond the standard model; from theory to experiment*, eds. I. Antoniadis, L.E. Ibanez and J.W.F. Valle, World Scient. 1998, hep-th/9712082.
21. A. Connes, M.R. Douglas and A. Schwarz, JHEP **02** (1998) 003, hep-th/9711162;
M.R. Douglas and C. Hull, JHEP **02** (1998) 008, hep-th/9711165;
N.A. Obers, B. Pioline and E. Rabinovici, Nucl. Phys. **B525** (1998) 163, hep-th/9712084.
22. V. Balasubramanian, D. Kastor, J. Traschen and K.Z. Win, *The spin of the M2-brane and spin-spin interactions via probe techniques*, hep-th/9811037;
W. Taylor and M. Van Raamsdonk, *Supergravity currents and linearized interactions for Matrix Theory configurations with fermionic backgrounds*, hep-th/9812239;
T. Sato, *Superalgebras in many types of M-brane backgrounds and various*

supersymmetric brane configurations, hep-th/9812014;
S. Hyun, Y. Kiem and H. Shin, *Non-perturbative membrane spin-orbit couplings in M/IIA theory*, hep-th/9901105.

23. P. Claus, R. Kallosh, J. Kumar, P. Townsend, A. Van Proeyen, JHEP **06** (1998) 004, hep-th/9801206.
24. M.J. Duff and K.S. Stelle, *Phys. Lett.* B2531991113.
25. R. Güven, *Phys. Lett.* B276199249.
26. G. Horowitz and A. Strominger, Nucl. Phys. **B360** (1991) 197.
27. R.R. Metsaev and A.A. Tseytlin, Nucl. Phys. **B533** (1998) 109, hep-th/9805028,
28. R. Kallosh, J. Rahmfeld and A. Rajaraman, JHEP **09** (1998) 002, hep-th/9805217.
29. R.R. Metsaev and A.A. Tseytlin, Phys. Lett. **B436** (1998) 281, hep-th/9806095.
30. M.J. Duff and C.N. Pope, *in* Supersymmetry and Supergravity '82, eds. S. Ferrara, J.G. Taylor and P. van Nieuwenhuizen (World Scientific, 1983), 183;
B. Biran, F. Englert, B. de Wit and H. Nicolai, Phys. Lett. **124B** (1983) 45; (E) 128B (1983) 461.
31. K. Pilch, P. van Nieuwenhuizen and P.K. Townsend, Nucl. Phys. **B242** (1984) 377.
32. B. de Wit, K. Peeters, J. Plefka and A. Sevrin, Phys. Lett. **B443** (1998) 153, hep-th/9808052.
33. G. Dall'Agata, D. Fabbri, C. Fraser, P. Fré, P. Termonia and M. Trigiante, *The* $Osp(8|4)$ *singleton action from the supermembrane*, hep-th/9807115.
34. P. Claus, *Super M-brane actions in* $AdS_4 \times S^7$ *and* $AdS_7 \times S^4$, hep-th/9809045.
35. D.G. Boulware and L.S. Brown, Ann. Phys. **138** (1982) 392.
36. R. Kallosh, *Superconformal actions in Killing gauge*, hep-th/9807206.

SO(32) SPINORS IN TYPE I STRING THEORY

ASHOKE SEN
Mehta Research Institute of Mathematics and Mathematical Physics
Chhatnag Road, Jhusi, Allahabad 211019, INDIA
E-mail: asen@thwgs.cern.ch, sen@mri.ernet.in

Duality between type I and SO(32) heterotic string theories in ten dimension requires that the type I string theory should contain in its spectrum states which transform in the spinor representation of SO(32). In this talk I describe explicit construction of these states following [hep-th/9808141].

1 Introduction and Summary

It has been conjectured that SO(32) heterotic string theory is dual to type I string theory [1,2,3,4]. Many tests of this conjectured duality have been performed, including comparison of the BPS spectrum in compactified string theories [5]. However, even in ten dimensions, the SO(32) heterotic string theory contains, in its perturbative spectrum, states which are stable but are not BPS. These are the states which transform as spinors of SO(32). The lightest states belonging to the spinor representation of SO(32) must be stable since they cannot decay into anything else. On the other hand they are not BPS states, since the theory in ten dimensions has only $N = 1$ supersymmetry and hence no central charge. Like BPS states, these stable non-BPS states must exist at all values of the coupling, although their mass is not determined by any known formula. Thus if the duality between the type I and SO(32) string theory is correct then we must find these states in the weakly coupled type I string theory as well. This is the problem we address in this talk. This talk will mostly contain the material covered in ref. [6].

In section 2 of the paper we discuss the description of the SO(32) spinor state of type I string theory as a tachyonic kink solution on the D-string anti-D-string pair. It is well known that on the world volume of the D-string anti-D-string pair there is a tachyonic mode coming from open string with one end on the D-string and the other end on the anti- D-string [8,9,10,11]. At the minimum of the tachyon potential, the negative contribution to the energy density from the tachyon potential exactly cancels the positive contribution from the tension of the D-string and the anti- D-string [7]. Since in type I theory the tachyon is a real field, and furthermore, the dynamics is invariant under a change of sign of the tachyon field, the minimum of the potential comes in pairs $\pm T_0$. Thus we can construct a tachyonic kink solution on the D-string anti- D-string pair interpolating between the degenerate vacua at $\pm T_0$. We show that this state carries SO(32) spinor quantum numbers. Furthermore its mass is finite since far away from the core of the soliton the energy density vanishes. Thus this is the ideal candidate for the state we are looking for.

The mass of this state can be easily shown to be proportional to the inverse coupling (g^{-1}) of the type I theory by a simple scaling argument. However computing the constant of proportionality directly requires the detailed knowledge of the tachyon potential which is not available to us. We use an indirect method for computing this coefficient by considering the case where the type I theory has been compactified on a circle of radius R, and the D-string anti- D-string pair is wrapped along that circle. We show that in order to get an SO(32) spinor state we must impose an anti-periodic boundary condition on the tachyon, and as a result the zero momentum mode of the tachyon is absent. As we reduce the radius of the circle, the contribution to the effective mass^2 of various modes from the momentum along the circle increases, and one finds that at a critical radius $R = (1/\sqrt{2})$ the tachyonic mode ceases to be tachyonic. In fact at this point the lowest momentum mode of the tachyon represents an exactly marginal deformation. We interprete this critical radius as the point where the SO(32) spinor state becomes degenerate with the trivial vacuum of the D-string anti- D-string pair where the tachyon vanishes. Since the mass of the wrapped D-string anti- D-string pair in the absence of tachyon vacuum expectation value (vev) can be easily computed in the weak coupling limit, this gives an indirect measurement of the mass of the spinor state. Although this gives us the mass only at the critical radius, we give a heuristic argument showing that the mass is independent of the radius and hence also represents the mass of the spinor state at infinite radius.

This 'target space viewpoint' gives an intuitive description of the SO(32) spinor state; however it does not give us an exact conformal field theory description of the system. This is the problem we address in sections 3 and 4. In section 3 we work at the critical radius $R = (1/\sqrt{2})$, and study the effect of switching on the marginal deformation associated with 'tachyon' condensation. Parametrizing the tachyon vev at this critical radius by an appropriate parameter α we compute the spectrum of open string states on the D-string anti- D-string system at arbitrary α. In particular we find that α is a periodic variable with periodicity 2. At any value of α other than zero (mod 2) the mode representing the freedom of separating the string anti-string pair becomes massive, showing that the pair is bound. In fact the mass of this mode is maximum at $\alpha = 1$ showing that at this point the pair is maximally bound. Also for all $\alpha \neq 0$ there is a zero mode representing the freedom of translating the solution along the compact direction, indicating the fact that the translation invariance of the original string anti-string configuration along the compact direction is broken as we switch on the tachyon vev.

At $R = (1/\sqrt{2})$ all values of α represent configurations of same mass, and hence we cannot unambiguously identify the point which corresponds to the spinor state, *i.e.* which will correspond to the minimum of the potential when we increase R beyond this critical radius. This is the problem we address in section 4 by studying the effect of switching on the perturbation in the closed string sector corresponding to the radius deformation. We find that the 'tachyonic mode' develops a one point function at all α except at $\alpha = 0$ and 1. Thus only these two points correspond to the extremum of the tachyon potential. Furthermore, the 'tachyonic mode' becomes tachyonic at $\alpha = 0$ as expected, but acquires positive mass^2 at $\alpha = 1$. This shows that the point $\alpha = 1$ represents a stable minimum of the potential, and hence represents the SO(32) spinor state that we have been looking for.

One can now examine the limit $R \to \infty$. This is best studied using a different set of world-sheet bosonic and fermionic fields, related to the original bosonic and fermionic fields representing cooordinate along the compact direction by a series of bosonization, fermionization and duality transformation. The net result is that in terms of the new bosonic coordinate, the soliton at $\alpha = 1$ can be represented as a D0-brane, with Dirichlet boundary condition in all directions, and the radius of the new boson goes to ∞ as $R \to \infty$. This gives an exact world-sheet description of the SO(32) spinor state of type I string theory and allows us to calculate its mass. This is same as the mass calculated in section 2. The R independence of the mass. which was a crucial assumption in the analysis of section 2, can be easily understood by noting that the mass of a D0-brane situated on a compact circle does not depend on the radius of the circle.

We conclude this introduction by mentioning some related developments which have taken place during the recent years. Various examples of tachyon condensation in string theory have been studied in ref. [13]. Other stable non-BPS states in various string theories have been discussed in ref. [14,15]. Stable non-BPS states in the context of supersymmetric field theory have been analyzed in ref. [16,17,18]. In particular, ref. [16] discusses non-BPS states carrying Z_2 quantum numbers similar to the states considered here. Other aspects of tachyons in non-supersymmetric string theories have been discussed in ref. [19].

2 SO(32) Spinor State as a Tachyonic Kink on the D-string Anti- D-string Pair

Let us consider type I string theory compactified on a circle S^1 of radius R and a D-string of type I wrapped on S^1. It is known that the world volume theory on the D-string is identical to that of the fundamental heterotic string in SO(32) heterotic string theory [4]. In particular the U(1) gauge field on the D-string is projected out by the requirement of world-sheet parity invariance. There is however a Z_2 subgroup of this U(1) which survives this projection, and hence we can introduce a Z_2 Wilson line on the S^1. Since the end of a fundamental string ending on the D-string is charged under the Z_2, in the presence of a Wilson line the wave function of an open string with one end on the D-string will pick up an extra $-$ sign under a $2\pi R$ translation along S^1.

The open string states with one end on the D-string and the other end on any of the 32 9-branes filling up the space-time provide 32 fermionic degrees of freedom on the D-string world-volume. Due to the possibility of putting the Z_2 Wilson line on the D-string, these fermions can satisfy either periodic or anti-periodic bounday condition along S^1. If we put anti-periodic boundary condition, then there are no fermion zero modes, and the states are in the scalar conjugacy class of SO(32). On the other hand if we put periodic boundary condition on these fermions, then there are zero modes of the 32 fermions, and the quantization of these zero modes give a ground state (and hence all excited states) in the spinor conjugacy class of SO(32).

Thus if we consider a system of a D-string – anti- D-string[a] pair of type I, both wrapped on S^1, and put Z_2 Wilson line on one of them, the state will belong to the spinor conjugacy class of SO(32), and will not carry any other conserved charge as the Ramond-Ramond (RR) two form charges of the D-string and the anti- D-string will cancel each other. Thus this state has the correct quantum numbers, and we expect the lowest mass state in this sector to represent the lightest SO(32) spinor state of type I string theory. Naively, the mass of this state is given by,

$$2 \cdot 2\pi R \cdot T_D \,, \tag{1}$$

where T_D represents the tension of a single D-string:

$$T_D = \frac{1}{2\pi g} \,, \tag{2}$$

g being the string coupling constant. (We are working in units where we have set $\alpha' = 1$. Also we shall view type I string theory as the result of modding out

[a]An anti- D-string is simply a D-string with opposite orientation.

type IIB theory by the world-sheet parity transformation Ω, and use type IIB units before projection under Ω to calculate all masses.) From eq.(1) we see that as $R \to \infty$, the mass of the D-string anti-D-string pair goes to infinity. Thus although this configuration has the right quantum numbers, it cannot represent the state we are looking for.

The situation improves upon noting that on the D-string anti- D-string world volume there is a tachyon field [8,9,10,11], representing the fact that we are actually sitting at the top of a potential well. Thus in order to find the lowest energy configuration with these quantum numbers we must allow the tachyon field to acquire a vacuum expectation value and go to the minimum of its potential. Let T denote the tachyon field living on the D-string anti-D-string world-volume, $g^{-1}V(T)$ denote the tachyon potential,[b] and T_0 be the point where $V(T)$ is at its minimum. It was argued in ref. [7] that,

$$g^{-1}V(T_0) + 2T_D = 0\,, \tag{3}$$

so that at the minimum of the tachyon potential the net energy per unit length of the D-string anti- D-string pair vanishes. Note that since the tachyon arises in the open string sector with one end on the D-string and the other end on the anti- D-string, it is odd under the Z_2 gauge transformation associated with either of the strings, and hence $V(T)$ is symmetric under $T \to -T$. Thus the locations of the minimum of $V(T)$ come in pairs $\pm T_0$.

Naively one would think that the ground state of the D-string anti- D-string system that we have been considering will correspond to a constant tachyon configuration $T(x) = T_0$, where x denotes the coordinate along S^1. However since we have put a Z_2 Wilson line on one of the D-strings, and since the tachyon is odd under this Z_2 gauge transformation, it must be anti-periodic under translation by $2\pi R$ along S^1. This shows that $T = T_0$ is not an allowed configuration. Instead, for large R, the minimum energy configuration will correspond to a kink solution such that[c]

$$\begin{aligned} T(x) &\to -T_0 \quad \text{for} \quad x << 0\,, \\ T(x) &\to T_0 \quad \text{for} \quad x >> 0\,, \end{aligned} \tag{4}$$

so that the tachyon satisfies the anti-periodic boundary condition:

$$T(\pi R) = -T(-\pi R)\,. \tag{5}$$

This configuration has been shown in Fig. 1 and has finite mass in the $R \to \infty$ limit as a consequence of (3) since the energy density vanishes far away

[b] We normalise various fields so that the world-volume action on the D-string has an overall factor of g^{-1} in front, and does not have any other g dependence.
[c] Here we have taken the core of the soliton to be located at $x = 0$.

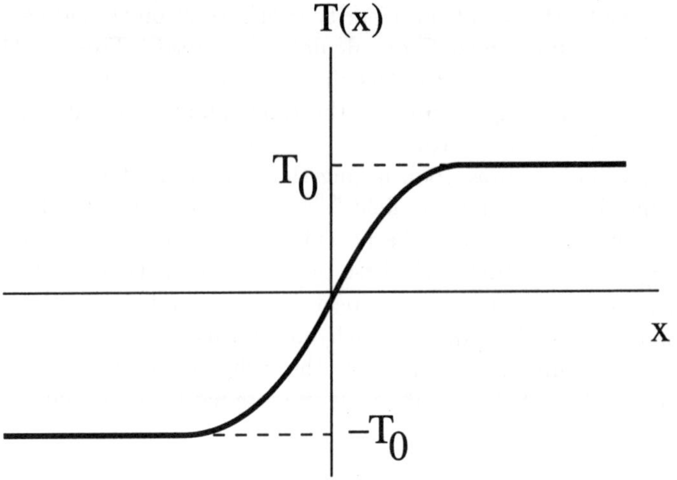

Figure 1. The minimum energy configuration of the tachyon subject to anti-periodic boundary condition.

from the core. In fact, far away from the core the D-string anti- D-string configuration is indistinguishible from the vacuum since it does not carry any net charge or energy density, so the configuration is localised near $x = 0$ and represents *a particle like state* in the ten dimensional type I string theory. The total mass of this state can be computed by integrating the energy density along x. Since the dependence on the string coupling g of this energy density is through an overall multiplicative factor of g^{-1}, we can immediately conclude that the mass of this soliton is given by

$$m_{spinor} = C/g\,, \tag{6}$$

where C is a numerical constant. Explicit computation of C requires a detailed knowledge of the tachyon potential, but we shall compute it through an indirect argument later.

Thus we see that *the SO(32) spinor state in type I string theory can be identified as the tachyonic kink on the D-string anti- D-string pair.* Although our argument based on circle compactification already shows that this state transforms in the spinor representation of SO(32), we can also see it directly without the circle compactification.[d] On the world-volume of the D-string

[d]The following investigation was suggested by N. Seiberg.

there are 32 left-moving Majorana fermions $\chi^i_{(1)}$, and on the world volume of the anti- D-string there are 32 right-moving Majorana fermions $\chi^i_{(2)}$ $(1 \leq i \leq 32)$. Since under the $Z_2 \times Z_2'$ gauge symmetry on the world volume of the D-string anti- D-string system the fields $\chi^i_{(1)}$, $\chi^i_{(2)}$ and T carry charges $(-1,1)$, $(1,-1)$ and $(-1,-1)$ respectively, the coupling between these fields must have the form:

$$L_{int} = if(T)\chi^i_{(1)}\chi^i_{(2)} + \cdots , \tag{7}$$

where $f(T)$ is an odd function of T. $\cdots$ in the above equation denote terms quartic and higher order in the fermion fields. Thus the relevant quadratic part of the fermionic lagrangian is given by:

$$L_{tot} = i[\chi^i_{(1)}(\partial_t - \partial_x)\chi^i_{(1)} + \chi^i_{(2)}(\partial_t + \partial_x)\chi^i_{(2)}] + if(T)\chi^i_{(1)}\chi^i_{(2)} . \tag{8}$$

In general the kinetic terms could be renormalized by even functions of T, but these could be absorbed into a redefinition of the fields $\chi^i_{(1)}$ and $\chi^i_{(2)}$. We shall now look for fermion zero modes by looking at the time independent equations of motion derived from the lagrangian (8):

$$\partial_x \begin{pmatrix} \chi^i_{(1)} \\ \chi^i_{(2)} \end{pmatrix} = f(T)\sigma_1 \begin{pmatrix} \chi^i_{(1)} \\ \chi^i_{(2)} \end{pmatrix} , \tag{9}$$

where σ_i denote the Pauli matrices:

$$\sigma_1 = \begin{pmatrix} 0 & 1 \\ 1 & 0 \end{pmatrix} , \qquad \sigma_2 = \begin{pmatrix} 0 & -i \\ i & 0 \end{pmatrix} , \qquad \sigma_3 = \begin{pmatrix} 1 & 0 \\ 0 & -1 \end{pmatrix} . \tag{10}$$

Let us now assume that $f(T_0) > 0$. Since $f(T)$ is an odd function of T, this would imply that $f(-T_0) < 0$. Using this we get the following normalizable solutions of eq.(9)

$$\begin{pmatrix} \chi^i_{(1)} \\ \chi^i_{(2)} \end{pmatrix} = \chi^i_0 \exp(\sigma_1 \int_0^x f(T(x'))dx') \begin{pmatrix} 1 \\ -1 \end{pmatrix} . \tag{11}$$

χ^i_0 are independent constants. If $f(T_0) < 0$, then the vector $\begin{pmatrix} 1 \\ -1 \end{pmatrix}$ in (11) is replaced by $\begin{pmatrix} 1 \\ 1 \end{pmatrix}$. In either case we get 32 fermionic zero modes χ^i_0 transforming in the vector representation of SO(32). Quantization of these zero modes gives states in the spinor as well as in the conjugate spinor representation of SO(32). However, since the zero modes are odd under the combined action of Z_2 and Z_2' generators which leave T and hence the background solution invariant, by demanding that the states are invariant under this diagonal Z_2 group leaves us only with states in the spinor representation of the SO(32) group. This shows that the tachyonic kink solution indeed has the right quantum numbers.

Let us now turn to the indirect argument determining the value of the numerical coefficient C appearing in the mass formula (6). For this we start with type I theory on a circle S^1 of radius R, and consider the original configuration, namely a D-string anti- D-string pair wrapped on S^1 with a Z_2 Wilson line on one of the strings. Since the tachyon field living on the world volume of this system must satisfy anti-periodic boundary condition along S^1, it admits a mode expansion of the form:

$$T(x,t) = \sum_{n\in Z} T_{n+\frac{1}{2}}(t) e^{i\frac{n+\frac{1}{2}}{R}x} . \qquad (12)$$

Note that the zero momentum mode is absent. The effective mass^2 of $T_{\pm(n+\frac{1}{2})}$ is given by

$$m_n^2 = \frac{(n+\frac{1}{2})^2}{R^2} - \frac{1}{2} , \qquad (13)$$

where $-1/2$ is the mass^2 of the tachyon field on the infinite D-string anti-D-string pair. From eq.(13) we see that $m_n^2 \geq 0$ for all n if

$$R \leq \frac{1}{\sqrt{2}} . \qquad (14)$$

In particular this shows that at $R = \frac{1}{\sqrt{2}}$ the modes $T_{\pm\frac{1}{2}}$ represent marginal deformation. In fact as will be shown in the next section, this mode actually represents an exactly marginal deformation. Thus we arrive at the following intuitive picture.

- For $R > (1/\sqrt{2})$ the total mass of the D-string anti- D-string pair before tachyon condensation is larger than the mass m_{spinor} of the spinor state, signalling the presence of a tachyonic mode in the D-string anti- D-string system:

$$4\pi R T_D > m_{spinor} \qquad \text{for} \qquad R > \frac{1}{\sqrt{2}} . \qquad (15)$$

- For $R < (1/\sqrt{2})$ the total mass of the D-string anti- D-string pair before tachyon condensation is smaller than the mass m_{spinor} of the spinor state, signalling the absence of a tachyonic mode in the D-string anti- D-string system:

$$4\pi R T_D < m_{spinor} \qquad \text{for} \qquad R < \frac{1}{\sqrt{2}} . \qquad (16)$$

This also shows that in this range of R the spinor state becomes unstable against decay into a D-string anti- D-string pair.

- For $R = (1/\sqrt{2})$ the total mass of the D-string anti- D-string pair before tachyon condensation is equal to the mass m_{spinor} of the spinor state, signalling the presence of a marginal deformation connecting the two states.

$$4\pi R T_D = m_{spinor} \qquad \text{for} \qquad R = \frac{1}{\sqrt{2}}. \tag{17}$$

This gives

$$m_{spinor} = \sqrt{2}/g\,. \tag{18}$$

Comparing with eq.(6) we get

$$C = \sqrt{2}\,. \tag{19}$$

The above analysis determines the value of m_{spinor} at $R = (1/\sqrt{2})$. Since we are interested in determining m_{spinor} for $R = \infty$, we have implicitly assumed that m_{spinor} does not depend on R, *i.e.* it does not get affected by compactification of the theory on a circle. We shall now give an indirect argument in favour of this, and leave a more detailed analysis of this question to section 4. In order to study the effect of compactification on the mass of the spinor state we can estimate the interaction energy between the spinor and its (infinite set of) images situated at distance $2\pi n R$ on the X-axis for all integer n. In particular the gravitational interaction in this case is of order

$$g^2 \cdot g^{-1} \cdot g^{-1} \sim 1\,, \tag{20}$$

where the g^2 factor arises from the Newton's constant and the factors of g^{-1} arise from the masses of each state in the interacting pair. Since the right hand side of (20) is much smaller than g^{-1} – the mass of the spinor state as computed earlier – we expect that the gravitational interaction between the spinor and its images will not significantly modify the mass of this state. Similar argument can be given for electromagnetic interactions as well. Thus we conclude that the mass of the spinor state is not significantly modified due to interaction with its images, and hence (18) represents the mass of this state in the infinite radius limit as well.

Finally, note that although we would have expected the classical soliton representing the SO(32) spinor state in (9+1) dimension to be invariant under the SO(9) rotation group, the soliton that we have constructed only has manifest SO(8) symmetry since the direction along which the original D-string anti- D-string pair lies is somewhat special. However, we shall see in section 4 that the boundary conformal field theory describing this state does have the full SO(9) rotational invariance.

3 Conformal Field Theory at the Critical Radius

Let X denote the scalar field corresponding to coordinate x along the S^1 and ψ, $\widetilde{\psi}$ its right- and left-moving fermionic partners on the world-sheet of the fundamental string. At this particular radius the conformal field theory of the bosonic field X is equivalent to that of a pair of right-moving fermionic fields ξ, η and a pair of left-moving fermion fields $\widetilde{\xi}$, $\widetilde{\eta}$ [20]. If we decompose X into its left- and right-moving parts as

$$X = X_L + X_R \,, \tag{21}$$

then the bose-fermi relation takes the form:

$$e^{i\sqrt{2}X_R} = \frac{1}{\sqrt{2}}(\xi + i\eta)\,, \qquad e^{i\sqrt{2}X_L} = \frac{1}{\sqrt{2}}(\widetilde{\xi} + i\widetilde{\eta})\,. \tag{22}$$

We can find another representation of the same conformal field theory by rebosonising the fermions as follows:

$$\frac{1}{\sqrt{2}}(\xi + i\psi) = e^{i\sqrt{2}\phi_R}\,, \qquad \frac{1}{\sqrt{2}}(\widetilde{\xi} + i\widetilde{\psi}) = e^{i\sqrt{2}\phi_L}\,. \tag{23}$$

ϕ represents a free bosonic field with radius $1/\sqrt{2}$. There is a third representation in which we use a slightly different rebosonization:

$$\frac{1}{\sqrt{2}}(\eta + i\psi) = e^{i\sqrt{2}\phi'_R}\,, \qquad \frac{1}{\sqrt{2}}(\widetilde{\eta} + i\widetilde{\psi}) = e^{i\sqrt{2}\phi'_L}\,, \tag{24}$$

where ϕ' is another scalar field of radius $1/\sqrt{2}$. For later use we list here the operator product expansions, and the relations between the currents of free fermions and bosons:

$$\psi(z)\psi(w) \simeq \xi(z)\xi(w) \simeq \eta(z)\eta(w) \simeq \frac{i}{z-w}\,, \tag{25}$$

$$\begin{aligned}\partial X_R(z)\partial X_R(w) &\simeq \partial\phi_R(z)\partial\phi_R(w)\\ &\simeq \partial\phi'_R(z)\partial\phi'_R(w) \simeq -\frac{1}{2(z-w)^2}\,,\end{aligned} \tag{26}$$

$$\psi\xi = i\sqrt{2}\partial\phi_R\,, \qquad \eta\xi = i\sqrt{2}\partial X_R\,, \qquad \psi\eta = i\sqrt{2}\partial\phi'_R\,. \tag{27}$$

Here $\simeq$ denotes equality up to non-singular terms. There are also similar relations involving the left-moving currents.

Since X satisfies Neumann boundary condition $X_L = X_R$ at the boundary of the world sheet, this translates to the Neumann boundary condition on the fermions:[e]

$$\xi_B = \widetilde{\xi}_B\,, \qquad \eta_B = \widetilde{\eta}_B\,, \tag{28}$$

where the subscript B denotes the boundary values of the fields. On the other hand in the NS sector, ψ satisfies the boundary condition

$$\psi_B = \widetilde{\psi}_B\,, \tag{29}$$

at both boundaries. Thus we see from (23), (24) that ϕ and ϕ' both satisfy Neumann boundary condition at both ends. However, in the Ramond sector, at one of the two ends of the open string, ψ satisfies:

$$\psi_B = -\widetilde{\psi}_B\,. \tag{30}$$

This translates to the fact that ϕ and ϕ' satisfy Dirichlet boundary condition at one end and Neumann boundary condition at the other end.

Up to overall numerical factors, the vertex operators corresponding to the states $T_{\pm\frac{1}{2}}$ are given in the -1 picture as [21]:

$$V_{\pm}^{(-1)} = \mp i e^{-\Phi_B} e^{\pm(i/\sqrt{2})X_B} \otimes \sigma_1\,, \tag{31}$$

and in the zero picture as

$$V_{\pm}^{(0)} = \psi_B e^{\pm(i/\sqrt{2})X_B} \otimes \sigma_1\,. \tag{32}$$

Φ denotes the bosonized ghost [21]. σ_1 is the 2×2 Chan-Paton factor representing that the tachyon originates in the open string sector with two ends of the open string on two different strings. The overall multiplicative factor of ∓ 1 in (31) is a matter of convention. Let

$$V_T = \frac{1}{\sqrt{2}}(V_+ + V_-)\,, \tag{33}$$

represent vertex operators for the real component of $T_{1/2}$.[f] Using eqs.(22)-(33) we get

$$V_T^{(-1)} = e^{-\Phi_B}\eta_B \otimes \sigma_1, \qquad V_T^{(0)} = \psi_B \xi_B \otimes \sigma_1\,. \tag{34}$$

---

[e] These boundary conditions are written in the coordinate system where the open string world sheet is represented as the upper half plane.

[f] We could have chosen to work with any particular direction in the complex $T_{1/2}$ plane; we choose the real direction for definiteness.

We can now take ϕ, η and $\tilde{\eta}$ as independent fields, and represent the tachyon vertex operator in the zero picture as:

$$V_T^{(0)} = \frac{i}{\sqrt{2}} \partial\phi_B \otimes \sigma_1 \,, \tag{35}$$

for Neumann boundary condition on ϕ. ∂ denotes tangential derivative along the boundary. This shows that switching on the tachyon vev corresponds to switching on a U(1) Wilson line along the bosonic direction ϕ. On the other hand, for Dirichlet boundary condition on ϕ, we may rewrite (34) as

$$V_T^{(0)} = \frac{i}{\sqrt{2}} (\partial\phi_R - \partial\phi_L)_B \otimes \sigma_1 \tag{36}$$

Switching on the tachyon vev now corresponds to changing the boundary condition from $\phi = 0$ to ϕ =constant. Clearly, both (35) and (36) represent marginal deformation.

Before we study the effect of switching on such a tachyon vev on the spectrum of open strings, let us study the effect of various projections on the open string spectrum before switching on the tachyon vev. There are several discrete transformations under which the open string states are required to be invariant. Each such transformation includes transformation on the fields at the boundary, on the SL(2,R) invariant vacuum in the NS sector, as well as on the Chan-Paton factor which we shall represent by a 2×2 matrix Λ. These are as follows:

$$\begin{aligned} (-1)^F : X_B \to X_B, \quad & \psi_B \to -\psi_B, \quad X_B^\mu \to X_B^\mu, \\ & \psi_B^\mu \to -\psi_B^\mu, \quad |0\rangle \to -|0\rangle, \quad \Lambda \to \sigma_3 \Lambda \sigma_3, \\ h : X_B \to X_B + \frac{2\pi}{\sqrt{2}}, \quad & \psi_B \to \psi_B, \quad X_B^\mu \to X_B^\mu, \\ & \psi_B^\mu \to \psi_B^\mu, \quad |0\rangle \to |0\rangle, \quad \Lambda \to \sigma_3 \Lambda \sigma_3, \end{aligned} \tag{37}$$

where x^μ $(1 \le \mu \le 8)$ denote the non-compact directions transverse to the string. In $(-1)^F$ the conjugation by σ_3 reflects the fact that the open strings with two ends on two different strings, represented by off diagonal Chan Paton matrices, have opposite GSO projection compared to open strings with both ends on the same string. The $-$ sign in the transformation law of the vacuum represents the fact that the NS sector ground state for open strings with both ends on the same string is odd under $(-1)^F$. h represents translation by $2\pi/\sqrt{2}$ along x, and the conjugation of Λ by σ_3 in the transformation law of h is a reflection of the Z_2 Wilson line present along one of the strings which make open strings with off-diagonal Chan-Paton factors anti-periodic along x instead of peridic. Using eqs.(22) we see that the action of these transformations on the fields ξ_B and η_B at the boundary are given by:

$$\begin{aligned}(-1)^F &: \quad \xi_B \to \xi_B, \quad \eta_B \to \eta_B, \\ h &: \quad \xi_B \to -\xi_B, \quad \eta_B \to -\eta_B\end{aligned} \tag{38}$$

and hence, using eq.(23), (24) we get the following action on ϕ, ϕ':

$$\begin{aligned}(-1)^F &: \quad \phi_B \to -\phi_B, \quad \phi'_B \to -\phi'_B, \\ h &: \quad \phi_B \to \frac{2\pi}{\sqrt{2}} - \phi_B, \quad \phi'_B \to \frac{2\pi}{\sqrt{2}} - \phi'_B.\end{aligned} \tag{39}$$

Besides $(-1)^F$ and h, the states are also required to be invariant under the world-sheet parity transformation Ω which we have not listed here.

Let us now study the effect of switching on the tachyon vev. Since we have seen that the Ramond sector of the open strings correspond to putting Neumann boundary condition at one end and Dirichlet boundary condition at the other, and hence are forced to carry zero momentum and winding along ϕ, we see immediately that the tachyon vev, representing translation along ϕ at the Dirichlet end, and Wilson line along ϕ at the Neumann end, does not affect the Ramond sector states. On the NS sector states with Neumann boundary condition at both ends, the tachyon vev corresponds to switching on a Wilson line along ϕ. Let us parametrize this Wilson line by

$$\exp(i\int_0^{2\pi/\sqrt{2}} A_\phi d\phi) = \exp(i\alpha\pi\sigma_1/2), \tag{40}$$

corresponding to

$$A_\phi = \frac{\alpha}{2\sqrt{2}}\sigma_1. \tag{41}$$

α is a parameter labelling the strength of the tachyon vev. Typically the effect of switching on the Wilson line is to shift the quantization rule for the momentum along ϕ for states which are charged under the corresponding gauge field. In particular, for a state carrying q units of charge under σ_1, the ϕ momentum gets shifted by an amount

$$\frac{\alpha}{2\sqrt{2}}q. \tag{42}$$

Since the 2×2 identity matrix I and the matrix σ_1 commute with the Chan-Paton factor σ_1 associated with the tachyon vertex operator, we see that states with Chan Paton factors proportional to I or σ_1 are neutral under the Wilson line and hence their masses are not affected by the tachyon vev. This leaves us with the NS sector states with Chan Paton factors σ_3 and σ_2 respectively. To study the effect of the tachyon vev on states from these sectors, let us note that the Chan Paton factors:

$$\sigma_3 \mp i\sigma_2 \tag{43}$$

carry charges ± 2 respectively under the generator σ_1. In the presence of the tachyon vev (40) the ϕ momentum quantization rule of these states get shifted by an amount:

$$\pm \frac{\alpha}{\sqrt{2}} \,. \tag{44}$$

Since ϕ has radius $(1/\sqrt{2})$, this shows that α is a periodic variable with period 2. It will be more convenient for our analysis to use the fermionic language where we use the fermionic fields ψ_B, ξ_B and η_B as independent boundary fields. Defining

$$\chi_B = \frac{1}{\sqrt{2}}(\xi_B + i\psi_B) = e^{i\phi_B/\sqrt{2}} \,, \tag{45}$$

we can represent the effect of the tachyon vev in the sectors $\sigma_3 \pm i\sigma_2$ by taking the mode expansions of χ_B, $\chi_B^\dagger$ as follows:

$$\begin{aligned} \chi_B &= \sum_{n\in Z} \chi_{n+\frac{1}{2}\pm\alpha} e^{-i\pi(n+\frac{1}{2}\pm\alpha)\tau} \\ \chi_B^\dagger &= \sum_{n\in Z} \chi^\dagger_{n+\frac{1}{2}\mp\alpha} e^{-i\pi(n+\frac{1}{2}\mp\alpha)\tau} \,. \end{aligned} \tag{46}$$

Note that in our notation

$$(\chi_{n+\frac{1}{2}\pm\alpha})^\dagger = \chi^\dagger_{-n-\frac{1}{2}\mp\alpha} \,. \tag{47}$$

The non-trivial anti-commutators are:

$$\{\chi_{n+\frac{1}{2}\pm\alpha}, \chi^\dagger_{-m-\frac{1}{2}\mp\alpha}\} = \delta_{mn} \,. \tag{48}$$

In the Fock space built up from these oscillators, we need to project onto states invariant under $(-1)^F$, h and Ω. From (37) we see that both $(-1)^F$ and h exchange the Chan Paton factors $\sigma_3 \mp i\sigma_2$. Since Ω acts on the Chan Paton factor by transposition, it also exchanges the two sectors. Due to this reason it is more convenient to take the independent generators to be $(-1)^F$, $(-1)^F h$ and $(-1)^F\Omega$. Since $(-1)^F h$ and $(-1)^F\Omega$ leave the Chan Paton factors untouched, we can impose invariance under these transformations on the Fock space states, and then take appropriate linear combinations of the states from the two sectors to get states invariant under $(-1)^F$. Let us for definiteness focus on the sector with Chan Paton factor $\sigma_3 + i\sigma_2$. The simplest way to implement invariance under $(-1)^F h$ and $(-1)^F \Omega$ is to start from states at $\alpha = 0$ satisfying these projections and then simply follow them as we continuously deform α. During this process the oscillators $\chi_{n+\frac{1}{2}}$ and $\chi^{\dagger}_{n+\frac{1}{2}}$ get deformed to $\chi_{n+\frac{1}{2}+\alpha}$ and $\chi^{\dagger}_{n+\frac{1}{2}-\alpha}$ respectively, and the Fock vacuum $|0\rangle$ get deformed to $|0\rangle_\alpha$ satisfying

$$\chi_{n+\frac{1}{2}+\alpha}|0\rangle_\alpha = 0, \qquad \chi^{\dagger}_{n+\frac{1}{2}-\alpha}|0\rangle_\alpha = 0, \qquad \text{for} \quad n \geq 0\,. \tag{49}$$

Notice that as α passes through $(1/2)$, the state $|0\rangle_\alpha$ is no longer the state with lowest L_0 eigenvalue. Instead for $(3/2) > \alpha > (1/2)$ the state with lowest L_0 eigenvalue, denoted by $|G\rangle$, is related to $|0\rangle_\alpha$ by the relation:

$$|0\rangle_\alpha = \chi^{\dagger}_{\frac{1}{2}-\alpha}|G\rangle\,, \tag{50}$$

Note that $|G\rangle$ has opposite $(-1)^F h$ charge compared to $|0\rangle$ since χ, $\chi^\dagger$ are odd under $(-1)^F h$.

Once we have determined the spectrum in each sector invariant under $(-1)^F h$ and $(-1)^F \Omega$, we can find $(-1)^F$ invariant combination of states on the two sectors by noting that under this transformation:

$$\begin{aligned}(\sigma_3 + i\sigma_2) &\to (\sigma_3 - i\sigma_2)\\ |0\rangle_\alpha &\to -|0\rangle_{-\alpha}\\ \chi_{n+\frac{1}{2}+\alpha} &\to \chi^{\dagger}_{n+\frac{1}{2}+\alpha}\\ \chi^{\dagger}_{n+\frac{1}{2}-\alpha} &\to \chi_{n+\frac{1}{2}-\alpha}\,.\end{aligned} \tag{51}$$

We now note some important features of the spectrum.

- At $\alpha = 1$ the fermions χ, $\chi^\dagger$ are again half integer moded. However, since the ground state $|G\rangle$ has opposite $(-1)^F h$ charge compared to the vacuum $|0\rangle$ at $\alpha = 0$, this point is not equivalent to $\alpha = 0$. In particular, if we denote by X^μ, ψ^μ $(1 \leq \mu \leq 8)$ the bosonic coordinates and their fermionic partners transverse to the D-string, then the state $\psi^\mu_{-\frac{1}{2}}|G\rangle$ is odd under $(-1)^F h$ and is projected out. On the other hand, if we start from the state $\psi^\mu_{-\frac{1}{2}}|0\rangle$ at $\alpha = 0$ and follow it as α changes from 0 to 1, this state gets mapped to $\psi^\mu_{-\frac{1}{2}}\chi^{\dagger}_{-\frac{1}{2}}|G\rangle$, and has dimension 1. Thus there are no zero modes corresponding to separating the D-string and the anti-

D-string away from each other. This shows that at $\alpha = 1$ (as well as at other values of α which are not equivalent to $\alpha = 0$) the D-string anti-D-string system is bound.

χ, $\chi^\dagger$ have half integer mode expansion again at $\alpha = 2$. Following an identical analysis one can easily verify that this time the new ground state $|G'\rangle$ has the same quantum numbers under various projection operators as in the case of $\alpha = 0$ and hence this point is equivalent to $\alpha = 0$. In particular the set of states $\psi^\mu_{-\frac{1}{2}}|G'\rangle$ survive all projections and represent the freedom of separating the D-string - anti-D-string pair away from each other. This also shows that the D-string anti- D-string system is most tightly bound (*i.e.* the mode corresponding to the freedom of separating the pair has maximum mass) at $\alpha = 1$. Thus we might expect that the $\alpha = 1$ point will represent the true minimum of the tachyon potential when we increase the radius of X away from $1/\sqrt{2}$. We shall see this more explicitly in section 4.

- Note also that at $\alpha = 1$, the ground state $|G\rangle$ is invariant under $(-1)^F h$. This would correspond to a tachyonic mode. However, as we shall show now, this state is projected out by the $(-1)^F \Omega$ projection. First note that since this state can be written as $\chi_{1/2}|0\rangle_{\alpha=1}$, it is the image of the state $\chi_{-1/2}|0\rangle$ at $\alpha = 0$. The $(-1)^F$ invariant combination at $\alpha = 0$, as seen from (51), is given by

$$\begin{aligned}&\frac{1}{\sqrt{2}}\{\chi_{-\frac{1}{2}}|0\rangle \otimes (\sigma_3 + i\sigma_2) - \chi^\dagger_{-\frac{1}{2}}|0\rangle \otimes (\sigma_3 - i\sigma_2)\}\\ &= i(\psi_{-\frac{1}{2}}|0\rangle \otimes \sigma_3 + \xi_{-\frac{1}{2}}|0\rangle \otimes \sigma_2)\,. \end{aligned} \tag{52}$$

The first term on the right hand side of this equation represents the Wilson line of the U(1) gauge field on the D-string along the compact circle, and is known to have the wrong Ω projection [4]. The second state represents a state carrying momentum along x, but also has wrong Ω projection since the Chan Paton factor is anti-symmetric. Since by construction these states are even under $(-1)^F$, we conclude that both are odd under $(-1)^F\Omega$, and hence are projected out. Thus there is no tachyonic mode of the system at $\alpha = 1$.

This analysis also shows us that if we consider an identical configuration of D-string anti- D-string pair in type IIB string theory, then the $\alpha = 1$ state *does* have a tachyonic mode since there is no Ω projection in this case. This tachyonic instability can be traced to the fact that on (infinitely long) D-string anti- D-string pair in type IIB string theory the tachyon represents a complex field on the world volume, and hence the minimum of the potential lies along a circle $T = T_0 e^{i\theta}$ instead of at two points $\pm T_0$. Since $\pi_0(S^1)$ is trivial, the kink solution is no longer topologically stable (although it does represent a solution of the equations of motion) and hence has a tachyonic mode representing decay into the trivial solution $T = T_0$.

- Once we switch on the tachyon vev we break translational invariance along x, and hence there must be a zero mode representing the freedom of translating the soliton along x. The vertex operator for this zero mode is constructed as follows. Let V_x denote the vertex operator associated with this mode. If δ_ϵ denotes the transformation $x \to x + \epsilon$, then, since the translation invariance is broken by the tachyon vev, we would expect,

$$\delta_\epsilon V_T \propto \epsilon V_x \, . \tag{53}$$

V_T is the tachyon vertex operator defined in eqs.(31)-(33). This gives,

$$V_x^{(-1)} = -e^{-\Phi_B}\xi_B \otimes \sigma_1, \qquad V_x^{(0)} = \psi_B \eta_B \otimes \sigma_1 \, . \tag{54}$$

This represents a marginal operator for all values of tachyon vev α.

- Note that the sectors with Chan Paton factors I and σ_1 contain Fock space states carrying $((-1)^F, (-1)^F h)$ values (1,1) and $(-1,1)$ respectively, with the SL(2,R) invariant vacuum assigned quantum numbers $(-1,-1)$. On the other hand, after switching on the tachyon vev, the combined spectrum from the sectors with Chan Paton factors $\begin{pmatrix} 1 & \pm 1 \\ \mp 1 & -1 \end{pmatrix}$ contains Fock space states carrying $((-1)^F, (-1)^F h)$ values $(\pm 1, -1)$ *if we continue to assign quantum numbers* $(-1,-1)$ *to the SL(2,R) invariant vacuum.* Thus once we combine the states from all four Chan Paton sectors, all Fock space states are present in the spectrum before projection under Ω.

Besides the modes discussed here, there are other open string states with one end lying on the (anti-) D-string and the other end lying on the 9-brane. The effect of tachyon vev on the spectrum of these states can be analysed by very similar method, but we shall not do it here.

4 Deforming Away from the Critical Radius

In this section we shall consider the effect of switching on the perturbation that deforms the radius away from the critical radius. In particular we shall show that:

- In the presence of this perturbation the tachyon develops a tadpole except at $\alpha = 0$ and at $\alpha = 1$. Thus these are the only two points which correspond to solutions of equations of motion.

- As we increase the radius of the circle in the x direction, the tachyon represents a state with positive mass^2 if $\alpha = 1$. Thus $\alpha = 1$ represents a stable solution for $R > (1/\sqrt{2})$.

- In the $R \to \infty$ limit we can represent the $\alpha = 1$ solution as a D0-brane in type I string theory.

We begin by analysing the effect of radius deformation on the one point function of the tachyon at an arbitrary value α of the background tachyon field. If V_r denotes the closed string vertex operator associated with the radius deformation, then the effect of switching on this perturbation on the tachyon one point function is proportional to the two point function $\langle V_r V_T \rangle_\alpha$ on the disk, with V_T inserted at the boundary of the disk and V_r inserted in the interior of the disk. For definiteness we shall take V_T in the -1 picture and V_r in the $(-1, 0)$ picture for this calculation.[g] $V_T^{(-1)}$ is given in eq.(34), whereas using eqs.(21)-(27) we may express $V_r^{(-1,0)} \propto e^{-\widetilde{\Phi}} \partial X_R \widetilde{\psi}$ as

$$V_r^{(-1,0)} \propto e^{-\widetilde{\Phi}} \eta \Big(e^{i\sqrt{2}(\phi_L + \phi_R)} + e^{i\sqrt{2}(\phi_L - \phi_R)} - e^{-i\sqrt{2}(\phi_L - \phi_R)} - e^{-i\sqrt{2}(\phi_L + \phi_R)} \Big), \tag{55}$$

$\widetilde{\Phi}$ being the left-moving component of the bosonized ghost [21]. The vertex operators $e^{\pm i\sqrt{2}(\phi_L + \phi_R)}$ carry net ϕ momentum. Since neither the tachyon vertex operator nor the background tachyon field carry any ϕ momentum, and since the Neumann boundary condition on ϕ conserves ϕ momentum, the correlation function involving these two terms in $\langle V_r^{(-1,0)} V_T^{(-1)} \rangle_\alpha$ vanish. Thus the matter part of the correlation function is proportional to:

$$\langle \eta(P) \Big(e^{-i\sqrt{2}(\phi_L - \phi_R)} - e^{i\sqrt{2}(\phi_L - \phi_R)} \Big)(P) \eta_B(Q) \\ Tr \Big(\sigma_1 e^{\frac{i}{2\sqrt{2}} \alpha \sigma_1 \oint dt \partial_t \phi_B} \Big) \rangle, \tag{56}$$

where P denotes a point in the bulk, Q denotes a point on the boundary and t parametrizes the boundary. Note that we have included the effect of the tachyon background by including the exponential factor inside the trace over

---

[g] $(-1, 0)$ picture means that we use -1 picture for the left-moving sector and 0 picture for the right-moving sector.

the Chan Paton factor, so $\langle\ \rangle$ in the above equation represents correlation function in the theory without any tachyon vev.[h] The trace over the Chan Paton factors can be easily evaluated and the contribution may be written as

$$i\langle\eta(P)\eta_B(Q)\Big(e^{i\sqrt{2}(\phi_L-\phi_R)}-e^{-i\sqrt{2}(\phi_L-\phi_R)}\Big)(P)$$
$$\sin(\frac{1}{2}\alpha\pi w_\phi)\rangle\,, \tag{57}$$

where

$$w_\phi=\frac{\sqrt{2}}{2\pi}\oint dt\partial_t\phi_B\,, \tag{58}$$

measures the total ϕ winding number carried by the operators in the bulk. Since $e^{\pm i\sqrt{2}(\phi_L-\phi_R)}$ carry ϕ winding number ± 2 we see that this correlation function is proportional to

$$i\sin(\alpha\pi)\langle\eta(P)\eta_B(Q)\Big(e^{-i\sqrt{2}(\phi_L-\phi_R)}+e^{i\sqrt{2}(\phi_L-\phi_R)}\Big)(P)\rangle\,. \tag{59}$$

The correlation function $\langle\eta(P)\eta_B(Q)\rangle$ is non-vanishing on the disk. On the other hand, due to Neumann boundary condition on ϕ, the expectation values of $e^{\pm i\sqrt{2}(\phi_L-\phi_R)}(P)$ on the disk are non-vanishing and equal. Thus we see that the one point function of the tachyon in the presence of the radius perturbation is proportional to $\sin(\alpha\pi)$. This vanishes only at $\alpha=0$ and $\alpha=1$ (other integer values of α being equivalent to 0 or 1). Thus only at these two values of α we get solutions of the equations of motion.

Next we analyse the two point function of the tachyon field at $\alpha=1$. This time we can take the two tachyon vertex operators to be in the -1 picture and V_r in the (0,0) picture, given by:

$$V_r^{(0,0)}=\partial X_R\partial X_L=-\frac{1}{2}\xi\eta\tilde{\xi}\tilde{\eta}$$
$$=\frac{1}{4}\eta\tilde{\eta}\Big(e^{-i\sqrt{2}(\phi_L-\phi_R)}+e^{i\sqrt{2}(\phi_L-\phi_R)}$$
$$+e^{-i\sqrt{2}(\phi_L+\phi_R)}+e^{i\sqrt{2}(\phi_L+\phi_R)}\Big)(P)\,. \tag{60}$$

---

[h]Since the tachyon vertex operator connects a D-string with Z_2 Wilson line to an anti-D-string without a Z_2 Wilson line, there is some subtlety in writing the contribution of the operators at the boundary as a trace over Chan Paton factors. However, since the D-string and the anti- D-string differ in their coupling to the RR sector states, and also in their coupling to the states carrying odd winding number along the x direction (see, for example ref. [12]), this subtlety does not affect the computation when the operator in the bulk is in the NSNS sector and carries zero winding number along x.

Thus the matter part of the relevant correlation function is of the form:

$$\begin{aligned}\langle\eta(P)\tilde{\eta}(P)\Big(&e^{-i\sqrt{2}(\phi_L-\phi_R)}+e^{i\sqrt{2}(\phi_L-\phi_R)}\\&+e^{-i\sqrt{2}(\phi_L+\phi_R)}+e^{i\sqrt{2}(\phi_L+\phi_R)}\Big)(P)\\&\eta_B(Q_1)\eta_B(Q_2)Tr\Big(\sigma_1\sigma_1 e^{\frac{i}{2\sqrt{2}}\sigma_1\oint dt\partial_t\phi_B}\Big)\rangle\,,\end{aligned}\tag{61}$$

where again P denotes a point in the interior of the disk and Q_1 and Q_2 denote points on the boundary of the disk. The trace over the Chan Paton factors gives $\cos(\pi w_\phi/2)$. As in the case of one point function, the contribution to the correlator from the terms involving $\exp(\pm i\sqrt{2}(\phi_L+\phi_R))$ vanishes due to ϕ momentum conservation. For the other two terms carrying $w_\phi=\pm 2$, $\cos(\pi w_\phi/2)$ gives a contribution of -1. Thus the net effect of the tachyon vev corresponding to $\alpha=1$ is to change the sign of this three point function. Since we already know that in the absence of tachyon vev the tachyon mass^2 *decreases* as we increase the radius, we conclude that for $\alpha=1$ the tachyon mass^2 *increases* as we increase the radius. Thus $\alpha=1$ represents a stable ground state of the system for $R>(1/\sqrt{2})$. This analysis also shows that for $R<(1/\sqrt{2})$ the tachyon mass^2 becomes negative at $\alpha=1$, indicating that the spinor state becomes unstable in this range of values of R.

We can generalise this analysis to include arbitrary number of tachyon vertex operators on the boundary and arbitrary number of insertions of $V_r^{(0,0)}$ in the interior. For $\alpha=1$, the effect of the tachyon vev is to introduce a factor of

$$\exp(\frac{i}{2}\pi w_\phi\sigma_1)=\cos(\pi w_\phi/2)+i\sigma_1\sin(\pi w_\phi/2)\,,\tag{62}$$

at the boundary. Since each term in $V_r^{(0,0)}$ carries even w_ϕ charge, the total w_ϕ carried by all the closed string vertex operators is always even, and hence the contribution from the $\sin(\pi w_\phi/2)$ term vanishes. The effect of the $\cos(\pi w_\phi/2)$ term is to change the sign of every term in each $V_r^{(0,0)}$ carrying odd $(w_\phi/2)$. Using expression (60) for $V_r^{(0,0)}$ we see that it gets transformed to:

$$\begin{aligned}&\frac{1}{4}\eta\tilde{\eta}\Big(-e^{-i\sqrt{2}(\phi_L-\phi_R)}-e^{i\sqrt{2}(\phi_L-\phi_R)}\\&\qquad+e^{-i\sqrt{2}(\phi_L+\phi_R)}+e^{i\sqrt{2}(\phi_L+\phi_R)}\Big)(P)\\&=-\frac{1}{2}\eta\tilde{\eta}\psi\tilde{\psi}=-\partial\phi'_R\partial\phi'_L\,.\end{aligned}\tag{63}$$

Thus tachyon vev corresponding to $\alpha=1$ transforms the vertex operator $\partial X_R\partial X_L$ to $-\partial\phi'_R\partial\phi'_L$. In other words, increasing the radius of the x coordinate in the presence of the tachyon vev corresponds to decreasing the radius of ϕ' coordinate with no tachyon vev. If the x radius is scaled up by a factor of L, this would correspond to scaling down the ϕ' radius by a factor of L. Thus an x radius of $L/\sqrt{2}$ will correspond to a ϕ' radius of $1/(\sqrt{2}L)$. Since the tachyon vertex operator carries ϕ' momentum (as can be easily verified from eqs.(24) and (34)) a decrease in the ϕ' radius will increase the mass^2 of the tachyon. This confirms our previous perturbative result.

The analysis can be easily extended to all NS sector vertex operators with Chan Paton factors proportional to σ_1 or I. These Chan Paton factors commute with the Chan Paton factors associated with the tachyon. Furthermore, since for correlation functions involving NS sector states on the disk we have Neumann boundary condition on ϕ ($\phi_L = \phi_R$) everywhere on the boundary, the vertex operators inserted at the boundary may carry ϕ momentum but no ϕ winding. Hence insertion of these vertex operators do not create any discontinuity in ϕ_B on the boundary, and $\oint \partial_t \phi_B$ along the boundary can be identified as $\sqrt{2}\pi w_\phi$, with w_ϕ measuring the total ϕ winding number of all the operators inserted in the interior of the disk. Thus the effect of a tachyon vev corresponding to $\alpha = 1$ may again be represented by (62). We can now repeat the argument of the previous paragraph to show that the effect of increasing the radius of the x coordinate in the presence of the tachyon vev is to effectively decrease the radius of the ϕ' coordinate.

Next we consider the case where on the boundary we have vertex operators with Chan Paton factors proportional to $\sigma^\pm = \sigma_3 \pm i\sigma_2$. For illustration we consider the case where we have two vertex operators at the boundary – one proportional to σ^+ and the other to σ^- – but this construction can be easily generalized for any correlation function with arbitrary insertions of NS sector open string vertex operators at the boundary. For the two point correlator the typical operator insertions at the boundary take the form:

$$Tr\Big(\sigma^- e^{ik_1\phi_B(Q_1)} e^{\frac{i}{2\sqrt{2}}\sigma_1 \int_{Q_1}^{Q_2} \partial_t\phi_B dt} \sigma^+ e^{ik_2\phi_B(Q_2)} e^{\frac{i}{2\sqrt{2}}\sigma_1 \int_{Q_2}^{Q_1+2\pi} \partial_t\phi_D dt}\Big), \tag{64}$$

where t denotes the coordinate on the boundary of the disk with periodicity 2π, Q_1 and Q_2 represent the points on the boundary where the vertex operators have been inserted, and k_1 and k_2 are the ϕ momenta carried by the vertex operators. Using the commutation relations between the Pauli matrices, and the relations

$$\int_{Q_1}^{Q_2} \partial_t\phi_B dt = \phi_B(Q_2) - \phi_B(Q_1),$$

$$\int_{Q_2}^{Q_1+2\pi} \partial_t \phi_B dt = \phi_B(Q_1) - \phi_B(Q_2) + \sqrt{2}\pi w_\phi \,, \tag{65}$$

one can bring (64) to the form:

$$Tr\Big(\sigma^- e^{i(k_1+\frac{1}{\sqrt{2}})\phi_B(Q_1)} \sigma^+ e^{i(k_2-\frac{1}{\sqrt{2}})\phi_B(Q_2)} e^{i\pi w_\phi \sigma_1/2}\Big) \,. \tag{66}$$

Thus we see that the effect of switching on the tachyon vev is to shift the ϕ momenta in vertex operators associated with $\sigma^\mp$ by $\pm 1/\sqrt{2}$ – a fact which has already been taken into account in section 3 – and simultaneously introduce a factor of $\exp(i\pi w_\phi \sigma_1/2)$ at the boundary. The later factor transforms the operators $\partial X_R \partial X_L$ inserted in the interior of the disk to $-\partial\phi'_R\partial\phi'_L$ as before. This shows that for all correlation functions involving NS sector open string vertex operators, the effect of deforming the x-radius after switching on the tachyon vev is to change the ϕ' radius in the opposite direction.

From (54), and the bosonisation rules (27) we see that the vertex operator V_x in the zero picture is proportional to

$$i\partial\phi'_B \otimes \sigma_1 \,, \tag{67}$$

and hence represents a Wilson line along the ϕ' coordinate. This clearly remains a marginal deformation even when we change the ϕ' radius $R_{\phi'}$. This shows that when we change the radius of the x coordinate in the presence of tachyon vev, the translational zero mode continues to represent a marginal deformation. Furthermore, this zero mode, being a Wilson line along ϕ', represents a periodic coordinate with periodicity proportional to $1/R_{\phi'}$. Since $R_{\phi'}$ is proportional to $(1/R)$, R being the radius of the x coordinate, we see that the periodicity is proportional to R, as is expected of a zero mode representing translation along a circle of radius R. In particular, this shows that in the $R \to \infty$ limit the zero mode takes value on a real line, representing the position of the soliton along the x axis.

In calculating the spectrum in the $R \to \infty$ limit we can simplify the analysis by making a T-duality transformation in the coordinate ϕ'. Since (after taking into account the effect of the tachyon vev) the coordinate ϕ' has radius $1/(2R)$, the T-duality will give rise to dual coordinate ϕ'_D of radius $2R$, and at the same time transform the Neumann boundary condition on ϕ' to Dirichlet boundary condition on ϕ'_D. In the limit $R \to \infty$, ϕ'_D has infinite radius, and we recover an extra SO(9) symmetry that mixes the pair (ϕ'_D, ξ) with (X^μ, ψ^μ) for $1 \le \mu \le 8$.[i] Thus we see that although to start with we had broken the SO(9) symmetry of the problem by identifying one special

[i] Although $(-1)^F$ and $h : X \to X + \sqrt{2}\pi$ projections act differently on (ϕ'_D, ξ) and (X^μ, ψ^μ), from the analysis of the previous section we have seen that once we combine the states from

direction along which the D-string anti- D-string pair lies, we recover the full SO(9) symmetry of the spectrum at the end.

Next we turn to the fate of the Ramond sector states under radius deformation. We shall focus on two point function involving two Ramond sector vertex operators; this is sufficient for analysing the spectrum. Since the insertion of a Ramond sector vertex operators converts a Neumann boundary condition on ϕ to a Dirichlet boundary condition, we can no longer represent the tachyon vertex operator in the zero picture by (35) everywhere in the boundary. However we can still express it as

$$V_T^{(0)} = i\sqrt{2}\partial\phi_R \otimes \sigma_1 \,. \tag{68}$$

Since the Ramond sector vertex operators do not carry either momentum or winding along ϕ, ϕ_R is continuous across the Ramond sector vertex operators inserted at the boundary. Following the same arguments as in the case of NS sector states, we see that the effect of switching on the tachyon vev on the correlation functions is to

- give a factor of

$$\exp\left(\frac{i}{\sqrt{2}}\sigma_1 \oint dt\partial_t\phi_R\right), \tag{69}$$

  at the boundary, and

- introduce factors of

$$\exp(\mp i\sqrt{2}\phi_R)\,, \tag{70}$$

  at the location of Chan Paton factors $\sigma^{\pm}$. However, since the Ramond sector vertex operators are localized at $\phi = 0$, (70) becomes trivial.

Let us now switch on the radius deformation. This corresponds to adding a term proportional to integral of $\partial X_L \partial X_R$ to the two dimensional action, which in turn can be expressed as (60). Using the operator product expansion of scalar fields, it is easy to see that acting on the right hand side of

all four Chan Paton sectors, we have all the Fock space states carrying $(-1)^F = \pm 1$ and $h = \pm 1$. Thus the full spectrum is invariant under the exchange $(\phi'_D, \xi) \leftrightarrow (X^\mu, \psi^\mu)$ before Ω projection. On the other hand, one can show by a detailed analysis that the Ω projection also preserves the SO(9) symmetry of the spectrum. The main point is to note that Ω induces a transformation $\phi' \to -\phi'$, since otherwise the vertex operator $V_x^{(0)} \propto \partial_t\phi'_B \otimes \sigma_1$, involving tangential derivative of ϕ' along the boundary, could not be Ω invariant. This in turn implies that the dual coordinate ϕ'_D does not change sign under Ω, just as the X^μ's for $1 \le \mu \le 8$. Furthermore, since the vertex operators $\xi_B \otimes \sigma_1$ and $\psi^\mu \otimes I$ are both present in the spectrum in the -1 picture even after the Ω projection, the oscillators of ξ and ψ^μ transform under Ω in identical manner.

(60), (69) gives an overall factor of -1. Thus for this correlation function, the effect of the tachyon vev is to change the sign of the $\partial X_L \partial X_R$ term, *i.e.* to convert a perturbation corresponding to an increase in radius to one corresponding to a decrease in the radius. In particular in order to compute the spectrum of Ramond sector states at x radius $R = L/\sqrt{2}$ in the presence of a tachyon vev $\alpha = 1$, we simply need to compute the spectrum at a new x radius $R_x = 1/(\sqrt{2}L)$ in the absence of any tachyon vev. If we T-dualize in the x coordinate, and use the dual radius x_D as the independent bosonic variable, then the x_D radius goes to ∞ as $R \to \infty$, and the Neumann boundary condition on x corresponds to Dirichlet boundary condition on x_D. Thus we recover the full SO(9) invariance of the spectrum in this limit.[j]

We can now address the question of the dependence of the mass of the soliton on the radius. In the absence of tachyon vev, the mass of the D-string anti- D-string pair is expected to be proportional to R. Let us first see how this arises in the string calculation. For this we compute the interaction between a pair of such systems at large separation by computing the partition function of open strings with two ends lying on the two solitons, and equate it to

$$G_N M^2 / l^6 \,, \tag{71}$$

where G_N is the Newton's constant, M is the mass of the system, and l is the separation. G_N in the (8+1) dimensional theory obtained by compactifying the x direction is proportional to $(1/R)$. On the other hand, for large separation l between the two systems, the leading contribution to the open string partition function comes from the states with large x momentum. The density of such states is proportional to R. Thus we get, $(M^2/R) \propto R$. This gives $M \propto R$.

In the presence of tachyon vev, G_N is still proportional to $(1/R)$. The open string states in NS (R) sector are now better regarded as states carrying definite momentum along ϕ' (x), with ϕ' (x) having radius proportional to

[j] Again in this case we need to ensure that various projections do not destroy the symmetry under the exchange of x_D with one of the transverse coordinates x^μ. To see this, let us first note that in this case the h projection simply removes all states in Chan Paton sectors σ_1 and σ_2, since states in this sector carry odd unit of x momentum, and hence become infinitely massive in the limit when the effective x-radius goes to zero. Thus we are left with Chan Paton factors I and σ_3. The action of Ω on Ramond sector states with Chan Paton factors $I \pm \sigma_3$, representing wrapped D-branes and anti- D-branes respectively, can be shown to differ by a factor of -1 using the results of ref. [22]. Thus when we combine the set of Fock space states from both sectors, we can simply ignore the Ω projection. Finally, since $(-1)^F$ changes the sign of ψ^μ as well as of ψ, projection under $(-1)^F$ does not destroy SO(9) invariance.

$1/R$. For large l the leading contribution comes from states with large ϕ' (x) momentum whose density of states is proportional to $R_{\phi'}$ (R_x) and hence to R^{-1}. Thus we get $(M^2/R) \propto (1/R)$. This shows that M^2 is independent of R as has been advertised earlier.

Let us now summarise the main result of this section. Starting from the description of the SO(32) spinor of type I as a tachyonic kink solution on the D-string anti- D-string pair, we have shown that there is a description of these states in terms of an exact boundary conformal field theory. In this description these states are regarded as non-supersymmetric D0-branes of type I string theory. There is no GSO projection on the open string states[k]. The NS sector ground state of the open string represents a tachyonic mode, but this is odd under the world sheet parity transformation Ω and is projected out. Finally, the extra factor of $\sqrt{2}$ in the mass formula (eq.(18)) compared to the mass of an ordinary D0-brane can be understood as follows. The open string partition function of an ordinary D0-brane contains a projection operator $\frac{1}{2}(1+(-1)^F)$; of this the first term represents NSNS exchange diagram, whereas the second term represents the RR exchange diagram in the closed string channel. In the absence of this projection operator the NSNS exchange diagram doubles, whereas the RR exchange diagram is absent. This gives a net enhancement factor of $\sqrt{2}$ in the mass formula.[l]

References

1. E. Witten, Nucl. Phys. **B443** (1995) 85 [hep-th/9503124].
2. A. Dabholkar, Phys. Lett. **B357** (1995) 307 [hep-th/9506160].
3. C. Hull, Phys. Lett. **B357** (1995) 545 [hep-th/9506194].
4. J. Polchinski and E. Witten, Nucl. Phys. **B460** (1996) 525 [hep-th/9510169].
5. E. Gava, J. Morales, K. Narain and G. Thompson, [hep-th/9801128].
6. A. Sen, JHEP09(1998)023 [hep-th/9808141].
7. A. Sen, JHEP08(1998)012 [hep-th/9805170].
8. M. Green, Phys. Lett. **B329** (1994) 435 [hep-th/9403040].
9. T. Banks and L. Susskind, [hep-th/9511194].
10. M. Green and M. Gutperle, Nucl.Phys.**B476** (1996) 484 [hep-th/9604091].

[k]The possibility of existence of such D0-branes was first mentioned by Bergman [23].

[l]The mass is being measured in the variable of the type IIB theory before the Ω projection; hence in this calculation we do not take into account the extra factor of 1/2 coming from the Ω projection.

11. G. Lifschytz, Phys. Lett. **B388** (1996) 720 [hep-th/9604156];
V. Periwal, [hep-th/9612215].
12. A. Sen, JHEP08(1998)010 [hep-th/9805019].
13. J. Polchinski and A. Strominger, Phys. Lett. **B388** (1996) 736 [hep-th/9510227];
E. Gava, K. Narain and M. Sarmadi, Nucl.Phys. **B504** (1997) 214 [hep-th/9704006];
I. Antoniadis, E. Gava, K.Narain and T. Taylor, Nucl. Phys. **B511** (1998) 611 [hep-th/9708075];
M. Srednicki, [hep-th/9807138].
14. A. Sen, JHEP06(1998)007 [hep-th/9803194].
15. O. Bergman and M. Gaberdiel, Phys. Lett. **B441** (1998) 133 [hep-th/9806155].
16. M. Strassler, JHEP09(1998)017 [hep-th/9709081], [hep-th/9808073].
17. A. Smilga and A. Veselov, Nucl. Phys. **B515** (1998) 163 [hep-th/9710123]; [hep-th/9801142].
18. O. Bergman and B. Kol, Nucl. Phys. **B536** (1998) 149 [hep-th/9804160].
19. J. Blum and K. Dienes, Nucl. Phys. **B516** (1998) 83 [hep-th/9707160]; Nucl. Phys. **B520** (1998) 93 [hep-th/9708016];
O. Bergman and M. Gaberdiel, Nucl. Phys. **B499** (1997) 183 [hep-th/9701137].
20. P. Ginsparg, Nucl. Phys. **B295** (1988) 153; and references therein;
I. Antoniadis, C. Bachas and C. Kounnas, Nucl. Phys. **B289** (1987) 87.
21. D. Friedan, E. Martinec and S. Shenker, Nucl. Phys. **B282** (1987) 13.
22. E. Gimon and J. Polchinski, Phys. Rev. **D54** (1996) 1667 [hep-th/9601038].
23. O. Bergman, private communication.

KILLING SPINORS, SUPERSYMMETRIES AND ROTATING INTERSECTING BRANES

P.K. TOWNSEND
DAMTP, University of Cambridge,
Silver Street, Cambridge CB3 9EW, UK
E-mail: pkt10@damtp.cam.ac.uk

I review a recently proposed method for determining the symmetry superalgebra of a supergravity configuration from its Killing spinors, and its application to the 'near-horizon' limits of various rotating and intersecting branes.

A configuration of gravitational and matter fields is said to possess a symmetry if there is an infinitesimal coordinate transformation, associated with some non-zero vector field ξ, that leaves invariant all fields, including the metric. The invariance of the metric implies that ξ is Killing and the set of all such fields closes under commutation to define the symmetry algebra of the field configuration. For a supergravity theory one might analogously define a 'supersymmetric' configuration to be one for which there is an infinitesimal local supersymmetry transformation, asociated with a non-zero anticommuting spinor field ϵ, that leaves invariant all fields. In this case, however, it is far from clear how one extracts the symmetry superalgebra from the knowledge of these spinors. To begin with, there is no way to construct an ordinary vector field from anticommuting spinors. To circumvent this problem we may note that ϵ is defined by a set of *linear* differential equations, so a supersymmetric field configuration is naturally associated with a common zero mode of a set of differential operators. This zero mode is a *commuting* spinor, but it will generally have both a 'body' and a 'soul'. The body is an 'ordinary' commuting spinor, ζ, satisfying the same equations as ϵ but with all background fermions set to zero. For such purely bosonic backgrounds the supersymmetry variations of the bosons vanish identically so the equations determining ζ are found from the supersymmetry variations of the fermions. These are the Killing spinor equations, and the solutions are the Killing spinors. Thus, the number of supersymmetries of a purely bosonic background equals the dimension of the space of linearly independent Killing spinors.

Killing vectors may now be constructed as bilinears of Killing spinors. It is not difficult to verify that the vector field $\bar{\zeta}\Gamma^m\zeta'\partial_m$ is Killing if the spinor fields ζ and ζ' are. This suggests that the symmetry superalgebra is determined by the Killing spinor fields (up to purely bosonic factors) just as the bosonic symmetry algebra is determined by the Killing vector fields. One way to see how this works [1] starts by considering the supergravity field configuration of interest as a background for small fluctuations. Given a Killing spinor or vector field of the background it is then possible to construct a time-independent charge as a functional of the fluctuation fields. One thus finds a set of fermonic charges $Q_F(\zeta)$ and a set of bosonic charges $Q_B(\xi)$ for which the (anti)commutators may be computed via the canonical (anti)commutation relations of the fluctuation fields [1]. For the bosonic charges one has

$$[Q_B(\xi), Q_B(\xi')] = Q_B([\xi, \xi']) , \tag{1}$$

so that the algebra of bosonic charges is homomorphic to the algebra of Killing vector fields. For the mixed bosonic, fermionic commutator one has

$$[Q_B(\xi), Q_F(\zeta)] = Q_B(\mathcal{L}_\xi \zeta) , \tag{2}$$

where $\mathcal{L}_\xi$ is the Lie derivative with respect to ξ. Note that the Lie-derivative of a spinor field with respect to a vector field is defined only if latter is Killing, as is the case here. Finally, and most importantly, one has

$$\{Q_F(\zeta), Q_B(\zeta')\} = Q_B(\xi) \tag{3}$$

where

$$\xi^m = \bar{\zeta}\Gamma^m \zeta' . \tag{4}$$

What this shows [1] is that *the linear combination ξ of Killing vector fields associated to any pair of Killing spinor fields, ζ and ζ', is the same as the linear combination $Q_B(\xi)$ of bosonic charges appearing in the anticommutator of the fermionic charges $Q_F(\zeta)$ and $Q_F(\zeta')$*. In this way, the symmetry superalgebra may indeed be deduced, up to purely bosonic factors, from the Killing spinors.

In many cases this fact is not needed because a knowledge of the bosonic symmetry algebra and the number of supersymmetries suffices to determine the superalgebra. A simple example is the maximally supersymmetric $adS_2 \times S^2$ Bertotti-Robinson solution of N=2 D=4 supergravity, which arises as the near-horizon limit of the extreme Reissner-Nordström (RN) black hole [2,3]. This information is sufficient to determine the symmetry superalgebra; it can only be $su(1,1|2)$, for which the bosonic subalgebra is $su(1,1) \oplus su(2)$. Many higher-dimensional maximally supersymmetric adS_{p+2} spacetimes arise in a similar way as near-horizon limits of p-brane solutions of supergravity theories [4], and if one runs through the list of anti-de Sitter superalgebras [5] then one usually finds that there is only one candidate. An interesting exception occurs for certain adS_2 and adS_3 cases. As adS_3 superalgebras are direct sums of two

adS_2 superalgebras, we may concentrate on the adS_2 case. There is a one-parameter family of adS_2 superalgebras,

$$D(2|1;\alpha) \supset su(1,1) \oplus su(2) \oplus su(2) \,. \tag{5}$$

The parameter $0 < \alpha \leq 1$ is the relative weight of the the two $su(2)$ subalgebras. When $\alpha = 1$ we have the isomorphism

$$D(2|1;\alpha) \cong osp(4|2;\mathbb{R}) \,, \tag{6}$$

and it is convenient to define

$$D(2|1;0) = su(1,1|2) \oplus su(2) \,. \tag{7}$$

Clearly, the parameter α is not determined by a knowedge of the bosonic symmetry algebra and the number of supersymmetries. The simplest case in which this ambiguity arises is the D=5 extreme Tangherlini black hole, which is the D=5 generalization of the extreme RN black hole. This is a solution of D=5 supergravity with a maximally supersymmetric $adS_2 \times S^3$ near-horizon limit[6]. This near-horizon solution therefore has eight supersymmetries and is invariant under the isometry group of $adS_2 \times S^3$. The superalgebra must be $D(2|1;\alpha)$ for some α, but what is the value of α?

In this case there is an indirect argument that α must vanish. This comes from consideration of angular momentum. It will prove useful to first consider rotation in the context of D=4 black holes. The general stationary D=4 black hole solution depends on one angular momentum paramcter, in addition to its mass and charge parameters, but the requirement of supersymmetry forces the angular momentum to vanish. This could seem puzzling in view of the fact that angular momentum does not appear in the anticommutator of supersymmetry charges of the super-Poincaré algebra, because this fact implies that every non-rotating supersymmetric solution is a member of a class of rotating supersymmetric solutions. Indeed it is, but all the rotating solutions have naked singularities and so fail to qualify as 'black holes'. As an aside, I remark that angular momentum *does* appear in the anticommutator of supersymmetry charges of the adS superalgebra, so in this case the value of the angular momentum is of direct relevance to supersymmetry. In fact, it turns out[7] that asymptotically adS supersymmetric D=4 black holes are *necessarily* rotating; now it is the non-rotating supersymmetric solutions that are singular!

Returning to D=5, we first note that the general stationary asymptotically flat solution will depend on two angular momenta[8], corresponding to rotation in two independent 2-planes. When both vanish we have a static black hole with $SO(4)$ hyperspherical symmetry. When they are non-zero but equal the symmetry is reduced to $SO(3) \times SO(2)$ and when they are unequal it is further reduced to $SO(2) \times SO(2)$. In the context of D=5 supergravity these groups become

$$SU(2) \times SU(2) \supset SU(2) \times U(1) \supset U(1) \times U(1)\,. \tag{8}$$

It turns out that supersymmetry requires both angular momenta to be equal [9]. Let j be the one independent angular momentum parameter of the supersymmetric black hole solutions; for an appropriate normalization one finds that non-singularity on and outside the event horizon requires $|j| < 1$. When $j = 0$ we have the Tangherlini BH with its $adS_2 \times S^3$ near-horizon geometry. When $j \neq 0$ we have a rotating version with a near horizon geometry that is no longer a direct product. Let σ_i $(i = 1, 2, 3)$ be the three left-invariant 1-forms on S^3 satisfying $d\sigma_1 = \sigma_2 \wedge \sigma_3$ and cyclic permutations; then $d\Omega_3^2 = (1/4)(\sigma_1^2 + \sigma_2^2 + \sigma_3^2)$ is the $SU(2) \times SU(2)$ invariant metric on the unit sphere. In this notation the near-horizon metric of the rotating supersymmetric D=5 black hole is [10]

$$ds^2 = -(r^2 dt + \frac{j}{2}\sigma_3)^2 + r^{-2}dr^2 + d\Omega_3^2 \tag{9}$$

This metric is clearly $SU(2) \times U(1)$ invariant, and at least 1/2 supersymmetric, but its full isometry supergroup is far from evident. An analysis of the Killing spinor equations reveals that supersymmetry is again fully restored in the near-horizon limit [11]. This is remarkable because it suggests that rotation reduces the near-horizon isometry group without reducing its supersymmetry. This would be hard to understand unless the $SU(2)$ factor that is broken to $U(1)$ by the rotation is a purely bosonic factor of the isometry supergroup, and this is the case only if $\alpha = 0$. In fact, a direct determination [10] of the symmetry superalgebra using the method described above confirms that the near-horizon symmetry superalgebra of the D=5 supersymmetric black hole is $su(1,1|2) \oplus su(2)$, when $j = 0$, and $su(1,1|2) \oplus u(1)$ when $j \neq 0$. It also determines the $su(1,1)$ Killing vector fields to be[a]

$$\begin{aligned} h &= \partial_t \\ d &= r\partial_r - 2t\partial_t \\ k &= [(1 - j^2)r^{-4} + 4t^2]\partial_t - 4tr\partial_r + 2jr^{-2}m\,, \end{aligned} \tag{10}$$

where m is the conventionally normalized Killing vector field associated with the angular momentum parameter j. The remaining $su(2)$ Killing vector fields in the bosonic subalgebra of $su(1,1|2)$ are the right-invariant vector fields on S^3.

[a]The notation used here differs from that of ref. 10.

Supersymmetric D=5 black holes have various interpretations as intersecting M-branes. The most straightforward is as three intersecting[12], and possibly rotating [13], M2-branes, customarily denoted by $(0|M2, M2, M2)$. Compactification on the 6-dimensional space spanned by the M2-branes yields a 1/8 supersymmetric solution of T^6-compactified D=11 supergravity, and a subclass (the 'Güven' class) yields 1/2 supersymmetric solutions of the equations of pure D=5 supergravity. An instructive, although less direct, route begins from the dual $(1|MW, M2, M5)$ configuration. Wrapping the M2-brane on a circle and the M5-brane on a 4-torus yields a 1/8 supersymmetric string solution of T^5-compactified D=11 supergravity. The 'Güven' subclass now yields a 1/2 supersymmetric solution of the pure (1,1) supersymmetric D=6 supergravity for which the near-horizon geometry is $adS_3 \times S^3$, irrespective of the rotation [14] because its effects can be removed *in the near-horizon limit* by a coordinate transformation. The symmetry superalgebra of this near-horizon solution is

$$su(1,1|2)_L \oplus su(1,1|2)_R \,. \tag{11}$$

The total number of supersymmetries is thus 16, which is double that of the D=5 black hole, but only the 'left' supersymmetries survive further reduction on S^1. Moreover, the identification needed to perform this reduction is not preserved by the coordinate transformation needed to remove the effects of rotation in the near-horizon limit [1], so that the near-horizon geometry in D=5 depends on the rotation parameter, in agreement with the direct analysis in D=5. In fact, $SU(1,1|2)_R$ is broken to $SU(2)$ for zero rotation and to $U(1)$ for non-zero rotation. We thus recover the D=5 results deduced earlier.

The 1/8 supersymmetric $(1|MW, M2, M5)$ configuration of M-theory is a special case of the 1/8 supersymmetric $(1|MW, M2, M5, M5)$ configuration. For an appropriate choice of the various (generalized) harmonic functions associated with the M-Wave, M2-brane and M5-brane components, there is a non-singular near-horizon limit with $adS_3 \times S^3 \times S^3 \times \mathbb{E}^2$ geometry [15,16], with the ratio of the two S^3 radii equal to the ratio of the two M5-brane charges. The number of independent Killing spinors is 16, which represents a *fourfold* increase in supersymmetry relative to the full intersecting brane configuration (as was the case for the $(1|MW, M2, M5)$ configuration). The reason for this is that the M-Wave is irrelevant in the near-horizon limit because there are no local degrees of freedom of the gravitational field in adS_3.

The 1/2 supersymmetric $adS \times S^3 \times S^3 \times \times \mathbb{E}^2$ solution of D=11 supergravity can obviously be viewed as a 1/2 supersymmetric $adS_3 \times S^3 \times S^3$ solution of D=9 maximal supergravity. We now face the problem of determining the isometry superalgebra of this solution. Indirect arguments [16] suggest that it should be

$$D(2|1;\alpha) \oplus D(2|1;\alpha) \tag{12}$$

with α the ratio of the two S^3 radii. The removal of one M5-brane corresponds to taking this ratio to zero, and hence to $\alpha = 0$, which we earlier saw to be correct. However, these arguments are only suggestive and this example provides what is perhaps the best illustration of the need for a systematic method. An application of the method described earlier confirms [1] that the superalgebra is indeed $D(2|1;\alpha)$, with α the ratio of the two S^3 radii. Thus all values of the parameter α of the *adS* superalgebra $D(2|1;\alpha)$ have a physical realization within M-theory.

References

1. J.P. Gauntlett, R.C. Myers and P.K. Townsend, *Phys. Rev.* D **59**, 025001 (1999).
2. B. Carter, in *Black Holes*, eds. C. and B. DeWitt (Gordon and Breach, New York 1973).
3. G.W. Gibbons, in *Supersymmetry, Supergravity and Related Topics*, eds. F. Del Aguila, J.A. de Azcárraga and L.E. Ibañez (World Scientific, Singapore, 1985).
4. G.W. Gibbons and P.K. Townsend, *Phys. Rev. Lett.* **71**, 3754 (1993).
5. W. Nahm, *Nucl. Phys.* B **135**, 149 (1978).
6. A. Chamseddine, S. Ferrara, G.W. Gibbons and R. Kallosh, *Phys. Rev.* D **55**, 3647 (1977).
7. V.A. Kostelecky and M.J. Perry, *Phys. Lett.* B **371**, 191 (1996).
8. R.C. Myers and M.J. Perry, *Ann. Phys. (N.Y.)* **172**, 304 (1986).
9. J.C. Breckenridge, R.C. Myers, A.W. Peet and C. Vafa, *Phys. Lett.* B **391**, 93 (1997).
10. J.P. Gauntlett, R.C. Myers and P.K. Townsend, *Class. Quantum Grav.* **16**, 1 (1999).
11. R. Kallosh, A. Rajaraman and W.K. Wong, *Phys. Rev.* D **55**, 3246 (1977).
12. G. Papadopoulos and P.K. Townsend, *Phys. Lett.* B **380**, 273 (1996).
13. M. Cvetič and D. Youm, *Nucl. Phys.* B **499**, 253 (1997).
14. M. Cvetič and F. Larsen, *Nucl. Phys.* B **531**, 239 (1998).
15. P.M. Cowdall and P.K. Townsend, *Phys. Lett.* B **429**, 281 (1988); *erratum: Phys. Lett.* B **429**, 281 (1998).
16. H.J. Boonstra, B. Peeters and K. Skenderis, *Nucl. Phys.* B **533**, 127 (1988).

HOLOGRAPHY AND THE WEYL ANOMALY

MåNS HENNINGSON

Institute of Theoretical Physics, Chalmers University of Technology, S-412 96 Göteborg, Sweden
mans@fy.chalmers.se

KOSTAS SKENDERIS

Spinoza Institute, University of Utrecht, Leuvenlaan 4, 3584 CE Utrecht, The Netherlands
k.skenderis@phys.uu.nl

We review our calculation of the Weyl anomaly for d-dimensional conformal field theories that have a description in terms of a $(d+1)$-dimensional gravity theory.

1 Introduction

In the last few months, we have learned that a theory containing gravity and defined on an open $(d+1)$-manifold X can, in some cases, be equivalent to a d-dimensional conformal field theory defined on the boundary M of X [1], [2], [3]. The manifold X has a causal structure similar to that of $(d+1)$-dimensional anti-de Sitter space, i.e. the boundary M at conformal infinity is timelike. This means that the space X is not globally hyperbolic, i.e. it does not admit any Cauchy surfaces, since new information may come in through the boundary M. The action functional $S[\phi]$ or the equations of motion for the gravity theory on X therefore do not completely determine the dynamics and must be supplemented by appropriate boundary data $\phi_{(0)}$ on M for each field ϕ on X. The partition function is then a functional of these boundary data:

$$Z_{grav}[\phi_{(0)}] = \int_{\phi_{(0)}} \mathcal{D}\phi \exp(-S[\phi]),$$

where the subscript on the integration sign indicates that the functional integral is over field configurations ϕ that satisfy the boundary condition given by $\phi_{(0)}$. The precise relationship between this gravity theory on X and the conformal field theory on the boundary M can now be described as follows: There is a one-to-one correspondence between the fields ϕ on X and the primary operators $\mathcal{O}$ on M. The set of correlation functions of the latter are conveniently summarized by a generating functional:

$$Z_{CFT}[\phi_{(0)}] = \left\langle \exp \int_M d^d x \mathcal{O} \phi_{(0)} \right\rangle,$$

where $\phi_{(0)}$ is now regarded as a formal expansion parameter. The partition function of the gravity theory on X and the generating functional on M are then equal, regarded as functionals of $\phi_{(0)}$:

$$Z_{grav}[\phi_{(0)}] = Z_{CFT}[\phi_{(0)}].$$

This relationship is particularly useful when the gravity theory is weakly coupled. We can then calculate the partition function at tree level, where it is given by the exponential of the action functional evaluated for a field configuration $\phi^{cl}(\phi_{(0)})$ that solves the classical equations of motion and obeys the boundary conditions given by $\phi_{(0)}$:

$$Z^{tree}_{grav}[\phi_{(0)}] = \exp\left(-S[\phi^{cl}(\phi_{(0)})]\right).$$

A field of particular importance in a theory containing gravity is of course the metric $G_{\mu\nu}$. The corresponding operator in the boundary conformal field theory is the stress-energy tensor T_{ij}. The boundary data for the metric $G_{\mu\nu}$ is *not* a boundary metric $g_{(0)\,ij}$, but only a conformal structure $[g_{(0)}]$. (This is defined as an equivalence class of boundary metrics when two metrics that differ by a local rescaling are considered equivalent, i.e. $g_{(0)} \sim \exp 2\sigma(x) g_{(0)}$ for an arbitrary positive function $\sigma(x)$.) Indeed, the metric $G_{\mu\nu}$ on X has a second order pole at the boundary M and thus does not induce a metric there. To get a finite metric on M, we could multiply $G_{\mu\nu}$ by f^2, where f is some positive function on X with a simple zero on M and then restrict to M. The freedom to choose f means that only the conformal equivalence class of the metric so obtained is unambiguously defined. Conversely, a choice of conformal structure $[g_{(0)}]$ on M together with Einstein's equations on X suffice to determine an (up to diffeomorphisms) unique metric $G_{\mu\nu}$ on X [4]. We thus naturally get a *conformal* field theory on M. Indeed, to consider a more general theory on M, we would need a background metric. However, as we have seen, only the conformal structure of M is well-defined, so only conformal field theories are meaningful on M.

In the following, we will consider the effective action $W_{CFT}[g_{(0)}] = -\log Z_{CFT}[g_{(0)}]$ of the theory on M. This can be regarded as minus the logarithm of the zero-point function or partition function of the theory. (It is a zero-point function in the sense that there are no insertions of any operators). A priori, this is a functional of the metric $g_{(0)}$ on M, but by conformal invariance, it should actually only depend on the conformal equivalence class $[g_{(0)}]$. However, this invariance is sometimes broken by a conformal (or Weyl) anomaly. This means that $W_{CFT}[g_{(0)}]$ is not invariant under a conformal rescaling $\delta g_{(0)} = 2\delta\sigma g_{(0)}$ of the metric, but transforms as

$$\delta W_{CFT}[g_{(0)}] = \int_M d^d x \sqrt{\det g_{(0)}} \mathcal{A} \delta\sigma,$$

where $\mathcal{A}$ is the anomaly. On general grounds, the gravitational part of the Weyl anomaly vanishes when the dimension d of M is odd. When d is even, it is of the form

$$\mathcal{A} = E + I + D_i J^i,$$

where E is proportional to the d-dimensional Euler density, and I is a conformal invariant [5 6]. The total derivative term $D_i J^i$ is irrelevant, since it can be canceled by adding a local counterterm to the effective action. There is of course a unique Euler density in every even dimension d. The number of linearly independent conformal invariants grows with d, though. There are no such invariants for $d = 2$, one (the square of the Weyl tensor) for $d = 4$, three for $d = 6$, etcetera. We will outline how to calculate the conformal anomaly for d-dimensional conformal field theories that have a description in terms of a $(d+1)$-dimensional gravity theory as described above. Technical details (and some intermediate results not presented in [7]) are relegated to the appendix.

2 The calculation

According to the recipe above, $W_{CFT}[g_{(0)}]$ is given by minus the logarithm of the partition function of the gravity theory with a certain conformal structure $[g_{(0)}]$ induced by the metric $G_{\mu\nu}$ on the boundary. All other fields of the gravity theory should vanish on the boundary M, since we are not inserting any operators in the conformal field theory correlation function. If a tree-level computation is justified, so that we only need to consider field configurations on X that solve the classical equations of motion, this means that all fields expect the metric vanish everywhere on X. The theory in bulk is then reduced to pure gravity described by the Einstein-Hilbert action plus a cosmological constant term:

$$S_{bulk} = \frac{1}{16\pi G_N} \int_X d^{d+1}x \sqrt{\det G_{\mu\nu}} \, (R + 2\Lambda) \,. \tag{1}$$

On a manifold with boundary, we also have the term

$$S_{boundary} = \frac{1}{16\pi G_N} \int_M d^d x \sqrt{\det \tilde{g}_{ij}} 2K, \tag{2}$$

where $\tilde{g}_{ij}$ is the metric on the boundary and K is the trace of the second fundamental form. This boundary term is necessary in order to get an action that depends only on first derivatives of the metric[8]. As described in the first section, a choice of conformal structure $[g_{(0)}]$ on the boundary M determines a unique metric $G_{\mu\nu}$ in the bulk of X that solves the equations of motion

$$R_{\mu\nu} - \frac{1}{2} R G_{\mu\nu} = \Lambda G_{\mu\nu}. \tag{3}$$

(This is of course true only up to diffeomorphisms). However, the bulk action diverges when evaluated for such a field configuration because of the second order pole in $G_{\mu\nu}$ on the boundary. Furthermore, the boundary terms in the action are ill-defined, since $G_{\mu\nu}$ does not induce a finite metric $\tilde{g}_{ij}$ on M.

To regulate the theory in a manner consistent with general covariance, we pick a specific representative $g_{(0)}$ of the boundary conformal structure $[g_{(0)}]$. This determines a distinguished coordinate system (ρ, x^i), in which the metric on X takes the form [9]

$$G_{\mu\nu} dx^\mu dx^\nu = \frac{l^2}{4} \rho^{-2} d\rho d\rho + \rho^{-1} g_{ij} dx^i dx^j. \tag{4}$$

Here the tensor g has the limit $g_{(0)}$ as one approaches the boundary represented by $\rho = 0$. The length scale l is related to the cosmological constant Λ. Einstein's equations for $G_{\mu\nu}$ can then be solved order by order in ρ with the result that g is of the form

$$g = g_{(0)} + \rho g_{(2)} + \ldots + \rho^{d/2} g_{(d)} + \rho^{d/2} \log \rho \, h_{(d)} + \mathcal{O}(\rho^{d/2+1}). \tag{5}$$

The regularization procedure now amounts to restricting the bulk integral in the action to the domain $\rho > \epsilon$ for some cutoff $\epsilon > 0$ and evaluating the boundary integrals at $\rho = \epsilon$. The regulated action evaluated for this field configuration is then of the form

$$\begin{aligned} W[g_{(0)}] = \frac{1}{16\pi G_N} \int d^d x \sqrt{\det g_{(0)}} \Big(& \epsilon^{-d/2} a_{(0)} + \epsilon^{-d/2+1} a_{(2)} + \ldots \\ & + \epsilon^{-1} a_{(d-2)} - \log \epsilon \, a_{(d)} \Big) + W_{fin}[g_{(0)}]. \end{aligned} \tag{6}$$

The coefficients $a_{(0)}, a_{(2)}, \ldots, a_{(d)}$ are all given by covariant expressions in $g_{(0)}$ and its curvature tensor $R^i{}_{jkl}$ (see appendix). The divergences as ϵ goes to zero can thus be canceled by adding local counterterms to the action, so that we are left with a finite effective action $W_{fin}[g_{(0)}]$. To find the variation of $W_{fin}[g_{(0)}]$ under a conformal rescaling of the boundary metric $g_{(0)}$, we note that the regulated action $W[g_{(0)}]$ is invariant under the combined transformation $\delta g_{(0)} = 2\delta\sigma g_{(0)}$ and $\delta\epsilon = 2\delta\sigma\epsilon$ for a *constant* parameter $\delta\sigma$. The terms proportional to negative powers of ϵ are separately invariant, so the variation of $W_{fin}[g_{(0)}]$ must therefore equal minus the variation of the logarithmically divergent term. The latter is given by

$$\mathcal{A}\delta\sigma = \frac{1}{16\pi G_N}(-2a_{(d)})\delta\sigma,$$

since $\log\epsilon$ transforms with a shift whereas $a_{(d)}$ itself is invariant. Although this formula was derived under the assumption that $\delta\sigma$ is a constant, it follows from the general form of the conformal anomaly that it can be applied also for a non-constant $\delta\sigma$.

3 The results

The discussion in the preceding section shows that, up to a constant, the conformal anomaly of a theory related to a gravity theory only depends on the space-time dimension d. We will now evaluate the anomaly explicitly for the physically relevant cases $d = 2, 4, 6$.

3.1 $d = 2$ and the asymptotic symmetry algebra of adS_3

Here we get

$$\mathcal{A} = -\frac{c}{24\pi}R$$

with the central charge

$$c = \frac{3l}{2G_N}.$$

This agrees with a computation based on the asymptotic symmetry algebra of three-dimensional anti-de Sitter space [10]!

3.2 $d = 4$ and $\mathcal{N} = 4$ super Yang-Mills theory

Here we get

$$\mathcal{A} = -\frac{2l^3}{16\pi G_N}\left(-\frac{1}{8}R^{ij}R_{ij} + \frac{1}{24}R^2\right).$$

Inserting for example the values of l and G_N appropriate for the $adS_5 \times S^5$ geometry of N coincident $D3$-branes in type IIB string theory, we get N^2 ($\sim$ dim $SU(N)$) times the conformal anomaly of an $\mathcal{N} = 4$ supermultiplet (one vector, four chiral spinors and six scalars) [11]! The agreement between our strong coupling calculation and this weak coupling result indicates that there is a non-renormalization theorem for the conformal anomaly.

3.3 d = 6 and tensionless strings

Here we get

$$\mathcal{A} = -\frac{2l^5}{16\pi G_N}\left(-\frac{1}{128}RR^{ij}R_{ij} + \frac{3}{3200}R^3 + \frac{1}{64}R^{ij}R^{kl}R_{ijkl}\right.$$
$$\left. + \frac{1}{320}R^{ij}D_iD_jR - \frac{1}{128}R^{ij}D^kD_kR_{ij} + \frac{1}{1280}RD^iD_iR\right).$$

Inserting the values of l and G_N corresponding to the $adS_7 \times S^4$ geometry of N coincident five-branes in M-theory, we find that $\mathcal{A}$ is proportional to N^3 ($>>$ dim $GL(N)$). This agrees with considerations based on the entropy of the brane system [12], [13]. This provides information about the mysterious tensionless string theory that appears when $M5$-branes coincide.

4 Appendix

In this appendix we present some technical details of the calculation of the conformal anomaly.

Einstein's equation (3) for G given in (4) amount to

$$\begin{aligned} \rho\left(2g'' - 2g'g^{-1}g' + \mathrm{Tr}(g^{-1}g')g'\right) + l^2\mathrm{Ric}(g) - (d-2)g' - \mathrm{Tr}(g^{-1}g')g &= 0 \\ (g^{-1})^{jk}\left(\nabla_i g'_{jk} - \nabla_k g'_{ij}\right) &= 0 \\ \mathrm{Tr}(g^{-1}g'') - \frac{1}{2}\mathrm{Tr}(g^{-1}g'g^{-1}g') &= 0, \end{aligned} \tag{7}$$

where differentiation with respect to ρ is denoted with a prime, ∇_i is the covariant derivative constructed from the metric g and Ric(g) is the Ricci tensor[a] of g. We solve these equations iteratively in ρ.

The $g_{(k)}$ for $k < d$ in (5) are given by

$$\begin{aligned} g_{(2)ij} &= \frac{1}{d-2}\left(R_{ij} - \frac{1}{2(d-1)}Rg_{(0)ij}\right) \\ g_{(4)ij} &= \frac{1}{d-4}\left(-\frac{1}{8(d-1)}D_iD_jR + \frac{1}{4(d-2)}D_kD^kR_{ij}\right. \\ &\quad - \frac{1}{8(d-1)(d-2)}D_kD^kRg_{(0)ij} - \frac{1}{2(d-2)}R^{kl}R_{ikjl} \\ &\quad + \frac{d-4}{2(d-2)^2}R_i{}^kR_{kj} + \frac{1}{(d-1)(d-2)^2}RR_{ij} \end{aligned}$$

[a]Our conventions are as follows $R_{ijk}{}^l = \partial_i\Gamma_{jk}{}^l + \Gamma_{ip}{}^l\Gamma_{jk}{}^p - i \leftrightarrow j$ and $R_{ij} = R_{ikj}{}^k$.

$$+\frac{1}{4(d-2)^2}R^{kl}R_{kl}g_{(0)ij}-\frac{3d}{16(d-1)^2(d-2)^2}R^2g_{(0)ij}\Big). \quad (8)$$

Although the above formulas for $g_{(k)ij}$ do not make sense for $d=k$, their traces are smooth in the d goes to k limit and in fact give the correct values for $\mathrm{Tr}(g_{(0)}^{-1}g_{(k)})$ for $k\leq d$. Actually the easiest way to calculate the traces is to use the last equation in (7). By differentiating this equation with respect to ρ and then setting ρ to zero one obtains

$$\begin{aligned}\mathrm{Tr}\left(g_{(0)}^{-1}g_{(4)}\right) &= \frac{1}{4}\mathrm{Tr}\left((g_{(0)}^{-1}g_{(2)})^2\right)\\ \mathrm{Tr}\left(g_{(0)}^{-1}g_{(6)}\right) &= \frac{2}{3}\mathrm{Tr}\left(g_{(0)}^{-1}g_{(2)}g_{(0)}^{-1}g_{(4)}\right)-\frac{1}{6}\mathrm{Tr}\left((g_{(0)}^{-1}g_{(2)})^3\right)\end{aligned} \quad (9)$$

The above data are sufficient in order to obtain the anomaly up to $d=6$. In general, to determine the Weyl anomaly at dimension d one needs to determine the full metric up to order $d-2$. The trace of the metric at order d can be determined from metric at lower orders using the last equation in (7).

The explicit expression of $a_{(n)}$'s in (6) can be derived by inserting the general form of the metric (4) in (1), perform the ρ integration, and evaluate (2) at $\rho=\epsilon$. The coefficients $a_{(d)}$ (in d dimensions) of the logarithmic divergences, which are relevant for the calculation of the Weyl anomaly, receive contributions only from the bulk integral. Their explicit form for $d=2,3,6$, are given by

$$\begin{aligned}a_{(2)} &= l\,\mathrm{Tr}(g_{(0)}^{-1}g_{(2)})\\ a_{(4)} &= l^3\frac{1}{2}\left([\mathrm{Tr}(g_{(0)}^{-1}g_{(2)})]^2-\mathrm{Tr}[(g_{(0)}^{-1}g_{(2)})^2]\right)\\ a_{(6)} &= l^5\left(\frac{1}{8}[\mathrm{Tr}g_{(0)}^{-1}g_{(2)}]^3-\frac{3}{8}\mathrm{Tr}[g_{(0)}^{-1}g_{(2)}]\mathrm{Tr}[(g_{(0)}^{-1}g_{(2)})^2]\right.\\ &\quad\left.+\frac{1}{2}\mathrm{Tr}[(g_{(0)}^{-1}g_{(2)})^3]-\mathrm{Tr}[g_{(0)}^{-1}g_{(2)}g_{(0)}^{-1}g_{(4)}]\right).\end{aligned}$$

References

1. J.M. Maldacena, 'The large N limit of Superconformal Field Theories and Supergravity', `hep-th/9711200`.
2. S.S. Gubser, I.R. Klebanov and A.M. Polyakov, 'Gauge Theory Correlators from Non-Critical String Theory', `hep-th/9802109`.
3. E. Witten, 'Anti-de Sitter space and Holography', `hep-th/9802150`.
4. C.R. Graham and J.M. Lee, 'Einstein Metrics with Prescribed Conformal Infinity on the Ball', *Adv. Math.* **87** (1991) 186.

5. L. Bonora, P. Pasti and M. Bregola, 'Weyl Cocycles', *Class. Quant. Grav.* **3** (1986) 635.
6. S. Deser and A. Schwimmer, 'Geometric Classification of Conformal Anomalies in Arbitrary Dimensions', *Phys. Lett.* **B309** (1993) 279, `hep-th/9302047`.
7. M. Henningson and K. Skenderis, 'The Holographic Weyl Anomaly', *JHEP* **07** (1998) 023, `hep-th/9806087`.
8. G.W. Gibbons and S.W. Hawking, 'Action integrals and partition functions in quantum gravity', *Phys. Rev.* **D15** (1977) 2752.
9. C. Fefferman and C.R. Graham, 'Conformal Invariants', in *Elie Cartan et les Mathématiques d'aujourd'hui* (Astérisque, 1985) 95.
10. J.D. Brown and M. Henneaux, 'Central charges in the canonical realization of asymptotic symmetries: An example from three-dimensional gravity', *Commun. Math. Phys.* **104** (1986) 207.
11. M.J. Duff, 'Twenty years of the Weyl anomaly', *Class. Quant. Grav.* **11** (1994) 1387, `hep-th/9308075` and references therein.
12. S.S. Gubser and I.R. Klebanov, 'Absorption by Branes and Schwinger Terms in the World Volume Theory', *Phys. Lett.* **B413** (1997) 41, `hep-th/9708005`.
13. I.R. Klebanov and A.A. Tseytlin, 'Entropy of Near-Extremal Black *p*-branes', *Nucl. Phys.* **B475** (1996) 164, `hep-th/9604089`.

Black Holes and Super Black Holes as Chern Simons Theories in 2+1 Dimensions

Sharmanthie Fernando and Freydoon Mansouri
Physics Department, University of Cincinnati, Cincinnati, OH 45221, USA
E-mails: fernando@physung.phy.uc.edu, Mansouri@uc.edu

We study anti-de Sitter black holes in 2+1 dimensions in terms of Chern Simons gauge theory of the anti-de Sitter group coupled to a source. Taking the source to be an anti-de Sitter state specified by its Casimir invariants, we show how all the relevant features of the black hole are accounted for. Enlarging the gauge symmetry to super AdS group, we obtain a supermultiplet of AdS black holes. We give explicit expressions for masses and the angular momenta of the members of the multiplet.

1 Introduction

The AdS black hole in 2+1 dimensions is a solution of free Einstein's equations with a negative cosmological constant [1]. It is well known that the free Einstein theory in 2+1 dimensions with or without a cosmological constant can be formulated as a free Chern Simons theory [2,3]. In a free Chern Simons theory, the field strength vanishes identically, so that there are no local degrees of freedom. It is then somewhat surprising that a theory with no local degrees of freedom has a black hole solution which must have non-trivial degrees of freedom to account for its entropy! Presumably, M theory, once it is fully constructed, will provide us with the correct answer. In the meantime, one of the suggestions to resolve this issue is to modify the Chern Simons action with a WZW term [4,5]. Another suggestion is to make use of the AdS/CFT correspondence [6]. More recently, it has been suggested that the answer lies in the coupling of the Chern Simons theory to a source [7]. In this approach, the free Chern Simons theory is not only taken to be locally trivial. It is also taken to be globally trivial in accord with Mach's principle. Non-trivial topologies then arise as a result of coupling to sources. This is one of the issues which we will address in this work.

One of the notable advantages of this approach is that it allows us to express the asymptotic observables of the theory in terms of the properties of the sources. To implement this idea, we must identify a localized source (particle) with an irreducible representation of the gauge symmetry group [8]. For the present problem, this will amount to relating the asymptotic observables of the BTZ black hole to the Casimir invariants of an AdS state coupled to the

Chern Simons action. We will show that space-time will naturally emerge from such a gauge theory and will have all the ingredients necessary for the AdS black hole [1,9]. These include, in particular, the discrete subgroup underlying the identifications.

A second issue which we will address in this work relates to the manner in which black hole solutions fit into supersymmetric schemes. A conventional method of searching for signs of supersymmetry in black hole solutions is to look for Killing spinors. Many works along these lines already exist in the literature. We cite a representative few here [10-18], from which more references can be traced. One way to see whether a given black hole solution admits Killing spinors is identify it with the bosonic part of an appropriate supergravity theory [10-18]. Then, by requiring that the fermions in the theory as well as their variations vanish, one arrives at Killing spinor equation(s). The asymptotic supersymmetries depend on the number of non-trivial solutions of these equations consistent with the black hole topology. For example, in asymptotically flat space-times, a typical supermultiplet consists of a black hole and a number of ordinary particles all with the same mass. In contrast to the familiar situation in particle physics, where we have Supermultiplets consisting of particles only, in this approach there is no systematic way of looking for supermultiplets consisting of black holes only. We will show that in $2+1$ dimensions it is possible to construct a theory which permits macroscopic solutions consisting of all AdS black hole supermultiplets [19]. It involves the Chern Simons gauge theory of the (1,1) super AdS group coupled to a super AdS state (source). As we shall see below, to be able to accommodate the structure of the solution which emerges from such a theory it becomes necessary to broaden the standard notions of classical geometries to include some quantum mechanical elements.

2 Anti-de Sitter space and algebra

The anti-de Sitter space in 2+1 dimensions can be viewed as a subspace of a flat 4-dimensional space with the line element

$$ds^2 = dX_A dX^A = dX_0^2 - dX_1^2 - dX_2^2 + dX_3^2 \tag{1}$$

It is determined by the constraint

$$(X_0)^2 - (X_1)^2 - (X_2)^2 + (X_3)^2 = l^2 \tag{2}$$

where l is a real constant . The set of transformations which leave the line element invariant form the anti-de Sitter group $SO(2,2)$. It is locally isomorphic to $SL(2,R) \times SL(2,R)$. From here on by anti-de Sitter group we shall mean its universal covering group.

With $a = 0,1,2$, we can write the AdS algebra in two more convenient forms:

$$\begin{aligned} \epsilon^{abc} J_c &= \epsilon^{abc}(J_c^+ + J_c^-) \\ l\Pi^a &= (J^{+a} - J^{-a}) \end{aligned} \tag{3}$$

where

$$\epsilon^{012} = 1; \qquad \eta^{ab} = (1,-1,-1) \tag{4}$$

Then, the commutation relations in, say, $J_a^\pm$ basis take the form[7]

$$[J_a^\pm, J_b^+] = -i\epsilon_{ab}^c J_c^\pm; \qquad [J_a^+, J_b^-] = 0 \tag{5}$$

The Casimir operators in this basis have the form

$$j_\pm^2 = \eta^{ab} J_a^\pm J_b^\pm \tag{6}$$

Alternatively, we can use a combination of these with eigenvalues corresponding to the parameters of the BTZ solution:

$$\begin{aligned} M &= l^2(\Pi^a \Pi_a + l^{-2} J^a J_a) = 2(j_+^2 + j_-^2) \\ J/l &= 2l\Pi_a J^a = 2(j_+^2 - j_-^2) \end{aligned} \tag{7}$$

Unless otherwise stated, we will use the same symbols for operators and their eigenvalues.

An irreducible representation of AdS group can be labeled by the eigenvalues of either the pair (M, J) or the pair (j_+, j_-). For application to black holes, it is also possible to label these representations by eigenvalues which are proportional to the horizon radii $r_\pm$ of the AdS black hole[1,7].

3 Connection and the Chern Simons action

To write down the Chern Simons action, we begin by writing the connection in $SL(2,R) \times SL(2,R)$ basis.

$$A_\mu = \omega_\mu^{AB} M^{AB} = \omega_\mu^a J_a + e_\mu^a \Pi_a = A_\mu^{+a} J_a^+ + A_\mu^{-a} J_a^- \tag{8}$$

where

$$A_\mu^{\pm a} = \omega_\mu^a \pm l^{-1} e_\mu^a \tag{9}$$

Eq. 9 should be viewed as definitions of e and ω in terms of the two $SL(2,R)$ connections. The covariant derivative can be written as

$$D_\mu = \partial_\mu - iA_\mu = \partial_\mu - iA_\mu^{+a} J_a^+ - iA_\mu^{-a} J_a^- \tag{10}$$

Then the components of the field strength are given by

$$[D_\mu, D_\nu] = -iF_{\mu\nu}^{+a} J_a^+ - iF_{\mu\nu}^{-a} J_a^- = -iF_{\mu\nu}^+[A^+] - iF_{\mu\nu}^-[A^-] \tag{11}$$

For a simple or a semi-simple group, the Chern Simons action has the form

$$I_{cs} = \frac{1}{4\pi} Tr \int_M A \wedge \left(dA + \frac{2}{3} A \wedge A \right) \tag{12}$$

where Tr stands for trace and

$$A = A_\mu dX^\mu = A^+ + A^- \tag{13}$$

We require the 2+1 dimensional manifold M to have the topology $R \times \Sigma$, with Σ a two- manifold. So, The Chern Simons action with $SL(2,R) \times SL(2,R)$ gauge group will take the form

$$I_{cs} = \frac{1}{4\pi} Tr \int_M \left[\frac{1}{a_+} A^+ \wedge \left(dA^+ + \frac{2}{3} A^+ \wedge A^+ \right) + \frac{1}{a_-} A^- \wedge \left(dA^- + \frac{2}{3} A^- \wedge A^- \tag{14}\right.\right.$$

Here the quantities $a_\pm$ are, in general, arbitrary coefficients, reflecting the semisimplicity of the gauge group. Up to an overall normalization, only their ratio is significant. As explained elsewhere [7], in the presence of a source (or of sources), any á priori choice of the coefficients $a_\pm$ reduces the class of allowed holonomies, so that even the classical theory coupled to sources will be affected by such a choice. For this reason, we will keep the coefficients $a_\pm$ as free parameters in the sequel, so that we can generate the correct holonomies for solutions both outside and inside the horizon.

Under infinitesimal gauge transformations

$$u_\pm = \theta^{\pm\, a} J_a^\pm \tag{15}$$

the gauge fields transform as

$$\delta A_\mu = -\partial_\mu u - i[A_\mu, u] \tag{16}$$

More specifically,

$$\delta A_\mu^{\pm\, a} = -\partial_\mu \theta^{\pm\, a} - \epsilon^a_{bc} A^{\pm\, b} \theta^{\pm\, c} \tag{17}$$

As we have stated, the manifold M has the topology $R \times \Sigma$ with R representing x^0. Then subject to the constraints

$$F_a^{\pm}[A^{\pm}] = \frac{1}{2}\eta_{ab}\epsilon^{ij}(\partial_i A_j^{\pm\, b} - \partial_j A_i^{\pm\, b} + \epsilon^b{}_{cd} A_i^{\pm\, c} A_j^{\pm\, d}) = 0 \tag{18}$$

the Chern Simons action for $SO(2,2)$ will take the form

$$\begin{aligned} 2\pi I_{cs} = & \frac{1}{a_+}\int_R dx^0 \int_\Sigma d^2x \left(-\epsilon^{ij}\eta_{ab} A_i^{+a}\partial_0 A_j^{+b} + A_0^{+a} F_a^+\right) \\ & + \frac{1}{a_-}\int_R dx^0 \int_\Sigma d^2x \left(-\epsilon^{ij}\eta_{ab} A_i^{-a}\partial_0 A_j^{-b} + A_0^{-a} F_a^-\right) \end{aligned} \tag{19}$$

where $i, j = 1, 2$.

4 $SO(2,2)$ States and Interaction with sources

Following the approach which has been successful in coupling sources to Poincaré Chern Simons theory [8], we take a source for the present problem to be an irreducible representation of anti-de-Sitter group characterized by Casimir invariants M and J (or r_+ and r_-). Within the representation, the states are further specified by the phase space variables of the source Π^A and q^A, $A = 0, 1, 2, 3$, subject to anti-de Sitter constraint given by Eq. 2. To allow for the possibility of quantizing the Chern Simons theory consistently, we must require that our sources be represented by unitary representations of the AdS group. The choice of relevant representations from among these were discussed in reference [7]. Here we note that since the AdS group in 2+1 dimensions can be represented in the $SL(2,R) \times SL(2,R)$ form, the unitary representations of $SO(2,2)$ can be constructed from those of $SL(2,R)$. The latter group has four series of unitary representations all of which are infinite dimensional. Of these, the appropriate representations turn out to be the discrete series bounded from below [7]. The states in an irreducible representation of $SL(2,R)$ are specified by the eigenvalues of its Casimir operator and one of the generators. In reference [7], the compact generator was chosen to label the states. This corresponds to inducing representations of $SL(2,R)$ via its maximal compact subgroup $SO(2)$. Thus, suppressing the superscripts $\pm$ which distinguish our two $SL(2,R)$'s, we can write

$$j^2|F, m> = F(F-1)|F, m>$$

$$J_0|F, m> = (F+m)|F, m>$$

In these expressions

$$F = real\ \ number \geq 0; \qquad m = 0, 1, 2, ... \tag{20}$$

So, for this series, the eigenvalues of the Casimir invariants of $SL(2,R) \times SL(2,R)$ can be written as,

$$j_\pm^2 = F_\pm^2 - F_\pm \tag{21}$$

It follows that the infinite set of states can, in a somewhat redundant notation, be specified as

$$|j_\pm^2, F_\pm + m_\pm>; \qquad m_\pm = 0, 1, 2, ... \tag{22}$$

Clearly, the integers $m_\pm$ are not necessarily equal. Using these states, we can construct the discrete series of the unitary representations of $SO(2,2)$. A typical state will have the following labels:

$$|M, J>= |j_+^2, j_-^2, F_+ + m_+, F_- + m_- > \tag{23}$$

To be able to identify the labels M and J with the corresponding labels in the AdS black hole, we must require that $F_\pm \geq 1$ [7]. It would then follow that $|J/l| \leq |M|$, as required for having a black hole solution.

The main advantage of diagonalizing a compact generator is that the maximal compact subgroup $SO(2) \times SO(2)$ of $SO(2,2)$ allows a parametrization of the AdS space in terms of circular functions so that the corresponding angular variable is periodic to begin with and will remain so in the black hole solution. As a result [7], the periodicity necessary to obtain the discrete identification group need not be imposed as an additional requirement as was done by BTZ [1]. However, to obtain the BTZ solution from this starting point, it becomes necessary [1] to perform a "Wick rotation" in the space of Casimir invariants to recover the hyperbolicity of the angular variable. An alternative to this procedure is to construct the SL(2,R) representations by diagonalizing a non-compact generator. In that case, the irreducible representations of $SO(2,2)$ can be constructed via induction from the subgroup $SO(1,1) \times SO(1,1)$. One way to do this is to make the formal replacement $J_0 \to iJ_2$ to go from the formalism in which the compact generator J_0 is diagonal to one in which the non-compact generator J_2 is diagonal . Obviously, the structure and the classification of the unitary representations will remain the same. The only difference is that in the latter case the angular variables to be used to parametrize the AdS space are hyperbolic to begin with. Then, as we shall see below, the 2π periodicity of the hyperbolic angular variable will arise from the requirement of compatibility with the topological properties of the Chern Simons theory coupled to a source.

With these preliminaries, we can couple a source to the AdS Chern Simons theory in the following way:

$$I_s = \int_C d\tau \left[\Pi_A \partial_\tau q^A - (A^{+a} J_a^+ + A^{-a} J_a^-) + \lambda \left(q^A q_A - l^2\right)\right]$$

$$+\int_C d\tau \left[\lambda_+ \left(J^{+a} J_a^+ - l^2 j_+^2\right) + \lambda_- \left(J^{-a} J_a^- - l^2 j_-^2\right)\right] \tag{24}$$

In this expression, C is a path in M, τ is a parameter along C, and $J_a^{\pm}$ play the role of c-number generalized angular momenta which transform in the same way as the corresponding generators which label the source. The quantities λ and $\lambda_{\pm}$ are Lagrange multipliers. The first constraint in this action ensures that $q^A(\tau)$ satisfy the AdS constraint. As explained in previous occasions[7,8,20], it is not the manifold M over which the gauge theory is defined but the space of q_A's which give rise to the classical space-time. The last two constraints identify the source being coupled to the Chern Simons theory as an anti-de Sitter state with invariants j_+ and j_-. These constraints are crucial in relating the invariants of the source to the asymptotic observables of the coupled theory via Wilson loops.

The total action for the theory is given by:

$$I = I_{cs} + I_s \tag{25}$$

It is easy to check that in this theory the components of the field strength still vanish everywhere except at the location of the sources. So, the analog of Eq. ? become

$$\epsilon^{ij} F_{ij}^{\pm\ a} = 2\pi a_{\pm} J^{\pm\ a} \delta^2(\vec{x}, \vec{x_0}) \tag{26}$$

In particular, fixing the gauge so that $SO(2,2)$ symmetry reduces to $SO(1,1) \times SO(1,1)$, we get

$$\epsilon^{ij} F_{ij}^{\pm\ 2} = 2\pi a_{\pm} F_{\pm} \delta^2(\vec{x}, \vec{x^0}) \tag{27}$$

where $F_{\pm}$ are the invariant labels of the state as in Eq. 23, but now they are associated with the non-compact generator J_2. All other components of the field strength vanish. We thus see that because of the constraints appearing in the action given by Eq. 24, the strength of the sources become related to their Casimir invariants. These invariants, in turn, determine the asymptotic observables of the theory. Since such observables must be gauge invariant, they are expressible in terms of Wilson loops, and a Wilson loop about our source can only depend on, e.g., j_+ and j_-.

¿From the data on the manifold M given above, one can determine the properties of the emerging space-time by solving Eqs 27. The only non-vanishing components of the gauge potential are given by[7]

$$A_\theta^{2\pm} = 2a_{\pm} F_{\pm} \tag{28}$$

where θ is an angular variable. Although these are components of a connection which is pure gauge, they give rise to non-trivial holonomies around the source.

More explicitly, we have

$$\omega[A^+] = exp^{\pi(r_- + r_+)J_2^+} \tag{29}$$

and

$$\omega[A^-] = exp^{\pi(r_- - r_+)J_2^-} \tag{30}$$

Since we are diagonalizing $J_2^{\pm}$ operators, their matrix representation is given by

$$J_2^{\pm}(\alpha) = \frac{1}{2}\begin{pmatrix} 1 & 0 \\ 0 & -1 \end{pmatrix} \tag{31}$$

Then, the above holonomies will take the form

$$\omega[A^+] = \begin{pmatrix} exp^{\pi(r_- + r_+)} & 0 \\ 0 & exp^{-\Pi(r_- + r_+)} \end{pmatrix} \tag{32}$$

$$\omega[A^-] = \begin{pmatrix} exp^{\pi(r_+ - r_-)} & 0 \\ 0 & exp^{-\Pi(r_+ - r_-)} \end{pmatrix} \tag{33}$$

It was shown in reference [7] how these holonomies lead to a discrete identification subgroup of $SO(2,2)$, indicating that the manifold M_q of the $0+1$ dimensional fields q^A has all the relevant features of the macroscopic AdS black hole solution. As we shall see below, the same holonomies, suitably interpreted, will play a crucial role in establishing the space-time structure of the supersymmetric theory discussed below.

5 The black hole space-time

To see how the space-time structure emerges from our anti-de Sitter gauge theory, we follow an approach which led to the emergence of space-time from Poincaré [8] and super Poincaré [20] Chern Simons gauge theories. We have emphasized that the manifold M is not to be identified with space-time. But the information encoded in M and discussed in the previous section is sufficient to fix the properties of the emerging space-time. To this end, let us consider a manifold $\hat{M}_q$ satisfying the AdS constraint

$$\hat{q}_0^2 - \hat{q}_1^2 - \hat{q}_2^2 + \hat{q}_3^2 = l^2 = -\Lambda^{-1} \tag{34}$$

where $\Lambda =$ cosmological constant. In fact, our $SL(2,R) \times SL(2,R)$ formulation allows us to take $\hat{M}_q$ to be the universal covering space of the AdS space. As we shall see, the emerging space-time is the quotient of $\hat{M}_q$ by the discrete subgroup Γ discussed in the previous section. Moreover, the source coupled to the Chern Simons action is an AdS state characterized by the Casimir invariants (M, J) or, equivalently, (r_+, r_-). To parametrize $\hat{M}_q$ consistent with the above constraint, consider a pair of 2-vectors,

$$\vec{\hat{q}}_\phi = (\hat{q}^0, \hat{q}^1) = (f cosh\phi, f sinh\phi) \tag{35}$$

$$\vec{\hat{q}}_t = (\hat{q}^2, \hat{q}^3) = \left(\sqrt{f^2 - l^2} cosh(t/l), \sqrt{f^2 - l^2} sinh(t/l)\right) \tag{36}$$

where $f = f(r)$, with r a radial coordinate which for an appropriate $f(r)$ will become the radial coordinate appearing in the line element for the BTZ black hole. As far the constraint given by Eq. 34 is concerned, the functional form of $f(r)$ is irrelevant. Computing the line element in terms of the parameters $(t/l, r, \phi)$, we get

$$ds^2 = (\frac{f^2}{l^2} - 1)dt^2 - \frac{f^{'\,2} dr^2}{(\frac{f^2}{l^2} - 1)} - f^2 d\phi^2 \tag{37}$$

where "prime" indicates differentiation with respect to r.

Anticipating the results to be given below, let us compare this line element with that for the BTZ black hole [1]:

$$ds^2 = [\frac{r^2}{l^2} - M + \frac{J^2}{4r^2}]dt^2 - \frac{dr^2}{[\frac{r^2}{l^2} - M + \frac{J^2}{4r^2}]} - r^2[d\phi - \frac{J^2}{2r^2}dt]^2 \tag{38}$$

If we identify the labels M and J with the Casimir invariants of an irreducible representation of the AdS group as discussed in the previous sections, we see that the line element given by Eq. 37 corresponds to an irreducible representation with $J = 0$ and $M = 1$. Such a parametrization cannot provide us with an AdS black hole with a continuous range of values for J and M. What we need is a space-time manifold M_q in which the coordinates q^A carry an arbitrary irreducible representation of the AdS group. We will construct this manifold explicitly by performing appropriate gauge transformations on $\hat{M}_q$. Although the original theory was invariant under $SL(2, R) \times SL(2, R)$ gauge transformations, we have already reduced this symmetry by choosing to work in a gauge in which the left over symmetry is just $SO(1, 1) \times SO(1, 1)$ generated, respectively, by $J_2^\pm$. It turns out to be more convenient to work with generators

$$J_2 = J_2^+ + J_2^-; \qquad l\Pi_2 = J_2^+ - J_2^- \tag{39}$$

As was the case with the compact generators J_0 and Π_0, the non-compact operators J_2 and Π_2 generate a $SO(1, 1) \times SO(1, 1)$ subgroup of the AdS group.

We identifying the parameters ϕ and t/l, respectively, with each $SO(1,1)$ and proceed in the same manner as we did for the compact generators in reference[7]. In particular, consider the local gauge transformation[7,8,20,21,22]

$$\vec{q}_{\phi'}(\phi, t/l) = e^{\left(\frac{r_+}{l}\phi - \frac{r_- t}{l^2}\right)J^2}\vec{\hat{q}}_{\phi}(\phi)$$
$$\vec{q}_{t'}(t/l, \phi) = e^{\left(\frac{r_-}{l}\phi - \frac{r_+ t}{l^2}\right)l\Pi^2}\vec{\hat{q}}_{t}(t/l) \tag{40}$$

It then follows that

$$\vec{q}_{\phi'}(\phi + 2\pi, t/l) = e^{2\pi \frac{r_+}{l} J^2}\vec{q}_{\phi'}(\phi, t/l)$$
$$\vec{q}_{t'}(t/l, \phi + 2\pi) = e^{2\pi \frac{r_-}{l} l\Pi^2}\vec{q}_{t}(t/l, \phi)$$
$$\vec{q}_{\phi'}(\phi + 2\pi, t/l + 2\pi) = e^{2\pi\left(\frac{r_+}{l} - \frac{r_-}{l}\right)J^2}\vec{q}_{\phi'}(\phi, t/l)$$
$$\vec{q}_{t'}(t/l + 2\pi, \phi + 2\pi) = e^{2\pi\left(\frac{r_-}{l} - \frac{r_+}{l}\right)l\Pi^2}\vec{q}_{t'}(\phi, t/l) \tag{41}$$

The factors by which these quantities change as $\phi \to (\phi + 2\pi)$ are reminiscent of the holonomies which we obtained in section 4. There we found that these holonomies in M led to a discrete identification group. To be consistent with this identification group, we must require that the hyperbolic variable ϕ be periodic. Thus, the periodicity of phi follows from the topology of M in the presence of a source. With this provision, the vector $(\vec{q}_{\phi'}, \vec{q}_{t'})$ transforms in the same way as the one which in section 4 was parallel transported around a loop in the manifold M. Calling the manifold to which such vectors belong M_q, we see that this manifold incorporates the same dynamics as the phase space variables in M, and we are justified in using the same letter q for both. Thus, we can parametrize the manifold M_q as follows[1,7,9]:

$$q^0 = f cosh\left(\frac{r_+}{l}\phi - \frac{r_- t}{l^2}\right)$$
$$q^1 = f sinh\left(\frac{r_+}{l}\phi - \frac{r_- t}{l^2}\right)$$
$$q^2 = \sqrt{f^2 - l^2} cosh\left(\frac{r_-}{l}\phi - \frac{r_+ t}{l^2}\right)$$
$$q^3 = \sqrt{f^2 - l^2} sinh\left(\frac{r_-}{l}\phi - \frac{r_+ t}{l^2}\right) \tag{42}$$

From these we can compute the line element. It is easy to show that it is the same as that given by Eq. 38 for $r > r_-$.

6 Supersymmetric Black Holes

The theory which we will describe below is the supersymmetric generalization of the theory which was discussed in the previous sections. We will see that the emerging macroscopic theory consists of a supermultiplet of ordinary space-times and, as a special case of this, a supermultiplet consisting of black holes only. The simplest way of obtaining a supersymmetric extension of the anti-de Sitter group is to begin with the AdS group in its $SL(2,R) \times SL(2,R)$ basis. The $N = 1$ supersymmetric extension of each $SL(2,R)$ factor is the supergroup $OSp(1|2;R)$. Thus, one arrives at the (1,1) form of the $N = 2$ super AdS group. Its algebra is given by

$$[J^{\pm}_a, J^{\pm}_b] = -i\epsilon_{ab}{}^{c} J^{\pm}_c; \quad [J^{\pm}_a, Q^{\pm}_{\alpha}] = -\sigma^{a}_{\alpha}{}^{\beta} Q^{\pm}_{\alpha}; \quad \{Q^{\pm}_{\alpha}, Q^{\pm}_{\beta}\} = -\sigma^{a}_{\alpha\beta} J^{\pm}_a$$
$$\{Q^{+}_{\alpha}, Q^{-}_{\beta}\} = 0; \quad [J^{+}, J^{-}] = 0 \tag{43}$$

The Casimir invariants are given by

$$C_{\pm} = j^2_{\pm} + \epsilon^{\alpha\beta} Q^{\pm}_{\alpha} Q^{\pm}_{\beta} \tag{44}$$

The spinor indices are raised and lowered by antisymmetric metric $\epsilon^{\alpha\beta}$ defined by $\epsilon^{12} = -\epsilon_{12} = 1$. The matrices $(\sigma^a)^{\beta}_{\alpha}$, $(a = 0, 1, 2)$, form a representation of $SL(2,R)$ and satisfy the Clifford algebra

$$\{\sigma^a, \sigma^b\} = \frac{1}{2}\eta^{ab} \tag{45}$$

More explicitly, we can take them to be of the form:

$$\sigma^0 = \frac{1}{2}\begin{pmatrix} 1 & 0 \\ 0 & -1 \end{pmatrix}; \quad \sigma^1 = \frac{1}{2}\begin{pmatrix} 0 & i \\ i & 0 \end{pmatrix}; \quad \sigma^2 = \frac{1}{2}\begin{pmatrix} 0 & 1 \\ -1 & 0 \end{pmatrix} \tag{46}$$

It is important to note that the supersymmetry generators of $OSp(1|2,R)$ do not commute with the Casimir invariant of its $SL(2,R)$ subgroup. That is,

$$[j^2_{\pm}, Q_{\alpha}] \neq 0 \tag{47}$$

Since super AdS group is semi-simple, we can construct its irreducible representations by first constructing the irreducible representations of $OSp(1|2,R)$ [13]. Depending on which $OSp(1|2,R)$ we are considering, the states within any such supermultiplet are the corresponding irreducible representations of $SL(2,R)$ Characterized by the Casimir invariants j_{+} or j_{-}, respectively. Based on the rationale given for the non- supersymmetric case [7], the irreducible representations of interest for the present case are those which can be obtained from the unitary discrete series of $SL(2,R)$ and which are bounded from below. To construct the supermultiplet corresponding to, say, the "plus" generators in Eq. 22, we can take the Clifford vacuum state $|\Omega^+>$ to be the $SL(2,R)$ state with the lowest eigenvalue of J_2^+. In the notation of Eq. 22, this corresponds to an $m=0$ state:

$$|F_+, m >= |F_+, m = 0 >= |F_+ >$$

Then, the superpartner of this state, again with $m = 0$, is the state $|F_+ + 1/2 >$ obtained by the application of one of the Q's. The corresponding values of j_+^2 are $F_+(F_+ - 1)$ and $(F_+ + 1/2)(F_+ - 1/2)$, respectively. The supermultiplet for the second $OSp(1|2,R)$ can be constructed in a similar way.

We are now in a position to construct the (1,1) super AdS supermultiplet as a direct product of the two $OSp(1|2,R)$ doublets. Altogether, there will be four states in the supermultiplet. They will have the following labels:

$$|F_+, F_- >; \quad |F_+ + 1/2, F_- >; \quad |F_+, F_- + 1/2 >; \quad |F_+ + 1/2, F_- + 1/2 > \tag{48}$$

¿From these, we can also obtain the expressions for the eigenvalues (M,J) of various states within the supermultiplet:

$$\begin{aligned} |M_1, J_1 >&= |M, J > \\ |M_2, J_2 >&= |M + 2F_+ - 1/2, J + 2F_+ - 1/2 > \\ |M_3, J_3 >&= |M + 2F_- - 1/2, J - 2F_- + 1/2 > \\ |M_4, J_4 >&= |M + 2(F_+ + F_-) - 1, J + 2(F_+ - F_-) > \end{aligned} \tag{49}$$

these states transform into one another under supersymmetry transformations.

The Chern Simons action for simple and semisimple supergroups has the same structure as that for Lie groups. The only difference is that the trace operation is replaced by super trace (Str) operation. So, in the $OSp(1|2,R) \times OSp(1|2,R)$ basis the Chern Simons action for the super AdS group has the same form as that given by Eq. 14. But now the expression for connection is given by

$$A^\pm = \left[A_\mu^{\pm\, a} J_a^\pm + \chi_\mu^{\pm\alpha} Q_\alpha^\pm\right] dx^\mu \tag{50}$$

Just as in the non-supersymmetric case, to have a non-trivial theory, we must couple sources to the Chern Simons action. To do this in a gauge invariant and locally supersymmetric fashion, we must take a source to be an irreducible representation of the super AdS group. As we saw above, such a supermultiplet consists of four AdS states. To couple it to the gauge fields, we must first generalize the canonical variables we used in the AdS theory to their supersymmetric forms [13]:

$$\Pi_A \rightarrow (\Pi_A, \Pi_\alpha) \qquad\qquad q_A \rightarrow (q_A, q_\alpha) \tag{51}$$

Then, the source coupling can be written as

$$I_s = \int_C \left[\Pi_A dq^A + \Pi_\alpha dq^\alpha + (A^+ + A^-) + constraints\right] \tag{52}$$

where again C is a path in M. The constraints here include those discussed for the AdS group and, in addition, those which relate the AdS labels of the Clifford vacuum to the Casimir eigenvalues of the super AdS group. The combined action

$$I = I_{cs} + I_s \tag{53}$$

leads to the constraint equations

$$\epsilon^{ij} F^{\pm\ a}_{ij} = 2\pi a_\pm J^\pm\, \delta^2(\vec{x}, \vec{x^0}); \qquad\qquad \epsilon^{ij} F^{\pm\alpha}_{ij} = 2\pi a_\pm Q^{\pm\alpha} \delta^2(\vec{x}, \vec{x^0}) \tag{54}$$

Up to this point, everything proceeds in parallel with the non- supersymmetric case. Differences begin to show up when one attempts to solve these equations by choosing a gauge again so that the gauge symmetry is reduced to $SO(1,1) \times SO(1,1)$:

$$\epsilon^{ij} F^{\pm\ 2} = 2\pi a_\pm J^\pm_2 \delta^2(\vec{x}, \vec{x^0}) \tag{55}$$

Although this equation is identical in form to Eq. 27 for the non-supersymmetric case, there is an essential difference in the underlying physics. In the supersymmetric case, the supermultiplet which we couple to the Chern Simons action consist of four $SO(2,2)$ states with different values of $F_\pm$. As a result, in the parallel transport of q^A around a close path analogous to the non-supersymmetric case, there will bc four sets of holonomies with different values of (j_+, j_-) or, equivalently, (r_+, r_-). Moreover, in the non-supersymmetric case, a single source with Casimir invariants (r_+, r_-) or, equivalently, (M, J) will give rise to an AdS black hole [7] for which the line element is characterized with the corresponding values of M and J and has the form given by Eq. ?. In the supersymmetric case, the source is a supermultiplet in which there are four states of differing (M, J) values. Then, depending on which set (M, J) that we choose, we will get a different BTZ solution. Since M and J are not invariant under supersymmetry transformations, these solutions are transformed into each other under supersymmetry. This makes it impossible for a single c-number line element of the type given by Eq. 38 to correspond to all the AdS states of a supermultiplet.

The situation here runs parallel to what was encountered in connection with super Poincaré Chern Simons theory [20]. There it was pointed out that standard classical geometries were not capable describing these structures and that one must make use of *nonclassical geometries.* Such geometries can be described in terms of three elements:

1. An algebra such as a Lie algebra or a Lie superalgebra.
2. A line element operator with values in this algebra.
3. A Hilbert space on which the algebra acts linearly.

For the problem at hand, the algebra of interest is the $N = (1,1)$ super AdS algebra in $2+1$ dimensions. The corresponding Hilbert space is the representation space of the superalgebra given by Eq. 48. Then, instead of the BTZ line element given, we begin with a line element operator with values in the $N = (1,1)$ superalgebra and assume that its diagonal elements depend on the algebra only through the Casimir operators $(\hat{M}, \hat{J})$ of its $SO(2,2)$ subalgebra. The "hats" on top of M and J are meant to distinguish the operators from the corresponding eigenvalues. Thus, we have

$$ds^2 = ds^2(\hat{M}, \hat{J})$$

The matrix element of this operator for each state of the supermultiplet will produce a c-number line element:

$$< M_k, J_k | ds^2(\hat{M}, \hat{J}) | M_k, J_k >= ds^2(M_k, J_k) \tag{56}$$

In other words, for each state of the supermultiplet, the nonclassical geometry produces a layer of classical space-time. The number of the layers is equal to the dimension of the supermultiplet. Supersymmetry transformations act as messengers linking different layers of this multilayered space-time

An equivalent way of constructing the operator line element is to begin with the parametrization of q^A given by Eq. 41. Then, replacing the Casimir eigenvalues in those expressions with the corresponding Casimir operators, we can proceed to compute the line element operator. The result will look like the line element given by Eq. 38, except that now Casimir eigenvalues are replaced with Casimir operators. This means that, for consistency, we must also interpret the quantities $J_2^{\pm}$ in Eq. 54 as operators. Acting on different states of the supermultiplet, they will give the corresponding $F_{\pm}$ eigenvalues. There will therefore be not one set but four sets of holonomies $W[A^+]$ and $W[A^-]$. Each set will produce the discrete identification subgroup in the corresponding layer of space-time.

Consider, next, the conditions under which every layer of the supermultiplet corresponds to an AdS black hole. For this to be true, we must have

$$M_k \geq 0; \qquad |J_k| \leq l M_k$$

This, in turn implies that

$$F_+ \geq F_- \geq 1$$

In the notation of Eq. 48, for $|J| = lM$, two layers of the supermultiplet become extreme AdS black holes. The only exception is in the limiting case when $M = J = 0$, in which case there will be three extremal black holes in the supermultiplet. It is also interesting to note that for an appropriate choice of M and J or, equivalently, F_+ and F_-, the same supersource which generates a black hole in one layer can generate a naked singularity in another.

Acknowledgment

This work was supported, in part by the Department of Energy under the contract number DOE-FGO2-84ER40153. We would like to thank the Organizing Committee of Johns Hopkins Workshop for their hospitality and for the opportunity to present this material. We have also benefited from the hospitality of Aspen Center for Physics, where part of this work was carried out.

References

1. M. Bañados, C. Teitelboim and J. Zanelli, Phys. Rev. Lett. **69** (1992) 1849; M. Bañados, M. Henneaux, C. Teitelboim and J. Zanelli, Phys. Rev. D **48** (1993) 1506.
2. A. Achucarro, P. Townsend, Phys. Lett.B **180** (1986) 35
3. E. Witten, Nucl. Phys **B311** (1988) 46; **B323** (1989) 113
4. S. Carlip, Phys. Rev. **D51** (1995) 632
5. A.P. Balachandran, L. Chandar, A. Momen, Nuc. Phys. **B461** (1996) 581; eprint archive gr-qc/9506006
6. A. Strominger
7. S. Fernando, F. Mansouri, eprint archive hep- th/9804147, Int. Jour. Mod. Phys. **A**, *in press*
8. F. Mansouri, M.K. Falbo-Kenkel, Mod. Phys Lett. **A8** (1993) 2503; F. Mansouri, Proceedings of XIIth Johns Hopkins Workshop, ed. Z. Horwath, World Scientific, 1994
9. S. Carlip, eprint archive gr-qc/9506079; Clas. Quan. Grav. **12** (1995) 2853
10. P.C. Aichelburg, R. Guven, Phys. Rev. **D 24** (1981) 2066

11. R. Kallosh, A. Linde, T. Ortin, A. Peet, A. Van Proyen Phys. Rev. **D 46** (1992) 5278
12. O. Coussaert, M. Henneaux, eprint archive hep- th/9310194; Phys. Rev. Lett. **72** (1993) 183
13. J.M. Izquierdo and P.K. Townsend, eprint archive gr-qc/9501018; Clas. Quan. Grav. **12** (1995) 895
14. R. Kallosh, D. Kaster, T. Ortin, T. Torma, Phys. Rev. **D 50** (1994) 6374
15. M.J. Duff, J. Rahmfeld, Phys. Lett. **B 345** (1995) 441; eprint archive hep-th/9605085; M.J. Duff, H. Lu, C.N. Pope, Phys. Lett. **B 409** (1997) 136
16. A.R. Steif, Phys. Rev. Lett. **69** (1995) 1849
17. R. Kallosh, eprint archive hep-th/9503029; Phys. Rev. **D 52** (1995) 1234
18. T. Ortin, eprint archive hep-th/9705095
19. S. Fernando, F. Mansouri, eprint archive hep- th/9809139
20. Sunme Kim, F. Mansouri, Phys. Lett. **B397** (1997) 81; F. Ardalan, S. Kim, F. Mansouri, Int. Jour. Mod. Phys., **A12** (1997) 1183
21. K. Koehler *et al*, Nucl. Phys. **B348** (1990) 373
22. C. Vaz and L. Witten, Phys. Lett. **B327** (1994) 29

BLACK HOLES AND QUANTUM MECHANICS

RENATA KALLOSH

Physics Department, Stanford University, Stanford, CA 94305-4060, USA
E-mail: kallosh@physics.stanford.edu

The motion of a particle near the RN black hole horizon is described by conformal mechanics. Models of this type have no ground state with vanishing energy. This problem was resolved in past by a redefinition of the Hamiltonian which breaks translational time invariance but gives a normalizable ground state. We show that this change of the Hamiltonian is a quantum mechanical equivalent of the change of coordinates near the black hole horizon removing the singularity. The new Hamiltonian of quantum mechanics is identified as an operator of a rotation between 2 time-like coordinates of the adS hypersurface which translates global time. Therefore conformal quantum mechanics may eventually help to resolve the puzzles of the classical black hole physics.

1. Recently a surprising connection [1] between black holes and conformal mechanics of De Alfaro, Fubini and Furlan [2] (DFF) and superconformal [3] mechanics have been established[a]. The dynamics of a (super)particle near the horizon of an extreme Reissner-Nordström black hole was shown to be governed by an action that reduces to a (super)conformal mechanics model in the limit of a large black hole charge. The Hamiltonian, and the rest of the generators of the conformal group were found in [1] to be equal to

$$H = \frac{p^2}{2f} + \frac{g}{2x^2 f}\,, \qquad K = f\frac{x^2}{2}\,, \qquad D = \frac{xp+px}{4}\,. \tag{1}$$

Here the function $f(x,p)$ in the limit of the large black hole charge Q tends to 1 and we find a complete agreement with the conformal model [2] in this limit. Such models upon quantization give the commutators of the conformal algebra

$$[H,D] = iH\ , \qquad [K,D] = -iK\ , \qquad [H,K] = 2iD\ . \tag{2}$$

This quantum mechanical model was shown in [2] to a have a continuous spectrum of energy eigenstates with energy eigenvalue $E > 0$, but there is no ground state at $E = 0$.

In ref. [1] a black hole interpretation of this model was suggested. The classical analog of an eigenstate of H is an orbit of a timelike Killing vector field k, equal to $\partial/\partial t$ in the region outside the horizon, and the energy is then

[a] Various aspects of black holes and ADS_2/CFT_1 duality were studied in [4].

the value of k^2. The absence of a ground state of H at $E = 0$ can now be interpreted as due to the fact that the orbit of k with $k^2 = 0$ is a null geodesic generator of the event horizon, which is not covered by the static coordinates adapted to ∂_t.

The procedure used by DFF to cure the problem of the absence of a ground state was to choose a different combination of conserved charges as the Hamiltonian. From the perspective of the quantum mechanics this shift of the Hamiltonian is unusual and motivated by the fact that when the theory does not have a ground state, an additional prescription may be used such that the new theory has a well defined vacuum. It has been realized that such additional prescription leads to the breakdown of time translational invariance in favor of maintaining a well-defined vacuum and an anti de Sitter group.

From the perspective of the particle motion near the black hole horizon, the procedure adopted in [2] is extremely well motivated. It was stated in [1] that it corresponds to a different choice of time, one for which the worldlines of static particles pass through the black hole horizon instead of remaining in the exterior spacetime.

2. We will discuss here the connection between the black hole horizon and absence of a normalizable lowest eigenstate of a quantum mechanical Hamiltonian in more detail. By studying the conformal quantum mechanics DFF reached the conclusion that the Hamiltonian of the conformal quantum mechanics is not acceptable for solving the problem of motion. They suggested to solve the problem of motion to use a compact operator L_0, which is not a Hamiltonian but a combination of the Hamiltonian H and conformal operator K,

$$L_0 = \frac{1}{2}\left(\frac{1}{a}K + aH\right) = \frac{a}{4}\left(\frac{x^2}{a^2} + p^2 + \frac{g}{x^2}\right) . \qquad (3)$$

This compact operator was shown to exhibit the pleasant features of having a discrete spectrum and normalizable eigenstates. Here a is a constant (with dimension of length) whose introduction breaks scale invariance. The action is scale invariant but the ground state is not, so that one may qualify this as a spontaneous breaking of scale invariance.

It has been conjectured in [2] that the constant a may have a fundamental meaning since it provides a sort of infra-red cut-off responsible for the good behaviour at large distances and the discrete spectrum. In fact it was found in [2] that from three independent operators H, K, D only one combination can be made which corresponds to a generator of a compact rotation. The other two operators,

$$L_{\pm} = \frac{1}{2}\left(\frac{1}{a}K - aH\right) \pm iD \tag{4}$$

correspond to hyperbolic non-compact transformations. The algebra of conformal generators H, K, D generators was replaced by the algebra of their linear combination,

$$[L_0, L_{\pm}] = \pm L_{\pm} \ , \qquad [L_+, L_-] = -2L_0 \ . \tag{5}$$

The eigenvalues and eigenstates of the operator L_0 have been found and $L_{\pm}$ were treated as raising and lowering operators. The eigenvalues of L_0 are given by a discrete series $r_n = r_0 + n$ where $r_0 = \frac{1}{2}(1 + \sqrt{g + 1/4})$.

The superconformal mechanics [3] faced the analogous problems of the absence of the well defined ground state and therefore in supersymmetric case the same approach as in non-supersymmetric case was developed, i.e. the shift of the Hamiltonian was performed. Moreover in the fermionic part it was necessary to find the appropriate combinations of supersymmetry and special conformal supersymmetry to provide the existence of the ground state with the vanishing energy.

The worrisome part of this procedure invented 22 years ago was the breakdown of time translational invariance which in a quantum mechanical system in the flat space is not clearly justified (apart from the fact that it helps to find the vacuum and the physical states of the theory).

DFF procedure of an unusual but successful treatment of conformal quantum mechanics comes out in a new light from the ADS/CFT perspective. The main difference with the old picture is that we start with the curved space of a black hole where it is not known how to describe quantum mechanically the motion of a particle near the horizon and in the black hole interior. However, the fact that the time translational symmetry is broken in the coordinate system where the horizon looks like a singularity is not surprising at all! Quite opposite, there is no translational symmetry in this situation in curved space since near the horizon the time coordinate is ill defined, as will be shown below in detail.

Note that in classical general relativity as different from quantum mechanics it is a standard procedure to demonstrate that the horizon is not a true singularity by changing the coordinate system.

3. Let us recall here the following [5] adS-type explanation as to why the extreme black hole horizon is not a true singularity. The metric of the extreme

RN metric in isotropic coordinates is

$$ds^2 = -\left(1 + \frac{|Q|L_P}{\rho}\right)^{-2} dt^2 + \left(1 + \frac{|Q|L_P}{\rho}\right)^2 \left[d\rho^2 + \rho^2 d\Omega^2\right], \tag{6}$$

where Q is the black hole charge, L_P is the Plank length ,$d\Omega^2 = d\theta^2 + \sin^2\theta\, d\varphi^2$ is the $SO(3)$-invariant metric on S^2, and $M = |Q|/L_P$ is the black hole mass. The near-horizon (or large charge) geometry is therefore

$$ds^2 = -\left(\frac{\rho}{QL_P}\right)^2 dt^2 + \left(\frac{QL_P}{\rho}\right)^2 d\rho^2 + (QL_P)^2 d\Omega^2\,, \tag{7}$$

which is the Bertotti-Robinson (BR) metric. It can be characterized as the $SO(1,2) \times SO(3)$ invariant conformally-flat metric on $adS_2 \times S^2$. The parameter QL_P may now be interpreted as the S^2 radius (which is also equal to the radius of curvature of the adS_2 factor). In horospherical coordinates $(t, \phi = \rho/QL_P)$ for adS_2, the 4-metric of the BR solution of Maxwell-Einstein theory is

$$ds^2 = -\phi^2 dt^2 + \frac{(QL_P)^2}{\phi^2} d\phi^2 + (QL_P)^2 d\Omega^2\,. \tag{8}$$

The metric is singular at $\phi = 0$, but this is just a coordinate singularity and $\phi = 0$ is actually a non-singular degenerate Killing horizon of the timelike Killing vector field $\partial/\partial t$. To see this one may define an adS_2 space as *a surface in a flat space with one more time variable and the* $SO(1,2)$ *metric*

$$X^{\hat{m}} \eta_{\hat{m}\hat{n}} X^{\hat{n}} + a^2 = -(X^0)^2 - X^+ X^- + a^2 = 0\,. \tag{9}$$

One may introduce the hypersurface coordinates (ϕ, t) by choosing

$$X^0 = \phi t\,, \qquad X^- = a\,\phi\,, \qquad X^+ = \frac{a^2 - t^2\phi}{a\,\phi}\,. \tag{10}$$

The metric induced on the surface is the BR metric (8) where the radius of the adS space a is related to the black hole mass and charge as follows:

$$a = |Q_{BH}|L_P = M_{BH} L_P^2\,. \tag{11}$$

The horospherical coordinates do not cover all of adS_2 but only a half of it, they cover $\frac{adS_2}{J}$ where J is the antipodal map. At $\phi = 0$ the metric (8) has a singularity related to the black hole horizon and one has to choose either $\phi > 0$ or $\phi < 0$. The coordinate $X^- = a\,\phi$ can be positive as well as negative, but if ϕ is restricted to be positive, this also means that X^- has to be positive, i.e. take half of the values which it is allowed when considered as defining a surface (9).

The coordinates X^0, X^+, X^- evaluated on the hypersurface are smooth, therefore ϕ remains a good coordinate at the horizon $\phi = 0$, while the time coordinate t becomes ill defined. This means that the original black hole metric near the horizon has to be described using different coordinates. Let us first introduce a complex variable T which is a complex combination of two time-like coordinates,

$$T = T_1 + iT_2 \ , \qquad T_1 \equiv \frac{X^+ + X^-}{2} \ , \qquad T_2 \equiv X^0 \ . \tag{12}$$

A space-like coordinate is

$$r \equiv \frac{X^+ - X^-}{2} \ . \tag{13}$$

In terms of coordinates (15) the surface is

$$-(T_1)^2 - (T_2)^2 + r^2 + a^2 = TT^* + r^2 + a^2 = 0 \ . \tag{14}$$

One may use a different solution to the surface constraint,

$$T = e^{\frac{i\tau}{a}} \left(a^2 + r^2\right)^{1/2} \tag{15}$$

In these coordinates the metric is

$$ds^2 = -\left(\frac{a^2 + r^2}{a^2}\right) d\tau^2 + \left(\frac{a^2}{a^2 + r^2}\right) dr^2 + a^2 d\Omega^2 \ . \tag{16}$$

4. The $SO(1,2)$ symmetry is linearly realized on the coordinates of the surface $X^{\hat{m}} \eta_{\hat{m}\hat{n}} X^{\hat{n}} + a^2 = 0$ as $\delta X^{\hat{m}} = \Lambda^{\hat{m}}{}_{\hat{n}} X^{\hat{n}}$. No translations are allowed, only rotations. The generators of this symmetry $J^{\hat{m}\hat{n}} = iX^{[\hat{m}} \partial^{\hat{n}]}$ form the $SO(1,2)$ algebra:

$$[J_{\hat{m}\hat{n}}, J_{\hat{p}\hat{q}}] = i(\eta_{\hat{m}[\hat{p}} J_{\hat{q}]\hat{n}} - \eta_{\hat{n}[\hat{p}} J_{\hat{q}]\hat{m}}) \ . \tag{17}$$

The $SO(1,2)$ symmetry consists of a compact rotation between two time-like coordinates, T_1 and T_2 and of the two boosts between the space-like coordinates r and two times. The rotation between two times can be easily exponentiated by considering it as a $U(1)$ transformation

$$T' = e^{i\alpha} T \tag{18}$$

This transformation is generated by the shift of the global time $\tau' = \tau + a\alpha$ so that

$$T' = e^{\frac{i\tau'}{a}} \left(a^2 + r^2\right)^{1/2} = T' = e^{i\alpha} T \ . \tag{19}$$

Two boosts

$$\delta T = \Lambda r \qquad \delta r = \frac{1}{2}(\Lambda T^* + \Lambda^* T) \tag{20}$$

generate a $U(1)$ transformation. The time translation generator is L_0 and the complex boost is generated by $L_\pm$.

In terms of $SO(1,2)$ transformations (17) the generators of the quantum mechanics with the well defined ground state are given by

$$J_{T_1T_2} = L_0 \qquad J_{T_1r} \pm iJ_{T_2r} = L_\pm \tag{21}$$

Thus we described an interpretation of the quantum mechanical operators $L_0, L_\pm$ starting with black holes and changing coordinates near the horizon to avoid a singularity.

The new Hamiltonian L_0 of the conformal quantum mechanics does generate translation of the time τ in the coordinate system describing the extreme black hole near the horizon, where the time t of the horospherical coordinate system is ill defined. Somehow the conformal quantum mechanics found the way to a correct description of the physics by breaking t-time translation symmetry which is now understood in the context of the black hole horizon.

The black hole interpretation suggested in [1] of the unusual features of a quantum mechanical system is clear. One may want to use this correspondence in the opposite direction, to resolve the problems of black holes of the classical general relativity by some kind of generalized quantum mechanics which may have somewhat unusual features as in the example of the conformal DFF model, described above. A quantum mechanics of such nature may tell us about the black hole interior.

I had clarifying discussions of the issues of black holes and quantum mechanics with G. Gibbons, A. Strominger, L. Susskind and P. Townsend. I am grateful to the organizers of the conference "Novelties in String Theories" in Gothenburg for the stimulating environment. This work is supported by the NSF grant PHY-9870115.

References

1. P. Claus, M. Derix, R. Kallosh, J. Kumar, P. Townsend, and A. Van Proeyen, *Black Holes and Superconformal Mechanics*, hep-th/9804177, Phys. Rev. Letters. **81** (1998) 4553.
2. V. de Alfaro, S. Fubini and G. Furlan, *Conformal Invariance in Quantum Mechanics*, Nuovo. Cimento. **34A** (1976) 569.

3. V.P. Akulov and I.A. Pashnev, *Quantum superconformal model in (1,2) space*, Theor. Math. Phys. **56** (1983) 862; S. Fubini and E. Rabinovici, *Superconformal Quantum Mechanics*, Nucl. Phys. **B245** (1984) 17.
4. J. Maldacena, *The Large N Limit of Superconformal Field Theories and Supergravity*, hep-th/9711200, Adv.Theor.Math.Phys. 2 (1998) 231-252; A. Strominger, *AdS_2 Quantum Gravity and String Theory*, hep-th/9809027; G. Gibbons, P. Townsend, *Black Holes and Calogero Models*, hep-th/9812034; T. Nakatsu, N. Yokoi, *Comments on Hamiltonian Formalism of AdS/CFT Correspondence*, hep-th/9812047; J. Maldacena, J. Michelson, A. Strominger, *Anti-de Sitter Fragmentation*, hep-th/9812073; J. Kumar, *Conformal Mechanics and the Virasoro Algebra*, hep-th/9901139.
5. M.J. Duff, G.W. Gibbons and P.K. Townsend, *Macroscopic superstrings as interpolating solitons*, Phys. Lett. **332 B** (1994) 321.

MULTI GRAVITON SCATTERING IN MATRIX THEORY

GABRIELE FERRETTI

Scuola Internazionale Superiore di Studi Avanzati (SISSA) Via Beirut 2, I-34014 Trieste, Italy

We briefly review the status of matrix theory and present a wilsonian method of computing multi particle scattering.

1 Introduction

In the October of 1996, Banks, Fischler, Shenker and Susskind proposed a formulation of M-theory in terms of a supersymmetric matrix model [1]. Such a proposal has become known as the "matrix theory conjecture" and has attracted much attention in the last two years. In its original form, it states that M-theory in the infinite momentum frame (IMF) is described by the large N limit of the quantum mechanical system obtained by reducing $9+1$-dimensional super Yang–Mills to $0+1$ dimensions.

This proposal was further sharpened by Susskind [2] who interpreted the finite N version of the same system as describing M-theory with a compactified light–like direction (DLCQ) of radius R. This conjecture is on firmer grounds for two reasons: firstly it avoids the subtleties of the large N limit and, secondly, it can be "derived" by simple scaling arguments and kinematics [3]. Of course, if one were convinced that this derivation is correct, there would be no need to perform further tests. However, as stressed in [4], there might be subtleties associated with the light-like limit and it is therefore of great interest to perform explicit comparisons in order to test the conjecture and its derivation.

The simplest phase of matrix theory is the one with nine non-compact transverse directions and no branes. This phase has only a finite number of degrees of freedom, yielding a genuine finite rank quantum mechanical matrix model. This is the phase that will be discussed in this note and this is where the agreement between matrix theory and classical supergravity is most impressive. The scattering of two [5] and three [6] gravitons has been exactly reproduced, including recoil effects [7]; the full kinematic structure for the scattering of gravitons and three-forms has been obtained [8], and even the scattering off ortientifolds [9] (another case with a finite number of degrees of freedom) has been successfully compared with the supergravity predictions.[a]

[a]Some of this evidence appeared after my talk at the meeting, reinforcing my conviction that

Things are not as clear cut for the compactified case. The most serious objection comes from the work of Douglas, Shenker and Ooguri [10], where it is argued that the finite N matrix model cannot reproduce the results for supergravity on K3. This puzzle still awaits a resolution that is likely to be very interesting, perhaps involving infrared effects of soft gravitons from the K3 background. Another challenge for matrix theory is the computation of quantum correction to the supergravity action, the motivation for which matrix theory was put forward in the first place! A first attempt at computing the coefficient of the $t_8 t_8 R^4$ term is still inconclusive [11].

In this talk, however, we shall ignore these problems and look at the bright side of the story, in particular three-graviton scattering in uncompactified space. The aim is not to present the full computation of [6], but to show how, by using an effective action approach and without doing much work, it is possible to isolate particular terms in the amplitude. We will focus in particular on one of the terms which was previously claimed not to be reproduced by matrix theory [12], and show how this term is correctly reproduced by an effective action approach[b] [13]. It should be clear that the techniques generalize to more complex situations.

2 Kinematic Setup

We work within the eikonal approximation and compute the phase shift associated with straight trajectories [15]. Let us label the three graviton trajectories in transverse space as $\mathbf{r}_i = \mathbf{v}_i t + \mathbf{b}_i$ $(i = 1, 2, 3)$ and their relative distances, velocities, and impact parameters by $\mathbf{r}_{ij}$, $\mathbf{v}_{ij}$, and $\mathbf{b}_{ij}$ respectively. We organize the computation in a wilsonian manner, integrating out stretched open strings according to their mass which we know as being proportional to the distance between branes in units of α'. Let us therefore assume that graviton number one is much further away than the other two, i.e.[c]

$$r_{12} \approx r_{31} << r_{23}. \tag{1}$$

Let us also denote the eleven-momentum as $\tilde{p}_i = (p_i^-, p_i^+, \mathbf{p}_i)$, where, because of DLCQ, $p_i^+ = N_i/R$, (R is the DLCQ radius), $\mathbf{p}_i \equiv p_i^+ \mathbf{v}_i$. The mass shell condition $\tilde{p}_i^2 = 0$ implies $p_i^- = p_i^+ v_i^2/2$. The product of two eleven-dimensional

matrix theory contains all of the dynamical information present in the classical supergravity lagrangian.

[b] A related argument in favor of matrix theory was given in [14]. From the following discussion it should be clear that the two arguments of [13] and [14] are not in contradiction with one other.

[c] Throughout the paper, $r_{ij} = |\mathbf{r}_{ij}|$, and so on.

momenta can then be expressed as

$$\tilde{p}_i \cdot \tilde{p}_j = \frac{1}{2} \frac{N_i N_j}{R^2} v_{ij}^2. \tag{2}$$

A typical scattering amplitude in the eikonal approximation involves terms of type

$$A = \frac{(\tilde{p}_1 \cdot \tilde{p}_2)(\tilde{p}_2 \cdot \tilde{p}_3)(\tilde{p}_3 \cdot \tilde{p}_1)}{q_{12}^2 q_{23}^2}, \tag{3}$$

$\mathbf{q}_{12}$ and $\mathbf{q}_{23}$ being the transverse transfer momenta. This term appears in the so called V-diagram [6,12,13], whereas terms where the momentum transfer appears at the numerator as well come from the so called Y-diagram and shall not be considered here. Finally, terms where the external momenta are contracted with the polarizations are suppressed in the eikonal limit.

After performing a Fourier transform and switching to a normalization appropriate to non relativistic scattering in ten dimensions, the above term takes the form

$$\tilde{A} = \frac{N_1 N_2 N_3}{R^5} \frac{v_{12}^2 v_{23}^2 v_{31}^2}{r_{12}^7 r_{23}^7}. \tag{4}$$

Since the presence of such term has generated some confusion in the recent literature, let me make the following elementary remarks: the trivial fact that $\mathbf{v}_{12} + \mathbf{v}_{23} + \mathbf{v}_{31} = 0$ combined with $SO(8)$ invariance, means that one can choose to represent the amplitude by using either the two magnitudes v_{12}, v_{23} and the "angle" $\mathbf{v}_{12} \cdot \mathbf{v}_{23}$, or the three magnitudes v_{12}, v_{23} and v_{31}. The full amplitude for the V-diagram was computed in [6] to be

$$\begin{aligned} \tilde{A}^{\text{tot}} &\approx \frac{N_1 N_2 N_3}{R^5} \frac{v_{12}^2 v_{23}^2 \mathbf{v}_{12} \cdot \mathbf{v}_{23}}{r_{12}^7 r_{23}^7} \\ &= \frac{N_1 N_2 N_3}{R^5} \frac{1}{2} \frac{v_{12}^2 v_{23}^2 (v_{31}^2 - v_{12}^2 - v_{23}^2)}{r_{12}^7 r_{23}^7}. \end{aligned} \tag{5}$$

Hence, the term (4) is indeed part of the V-amplitude (5), provided that we express the same amplitude in the appropriate kinematic variables. A final observation is that the term $O(v_1^4)$ in (5), that is, the velocity of the far away graviton in a given reference frame, is not present in (5), a fact first pointed out in [14]. This is not in contradiction with (4), it simply implies the presence of additional terms canceling the leading $O(v_1^4)$ dependence, precisely as in (5). The remainder of this paper is devoted to showing how the contribution (4) can be reproduced in matrix theory.

3 Computation of the Effective Action

In this section we follow closely [13] and sketch how to reproduce the previous results by an effective action approach to matrix theory. Matrix theory is defined by a supersymmetric $SU(N)$ Yang-Mills (YM) field theory in $1+0$ dimensions with the classical action

$$S(A_\mu, \Psi) = \int dt \left(-\frac{1}{2} \operatorname{Tr} F^{\mu\nu} F_{\mu\nu} + i \operatorname{Tr} \bar{\Psi} \Gamma \cdot D\Psi \right), \tag{6}$$

where $F_{0k} = \partial_t A_k + i[A_0, A_k]$, $F_{jk} = i[A_j, A_k]$ and $D_0\Psi = \partial_t \Psi + i[A_0, \Psi]$, $D_k\Psi = i[A_k, \Psi]$. [d]

In order to compare the three-graviton scattering in supergravity with finite-N matrix theory, one has to make a two-loop computation for $SU(3)$ YM in the presence of the following classical external field

$$B_\mu = (0, B_k); \quad B_k = \begin{pmatrix} 0 & 0 & 0 \\ 0 & r_2 - r_1 & 0 \\ 0 & 0 & r_3 - r_1 \end{pmatrix}_k, \tag{7}$$

which is a solution of the classical equations of motion because, as explained earlier, the graviton positions depend linearly on time. Notice that B_k contains an irrelevant decoupled $U(1)$ component that we retain for convenience, since it is well known that it only contributes to the motion of the center of mass.

We quantize in the background gauge by setting

$$A_\mu = B_\mu + A^f_\mu, \tag{8}$$

where A^f_μ is the quantum (fluctuating) field.

We use the standard gauge fixing defined by

$$D^\mu(B)\, A^f_\mu = \partial_t A^f_0 + i[B_k, A^f_k] = 0, \tag{9}$$

and perform the functional integration

$$e^{-F(B)} = \int dA^f_\mu d\Psi d\bar{C} dC e^{-S(B_\mu + A^f_\mu, \Psi) - \int [D^\mu(B)\, A^f_\mu]^2 - \int \bar{C} D^\mu(B) D_\mu(A) C}, \tag{10}$$

where C and $\bar{C}$ are the Faddeev-Popov ghost fields. The 2-loop action $F_{(2)}(B)$ is to be compared with the supergravity amplitude for the three-graviton scattering.

---

[d]Greek indices go from 0 to 9 and Latin ones from 1 to 9.

Let us write

$$A_\mu = \begin{pmatrix} a_1 & a_2^\dagger & a_3^\dagger \\ a_2 & Y_{22} & Y_{23} \\ a_3 & Y_{32} & Y_{33} \end{pmatrix}_\mu , \tag{11}$$

where the 2×2 matrix $Y_\mu = B_\mu + Y_\mu^f$ (by a slight abuse of notation, we also denote by B_μ the appropriate 2×2 sub-matrix of (7)); $a_{1,2,3}$ and Y^f are quantum fields.

For our choice of background (7), $a_{2\mu}, a_{3\mu}$ are the *heavy* modes since their mass terms are proportional to the large distances r_{12} and r_{13}. Similarly, for the fermions and the ghosts, we denote by $\psi_2, \psi_3, c_2, c_3, \bar{c}_2, \bar{c}_3$ the respective heavy modes, and by ϕ, $\mathcal{C}$, $\bar{\mathcal{C}}$ the remaining *light* fields.

We can then perform the functional integration in two steps, in a wilsonian manner. First, we compute at 1-loop the integral over the heavy modes:

$$e^{-W_{(1)}(Y^f, a_1, \phi, \bar{\mathcal{C}}, \mathcal{C}; B)} = \tag{12}$$
$$\int \mathrm{d}a_2 \mathrm{d}a_3 \mathrm{d}\psi_2 \mathrm{d}\psi_3 \mathrm{d}\bar{c}_2 \mathrm{d}\bar{c}_3 \mathrm{d}c_2 \mathrm{d}c_3 e^{-S(B_\mu + A_\mu^f, \Psi) - \int [D^\mu(B) A_\mu^f]^2 - \int \bar{C} D^\mu(B) D_\mu(A) C} .$$

To complete the computation one has then to integrate $W_{(1)}$ over the light modes:

$$e^{-F(B)} = \int \mathrm{d}Y^f \mathrm{d}a_1 \mathrm{d}\phi \mathrm{d}\bar{\mathcal{C}} \mathrm{d}\mathcal{C} e^{-W_{(1)}(Y^f, a_1, \phi, \bar{\mathcal{C}}, \mathcal{C}; B)} . \tag{13}$$

The effective action $W_{(1)}$ has the form

$$W_{(1)} = V_{(1)}(Y^f; B) + \cdots , \tag{14}$$

where the dots represent terms depending on a_1, derivative terms as well as fermionic and ghost terms. The quantity $V_{(1)}$ is the effective potential of the field Y^f.

This quantity was computed in [13], where it was found to be non zero for generic Y^f. In fact, let us take

$$Y_k^f = \begin{pmatrix} 0 & x \\ x^\dagger & 0 \end{pmatrix}_k , \tag{15}$$

where x_k is the gauge field component in the k direction having the light mass term r_{23}. We obtain that the part of $V_{(1)}$ quadratic in the field $\mathbf{x}$ is

$$V_{(1)}(Y^f; B) = \frac{|\mathbf{r}_{23} \cdot \mathbf{x}|^2 - r_{23}^2 x^2}{r_{13} r_{12} (r_{12} + r_{13})} . \tag{16}$$

Some comments are necessary at this stage. The presence of a potential in the effective action seems to clash with our expectations from the various non-renormalization theorems of [16]. In fact, the two things are perfectly consistent. The background (7) breaks the original $SU(3)$ symmetry down to $U(1)\times U(1)$. Those fields that commute with the background (the truly massless fields) are prevented from acquiring a potential. However, the field $\mathbf{x}$ in (15) does not commute with the background and it is therefore not protected by the non-renormalization theorems. The following remark makes our result even more plausible: although not yet rigorously established, we expect a non-renormalization theorem similar to [16] to hold for the non abelian case as well. If this is so, the potential (16) should vanish for those backgrounds that leave an unbroken $SU(2) \times U(1)$. The only way for this to happen, given the fact that we take graviton one to be far away, is if $\mathbf{r}_{23} = 0$ and, in such case, (16) vanishes identically.

We can now use the potential (16) to compute the contribution to the amplitude $F_{(2)}(B)$ and show that it yields a non zero $\tilde{A}$. We integrate over $\mathbf{x}$ using the propagator obtained in [13] and obtain

$$F_{(2)}(B) = \int dt \frac{1}{r_{12}r_{13}(r_{12}+r_{13})} \frac{(\dot{r}_{23})^6}{(r_{23})^{11}} + \cdots, \tag{17}$$

where the dots represent terms that are not needed for computing $\tilde{A}$. Notice that $F_{(2)}(B)$ is proportional to the sixth power of $\dot{r}_{23}$ and therefore is the first term in the expansion in velocities for which no supersymmetry-induced cancellation is expected.

To verify that the action (17) contains (4), one selects terms proportional to $v_{13}^2 v_{12}^2$, and performs the integration over t to get a contribution of the form (we use $b_{12} \approx b_{13}$)

$$\tilde{\mathcal{S}} \sim \frac{N_1 N_2 N_3}{R^5} \frac{v_{13}^2 v_{12}^2}{b_{12}^7} \int dt\, t^4 \frac{(v_{23})^6}{(v_{23}^2 t^2 + b_{23}^2)^{11/2}} \sim \frac{N_1 N_2 N_3}{R^5} \frac{v_{13}^2 v_{12}^2 v_{23}}{(b_{12})^7 (b_{23})^6} \,. \tag{18}$$

The appropriate powers of N_i in (18) can be deduced—following [5]— from the double-line notation in which the setting-sun diagram is of order N^3; this factor must be $N_1 N_2 N_3$ for the diagram to involve all three particles. The overall R dependence can also be easily fixed by a scaling argument.

We can now go back to the original expression for the amplitude (4) and check that eq. (18) reproduces the behavior of the supergravity result in the eikonal approximation:

$$\int dt\, \tilde{A} = \frac{N_1 N_2 N_3}{R^5} \int dt\, \frac{v_{23}^2 v_{13}^2 v_{12}^2}{(v_{23}^2 t^2 + b_{23}^2)^{7/2} (v_{12}^2 t^2 + b_{12}^2)^{7/2}}$$

$$\sim \frac{N_1 N_2 N_3}{R^5} \frac{v_{23} v_{13}^2 v_{12}^2}{(b_{12})^7 (b_{23})^6} = \tilde{\mathcal{S}}. \tag{19}$$

4 Conclusions

There is, by now, mounting evidence showing that *all* of the tree level supergravity lagrangian is reproduced by finite N matrix theory, at least in uncompactified space without branes. At this point the crucial test will consist in reproducing M-theory quantum corrections. In this case, the objections raised by Hellerman and Polchinski [4] is that the perturbative computation may not make sense in the DLCQ limit. This is what happens for all but a few very special quantum field theories. The objection is somewhat softened by the observation of Bilal [17] that string perturbation theory is better behaved in this regard. Not enough is known about M-theory to decide which of the above cases is closer to describing the present situation; all we can do is compute and see.

Even with the optimistic attitude that finite N matrix theory is able to reproduce the quantum corrections, clearly our way of computing using a semi-classical approximation is too naive [18]. The bound state at threshold describing the M-graviton is a highly complicated object and we need to take into account its "form-factor", that is, the distribution of D-partons inside it. This is likely to present a formidable computational task but it is, in principle, accessible with the techniques of [19].

A final pressing issue that needs to be better understood, even in the DLCQ approach, is the large N limit. Contrary to the original prescription of [1], in this case the decompactification limit is obtained by letting $R \to \infty$, $N \to \infty$, while keeping N/R *finite*. In matrix theory, this corresponds to a highly non-t'Hooftian limit where $g_{\mathrm{YM}} \sim N^{3/2}$ instead of $g_{\mathrm{YM}} \sim N^{-1/2}$, hence there are potentially diverging diagrams in the perturbative expansion. Because of the interpretation of N/R as the physical momentum given in the DLCQ approach, one cannot argue that the divergent term might resum to a finite quantity — in the DLCQ there should be no divergent term. This opens up the possibility of falsifying finite N matrix theory by exposing a diagram that does not converge in the above limit. The first candidate, however, is a *four* loop $O(v^8)$ diagram! This is because terms of lower order in the velocity have been shown to be protected by supersymmetry, and the three loop $O(v^8)$ contribution is convergent after taking into account the extra powers of $1/R$ needed to go from ten to eleven dimensional normalization of the amplitudes. The whole analysis should be performed in the framework of the $1/N$ expansion, in other words, it is not necessary that the dangerous diagrams vanish identically, only that their leading contribution(s) in $1/N$ vanish. This suggests a subtle interplay between supersymmetry and the large N limit. Given the technical difficulties in performing an explicit perturbative computation beyond two loops, it seems more promising to try to generalize the approach of [16] including the large N limit as an extra ingredient. Hopefully we will soon have the answer to some of these questions.

Acknowledgments

The results presented at this meeting were obtained together with Marco Fabbrichesi and Roberto Iengo. I also would like to express my gratitude to the members of the department of Theoretical Physics at Chalmers for their wonderful hospitality and for having organized the meeting in a brewery. This work was supported in part by the European Union TMR program CT960045.

References

1. T. Banks, W. Fischler, S.H. Shenker and L. Susskind, Phys. Rev. **D 55** (1997) 5112 `hep-th/9610043`.
2. L. Susskind, `hep-th/9704080`.
3. A. Sen, Adv. Theor. Math. Phys. **2** (1998) 51 `hep-th/9709220`.
   N. Seiberg, Phys. Rev. Lett. **79** (1997) 3577 `hep-th/9710009`.
4. S. Hellerman and J. Polchinski, `hep-th/9711037`.
5. K. Becker, M. Becker, J. Polchinski and A. Tseytlin, Phys. Rev. **D 56** (1997) 3174 `hep-th/9706072`.
6. Y. Okawa and T. Yoneya, `hep-th/9806108`.
7. Y. Okawa and T. Yoneya, `hep-th/9808188`.
8. J. Plefka, M. Serone and A. Waldron, `hep-th/9809070`.
9. U.H. Danielsson, M. Kruczenski and P. Stjernberg, `hep-th/9807071`.
   U.H. Danielsson, talk given at this conference.
10. M.R. Douglas, H. Ooguri and S.H. Shenker, Phys. Lett. `B` **402** (1997) 36 `hep-th/9702203`.
    M.R. Douglas and H. Ooguri, Phys. Lett. **B 425** (1998) 71 `hep-th/9710178`.
    H. Ooguri, talk given at this conference.
11. M. Serone, Phys. Lett. **B 422** (1998) 88 `hep-th/9711031`.
    E. Keski-Vakkuri and P. Kraus, Nucl. Phys. **B 529** (1998) 246 `hep-th/9712013`.
    K. Becker and M. Becker, Phys. Rev. **D 57** (1998) 6464 `hep-th/9712238`.

12. M. Dine, A. Rajaraman, Phys. Lett. **B 425** (1998) 77 hep-th/9710174.
13. M. Fabbrichesi, G. Ferretti and R. Iengo, J. High Energy Phys. **06** (1998) 002 hep-th/9806018.
    M. Fabbrichesi, G. Ferretti and R. Iengo, hep-th/9806166.
14. W. Taylor IV and M. Van Raamsdonk, hep-th/9806066.
15. M.R. Douglas, D. Kabat, P. Pouliot and S.H. Shenker, "D-branes and Short Distances in String Theory", hep-th/9608024.
16. S. Paban, S. Sethi and M. Stern, J. High Energy Phys. **06** (1998) 012 hep-th/9806028.
    S. Paban, S. Sethi and M. Stern, hep-th/9805018.
17. A. Bilal, Phys. Lett. **B 435** (1998) 312 hep-th/9805070.
18. D. Gross, unpublished as in [5].
19. S. Sethi and M. Stern, Commun. Math. Phys. **194** (1998) 675 hep-th/9705046.

M-theory graviton scattering in a curved background[a]

Ulf H. Danielsson
Institutionen för teoretisk fysik
Box 803
S-751 08 Uppsala
Sweden
E-mail: ulf@teorfys.uu.se

In this talk the scattering of gravitons off a five orientifold in eleven dimensions is reviewed. The supergravity result is compared with a two loop M(atrix) model calculation and exact agreement is found. The relevant supergravity calculation involves nonlinear three graviton effects.

1 Introduction

The M(atrix)-model[1] was originally introduced as a model for eleven dimensional M-theory based on D-particle quantum mechanics[2,3,4]. The theory is formulated using supersymmetric Yang-Mills theory with 16 supersymmetries and gauge group $SU(N)$ where N corresponds to the number of D-particles. D-particles are among the most important objects in Type IIA string theory. In [5,6] the type IIA string was obtained from M-theory through compactification on a circle. A D-particle is then a graviton with quantized momentum in the compact direction. To recover the full eleven dimensional theory we must make the circle very large and as argued in [1] this means that we must consider the limit $N \to \infty$. An even stronger conjecture was put forward in [7], where it was proposed that finite N makes sense as a lightlike compactification, the discrete light cone quantization (DLCQ). For reviews with extensive references, see for example [8-11].

Clearly these conjectures, if true, amounts to considerable progress in our understanding of string theory. The work of [12,13] is therefore extremely encouraging since it provides very strong arguments in favor of the DLCQ conjecture. Unfortunately the use of the M(atrix)-theory can nevertheless be quite limited if actual calculations in the model turns out to be inpractical. Several succesful tests have been performed but they seem all to depend on having enough supersymmetry. To understand this, let us take the DLCQ conjecture as a starting point and investigate more carefully the eleven dimensional limit.

The calculations that we have the means to perform in the M(atrix) model are perturbative. We must then find out when the perturbative expansion can

[a]The talk is based on work together with Martín Kruczenski and Pär Stjernberg.

be trusted and whether it conflicts in any way with the decompactification limit. The noncompact $D = 11$ supergravity is obtained by making sure that the compact lightlike direction is large enough. According to [14,15], this requires a large boost so that the physical dimensions of the system under study are much smaller than the compactification scale. This in turn implies taking N very large. To be more explicit, consider a system of characteristic mass M and characteristic length r. If the system is at rest we have $P_- = P_+ = M$. According to the argument above, we need $r << R$ (where R is the lightlike compactification radius) in order to have a reliable description. This implies $Mr << MR = P_- R = N$. Given r, the largest mass that we need to consider is the mass of a black hole of radius r. In eleven dimensions we have $M \sim r^8$ and we therefore conclude that $r << N^{1/9}$.

The M(atrix)-model calculation is a perturbative loop expansion and we must investigate whether its range of validity overlaps with supergravity. The M(atrix)-loop expansion is an expansion in N/r^3 which means that we must have $N/r^3 << 1$. Hence we find

$$N^{1/3} << r << N^{1/9} \tag{1}$$

It follows that there is no overlap between the M(atrix) regime and the supergravity regime. There is therefore no reason to expect the perturbative M(atrix) calculations to agree with supergravity and we can expect agreement only when there is a nonrenormalization theorem to rely on. This can turn out to be a severe limitation to the use of the M(atrix)-model.

The most studied example where there is nonrenormalization is the scattering of two gravitons. At one loop it can be argued from supersymmetry that the v^2 term vanishes and that the v^4 term is not renormalized and agrees directly with supergravity[4]. From the D-particle point of view this follows from the force being of the same form for short and long distances. The successful two loop calculation in [16,17] suggests that there is nonrenormalization of terms at least up to order v^6. The discussion of [18] suggests, however, possible problems at higher orders. Related are the successful one loop calculations for extended objects [19] represented by particular background matrices. An alternative representation for 5-branes is described in [20] through explicit addition of hypermultiplets to the M(atrix) theory where at one loop the supergravity result is again successfully reproduced. This is a parallel to the case of two D-particles, but now it is the v^2 which is protected with the force being of the same form for short and long distances.

Recently the three-graviton scattering[21] has been found to be in agreement using two loop calculations contrary to the first reported discrepancy in [22]. Several other papers have also appeared discussing various aspects of the three graviton scattering[23,24,25,26].

Other more elaborated examples include orientifolds. Some of these were discussed in [27]. Three cases are successful at the one loop level. These are R/Z_2 (see [28]), R^5/Z_2 and R^9/Z_2, e.g. those cases where you can find protected terms in the expansion according to the above prescription. For the other orientifolds, the situation is not as fortunate. In the case of R^8/Z_2, it is noted in [27] that a suitable term can be found at two loops but with the wrong N-dependence. In our view there is no reason to expect agreement. No nonrenormalisation theorem that might protect from higher loop corrections is expected.

What other tests can one perform? An obvious suggestion is to consider even more gravitons. However, in [29] some preliminary results have appeared suggesting possible problems with more than three gravitons. One might also consider replacing one of the gravitons with e.g. a five brane. That is, we consider the scattering of two gravitons in the background of a five brane. A variation on the same theme is to scatter a single graviton against a five orientifold; counting the mirror image of the graviton we have three objects. Compared to having a five brane and two gravitons the kinematical configuration is more restricted and the answer can be presented in a simple form. This, then, corresponds to graviton scattering in a curved background and will be the subject of this talk.

My review of graviton scattering in a R^5/Z^2 background is based on a paper written together with Martín Kruczenski and Pär Stjernberg, where further details may be found[30]. The result that we obtain is in excellent agreement with supergravity, where the contribution can be seen to be a three graviton effect involving the orientifold, the graviton and the mirror image. [30] therefore represents a further successful test of the M(atrix) model.

The outline of the talk is very simple. In section two we perform the supergravity calculation and in section three we perform the two loop M(atrix) model calculation. The conclusions follow in section four.

2 The supergravity calculation

From the supergravity point of view we need to calculate the scattering of two gravitons (a source and a probe) in the presence of an object with five brane charge. This is due to the fact that the Z_2 quotient produces an image for each particle, but also the 5-dimensional fixed plane becomes a source for the four form and the graviton[31].

The metric and four form corresponding to a five dimensional object charged under $F_{[4]}$ is

$$ds^2 = H^{-1/3}(dx_+dx_- + dx_i dx^i) + H^{2/3}dy_I dy^I \tag{2}$$

$$F_{I_1I_2I_3I_4} = 3k\epsilon_{I_1I_2I_3I_4I_5}\frac{y^{I_5}}{\rho_\perp^5}, \tag{3}$$

where $H = 1+\frac{k}{\rho_\perp^3}$, with k a parameter given by the five brane charge. $k = -1/2$ for an orientifold while for a five brane is $k = 1/2$ as follows from e.g. [20]. As explained in [27] the value of the charge for the orientifold comes from working on the covering space. $i = 1...4$, $I = 5...9$ and we have defined $\rho_\perp^2 = y_I y^I$ and $\rho_{||}^2 = x_i x^i$. We have introduced light-cone coordinates $x^\pm = x_{11} \pm t$ where x^+ plays the role of time $\tau = \frac{1}{2}x^+$. To the order that we will be working the field strength F will not be important.

The energy momentum tensor of the image graviton can be computed from the action for a massless particle

$$S = \frac{1}{2}\int d\tau\; e^{-1}g_{\mu\nu}\dot{x}^\mu\dot{x}^\nu, \tag{4}$$

where e is an arbitrary function of the parameter τ which we choose to be constant. The energy momentum tensor is then:

$$T_{\mu\nu}(x) = \frac{1}{\sqrt{-g}}\frac{\delta S}{\delta g^{\mu\nu}(x)} \tag{5}$$

$$= \frac{1}{2\sqrt{-g}}\int d\tau\; e^{-1}\delta(x - x(\tau))g_{\mu\alpha}g_{\nu\beta}\dot{x}^\mu\dot{x}^\nu. \tag{6}$$

Using

$$p_- = e^{-1}g_{-+}\dot{x}^+ = \frac{N}{R} \tag{7}$$

one finds

$$T^{\mu\nu} = p_-\delta^{(9)}(x^i, y^I)u^\mu u^\nu, \tag{8}$$

where u^μ is the velocity of the image graviton (i.e. the source graviton) and is given by

$$u^\mu = (2, -\frac{1}{8}v^2, 0, 0, 0, 0, 0, 0, -\frac{1}{2}v, 0, 0, 0), \tag{9}$$

i.e. it corresponds to a D-particle moving along y^2. The impact parameter will be given by $y^1 = -r$. The probe will be moving in the opposite direction with

$$s^\mu = (2, -\frac{1}{8}v^2, 0, 0, 0, 0, 0, \frac{1}{2}v, 0, 0, 0), \tag{10}$$

and impact parameter $y^1 = r$. The relative velocity of the probe and source gravitons is v. This means that the image particle produces the Aichelburg-Sexl metric

$$g^{AS}_{\mu\nu} = \eta_{\mu\nu} + \zeta^h_{\mu\nu}, \tag{11}$$

where

$$\zeta^h_{\mu\nu} = \frac{15N}{2R^2M^9(\rho_\perp^2 + \rho_{||}^2)^{\frac{7}{2}}} u_\mu u_\nu \equiv h_0 u_\mu u_\nu, \tag{12}$$

after averaging over the x^- direction.

What if the particle is in the background of the orientifold metric? Since the Einstein equations are nonlinear the result will not be just a superposition but we will also need to take interactions into account. We will now proceed to evaluate these corrections.

We expand the metric around a Minkowsky background according to

$$g_{\mu\nu} = \eta_{\mu\nu} + h_{\mu\nu}, \tag{13}$$

where the perturbation is given by

$$h_{\mu\nu} = \zeta^h_{\mu\nu} + \zeta^H_{\mu\nu} + \chi_{\mu\nu}. \tag{14}$$

The first two terms are the super imposed unperturbed backgrounds where the D-particle part is given by equation (12) above while the orientifold part is given by

$$\zeta^H_{-+} = -\frac{k}{6\rho_\perp^3} \quad \zeta^H_{||||} = -\frac{k}{3\rho_\perp^3} \quad \zeta^H_{\perp\perp} = \frac{2k}{3\rho_\perp^3} \tag{15}$$

$\chi_{\mu\nu}$ is the genuine three body contribution that will be obtained below.

The effective action consists of two parts.

$$S = S_D + S_g, \tag{16}$$

that we will now cinsider in turn.

- *The D-particle Lagrangian*

  We find a contribution

$$L_D = \frac{v^2}{8} + \frac{1}{2} h_{\mu\nu} s^\mu s^\nu (1 - h_{-\lambda} s^\lambda), \tag{17}$$

  which is obtained by using $L_D = -P_- \dot{x}^-$ with $m \to 0$ at fixed p_-. $g_{\mu\nu}\dot{x}^\mu \dot{x}^\nu = 0$ gives $\dot{x}^-$. With our ansatz for the metric and using e.g. $s^\mu u_\mu = -v^2/2$ we find

$$L_D = \frac{v^2}{8}(1 + \frac{k}{r^3}) + \frac{1}{8} h_0 v^4 (1 + \frac{5k}{6r^3}) + \frac{1}{2}\chi_{\mu\nu} s^\mu s^\nu. \tag{18}$$

  But this is not all.

- *The gravitational action*

We begin with

$$S_g = \frac{1}{2\kappa_{11}^2}\int d^{11}x\sqrt{-g}R. \tag{19}$$

Following [21] we write this symbolically as

$$S_g = \frac{1}{2\kappa_{11}^2}\int d^{11}x\left(\frac{1}{4}h\mathcal{D}h + \frac{1}{3}V_3hhh\right), \tag{20}$$

with the Einstein equations given by

$$-\frac{1}{2}\mathcal{D}h - V_3hh = \kappa_{11}^2T. \tag{21}$$

We plug this back in the action to obtain

$$\begin{aligned}\frac{1}{2\kappa_{11}^2}&\int d^{11}x\left(\frac{1}{4}h\mathcal{D}h + \frac{1}{3}h\left(-\frac{1}{2}\mathcal{D}h - \kappa_{11}^2T\right)\right)\\ &= \frac{1}{2\kappa_{11}^2}\int d^{11}x\left(\frac{1}{12}h\mathcal{D}h - \frac{1}{3}\kappa_{11}^2hT\right)\end{aligned} \tag{22}$$

We then use the ansatz

$$g_{\mu\nu} = \eta_{\mu\nu} + \zeta_{\mu\nu} + \chi_{\mu\nu} \tag{23}$$

with

$$-\frac{1}{2}\Delta\zeta_{\mu\nu} = \kappa_{11}^2T_{\mu\nu},$$

to obtain

$$\frac{1}{2\kappa_{11}^2}\int d^{11}x\left(\frac{1}{12}\left(\zeta_{\mu\nu}+\chi_{\mu\nu}\right)\left(\mathcal{D}\zeta_{\mu\nu}+\mathcal{D}\chi_{\mu\nu}\right) - \frac{1}{3}\kappa_{11}^2\left(\zeta_{\mu\nu}+\chi_{\mu\nu}\right)T^{\mu\nu}\right).$$

If we partial integrate twice and ignore terms quadratic in χ, the result is

$$-\frac{1}{4}\zeta_{\mu\nu}s^\mu s^\nu - \frac{1}{3}\chi_{\mu\nu}s^\mu s^\nu, \tag{24}$$

or

$$L = \frac{v^2}{8}(1+\frac{k}{2r^3}) + \frac{1}{16}h_0v^4(1+\frac{5k}{3r^3}) + \frac{1}{6}\chi_{\mu\nu}s^\mu s^\nu. \tag{25}$$

We must now determine $\chi_{\mu\nu}s^\mu s^\nu$ and to this end we need to consider the Einstein tensor. We find a first order piece given by

$$G^{(1)}_{\mu\nu} = \frac{1}{2}(-\partial^2 h_{\mu\nu} - \partial_\mu\partial_\nu h^\alpha_\alpha + \partial_\mu\partial_\alpha h^\alpha_\nu + \partial_\nu\partial_\alpha h^\alpha_\mu), \tag{26}$$

and a second order piece given by

$$\begin{aligned} G^{(2)}_{\mu\nu} = &-\frac{1}{2}h_{\alpha\beta}(\partial^\alpha\partial_\nu h^\beta_\mu + \partial^\alpha\partial_\mu h^\beta_\nu - \partial^\alpha\partial^\beta h_{\mu\nu} - \partial_\mu\partial_\nu h^{\alpha\beta}) \\ &+\frac{1}{2}\partial_\alpha h_{\beta\nu}\partial^\alpha h^\beta_\mu - \frac{1}{2}\partial_\alpha h_{\beta\nu}\partial^\beta h^\alpha_\mu + \frac{1}{4}\partial_\nu h_{\alpha\beta}\partial_\mu h^{\alpha\beta} \\ &-\frac{1}{2}(\partial_\mu h^\alpha_\nu + \partial_\nu h^\alpha_\mu - \partial^\alpha h_{\mu\nu})(\partial^\beta h_{\alpha\beta} - \frac{1}{2}\partial_\nu h^\alpha_\alpha) \\ &-\frac{1}{2}h_{\mu\nu}(-\partial^2 h^\alpha_\alpha + \partial_\alpha\partial_\beta h^{\alpha\beta}) - \frac{1}{2}\eta_{\mu\nu}R. \end{aligned} \tag{27}$$

We proceed by imposing the gauge choice

$$\partial^\mu(\chi_{\mu\nu} - \frac{1}{2}\eta_{\mu\nu}\chi^\lambda_\lambda) = 0, \tag{28}$$

and obtain $-\frac{1}{2}\partial^2\chi_{\mu\nu}$ from the first order term. The same gauge condition is also obeyed by the unperturbed background $\zeta_{\mu\nu}$. However, as opposed to [21] the background is not traceless since $\zeta^H_{\mu\nu}$ has a trace given by $\zeta^{H\lambda}_\lambda = \frac{4k}{3r^3}$. The relevant source terms obtained from the second order piece contain one ζ^h and one ζ^H, note that the third line vanishes due to the gauge condition. Furthermore, the energy momentum tensor (8) provides a source term for $\chi_{\mu\nu}$ when the indices are lowered by the orientifold metric. Following [21] we find the relevant contribution to be

$$\kappa^2_{11}(T^\alpha_\nu\zeta^H_{\alpha\mu} + T^\alpha_\mu\zeta^H_{\alpha\nu}). \tag{29}$$

This simple energy momentum tensor where all the non linearities are coming from the lowering of the indices is crucial for obtaining the M(atrix) model result both for [21] and in our case. As explained in [24] further non linearities are absorbed into the quantized lightlike momentum. We put

$$\chi = \chi_{\mu\nu}s^\mu s^\nu \tag{30}$$

and for convenience we redefine

$$h_{\mu\nu} = H^{2/3}\tilde{h}_{\mu\nu}. \tag{31}$$

The resulting equation for $\tilde{\chi}$ is

$$\Delta\tilde{\chi} + \frac{v^4}{4}\left(\frac{k}{\rho^3}\Delta_{||}h_0 + \nabla h_0 \cdot \nabla\frac{k}{\rho^3} + \frac{2\kappa_{11}^2 k}{3\rho^3}p_-\delta^{(9)}(x^i, y^I)\right)$$
$$+\frac{v^4}{4}\left(\frac{k}{\rho^3}\partial_y^2 h_0 - \frac{1}{6}\partial_y h_0 \partial_y \frac{k}{\rho^3} - \frac{1}{3}h_0\partial_y^2\frac{k}{\rho^3}\right) = 0, \qquad (32)$$

where $y = y^2$ is the direction along the motion of the graviton. We need not solve the equation in general since we are interested only in the value of χ at the position of the particle. It is in fact enough to consider the case $(y^1 = r, y^2 = \cdots y^4 = x^i = 0)$. This can be obtained using the 9-dimensional Green function. For convenience we define $\rho_\perp^2 = \rho^2 + z^2$ where $z = y^1$. In particular we find that the first of the source terms in equation (32) leads to the integral

$$\frac{225}{16}\int_{-\infty}^{\infty} dz \int_0^{\infty} d\rho_1 d\rho_2 \rho_\perp^3 \rho_{||}^3 \frac{1}{((z-r)^2 + \rho_\perp^2 + \rho_{||}^2)^{7/2}}$$
$$\times \frac{1}{(z^2+\rho_\perp^2)^{3/2}}\Delta_{||}\frac{1}{((z+r)^2+\rho_\perp^2+\rho_{||}^2)^{7/2}} = -\frac{105}{256 r^{10}} = -\frac{7k}{r^3}h_0. \qquad (33)$$

The integral has been evaluated using a change of variables according to $r_1^2 = (z-r)^2 + \rho_\perp^2 + \rho_{||}^2$, $r_2^2 = (z+r)^2 + \rho_\perp^2 + \rho_{||}^2$, and $r_3^2 = z^2 + \rho_\perp^2$. A similar integral gives $\frac{8k}{\rho^3}h_0$ for the second term of the first line while the last term of the first line immediately gives $\frac{k}{3\rho^3}h_0$. It can be verified that the rest of the terms do not contribute to the particular scattering amplitude that we are interested in. We can then conclude that

$$\tilde{\chi} = \frac{v^4}{3}h_0\frac{k}{r^3} \qquad (34)$$

and therefore

$$\chi = \frac{v^4}{2}h_0\frac{k}{r^3}. \qquad (35)$$

This allows us to obtain

$$L = \frac{v^2}{8}(1 + \frac{k}{2r^3}) + \frac{1}{16}h_0 v^4(1 + \frac{3k}{r^3})$$
$$= \frac{v^2}{8}(1 + \frac{k}{2r^3}) + \frac{15}{32}\frac{v^4}{(2r)^7}(1 + \frac{3k}{r^3}). \qquad (36)$$

This is however still not the whole story. We must also add the identical contribution from the image graviton action and the action of the five brane charged object (including gravitational actions). The latter is easily deduced without any further calculation. The integrated force on the orientifold must be equal and opposite the one from the two gravitons, it will therefore be precisely equal to the sum of the contributions proportional to k in the action of the the two gravitons. This effectively doubles the value of k and we obtain

$$L_{tot} = 2\Big(\frac{v^2}{8}(1+\frac{k}{r^3}) + \frac{15}{32}\frac{v^4}{(2r)^7}(1+\frac{6k}{r^3})\Big) = \\ 2\Big(\frac{v^2}{8}(1-\frac{1}{2r^3}) + \frac{15}{32}\frac{v^4}{(2r)^7}(1-\frac{3}{r^3})\Big), \tag{37}$$

where we have put $k = -1/2$. Note that the value for a five brane is $k = 1/2$ as follows from e.g. [20]. As explained in [27] the value for the orientifold comes from working on the covering space. At order v^2/r^3 the effect of the gravitational action is to cancel the five brane contribution explaining in the spirit of [21] why previous calculations, e.g. [4], where the gravitational action (and five brane action) have been neglected, nevertheless give the right answer.

More precisely, let us consider two interacting objects, e.g two gravitons or a graviton and a five brane. We get a contribution $K_1 + V - \frac{V}{2}$ from the first and a contribution $K_2 + V - \frac{V}{2}$ from the second. K_i is the kinetic energy while V is the potential. The last term in each case comes from the gravitaional action. Adding we find $K_1 + K_2 + V$. If the second object is at rest, $K_2 = 0$, this is the same result as if we only consider the first object and forgets the gravitational action.

3 The M(atrix) model calculation

We now turn to the M(atrix) theory description of the R^5/Z_2 orbifold with the aim of reproducing the supergravity result (37). To derive the action we start with a $U(2N)$ 0+1 dimensional supersymmetric Yang-Mills theory on the covering space with fields $X^i, : i = 1,\ldots,9$ in the adjoint representation, which describe N D-particles along with their mirror images. The fields are split into $X_\perp$ and $X_{||}$ corresponding to directions parallel and transverse to the orientifold plane. Keeping only states that are invariant under the combined action of orientation reversal and space-time reflection gives rise to the projection condition [28]

$$\begin{aligned} X_{||} &= M X_{||}^{\mathrm{T}} M^{-1}, \\ X_\perp &= -M X_\perp^{\mathrm{T}} M^{-1}, \\ \Theta &= \Gamma_\perp\, M \Theta^{\mathrm{T}} M^{-1}, \qquad\qquad \Gamma_\perp = \Gamma_5 \cdots \Gamma_9. \end{aligned} \tag{38}$$

These equations restricts the gauge group for the remaining states to be either $SO(2N)$ or $USp(2N)$, but consistency conditions enforces the choice $USp(2N)$ [32,33]. The action is given by [32]

$$S = \frac{1}{2}\int dt\ \mathrm{Tr}\Bigg\{\frac{1}{2\mathrm{R}}(\mathrm{D_t X_i})^2 + \frac{1}{2\mathrm{R}}(\mathrm{D_t Y_I})^2 + \frac{\mathrm{R}}{4}[\mathrm{X_i},\mathrm{X_j}]^2 + \frac{\mathrm{R}}{4}[\mathrm{Y_I},\mathrm{Y_J}]^2$$
$$+\frac{R}{2}[Y_I, X_j]^2 + S_a D_t S_a + S_{\dot a} D_t S_{\dot a} + 2iRX_i\sigma^i_{a\dot a}\{S_a, S_{\dot a}\}$$
$$-S_a\gamma_I[Y_I, S_a] + S_{\dot a}\gamma_I[Y_I, S_{\dot a}]\Bigg\}, \tag{39}$$

where R is the eleven dimensional radius and we have put the eleven dimensional Planck length $l_p = 1$. The indices take values $i = 1,\ldots,4$ and $I = 5,\ldots,9$ and $D_t = \partial_t - i[A_0,\cdot]$. An extra factor of $1/2$ has been included in front of the action in account for the fact that our system does not include the action of the image. This is clearly seen from the kinetic term which must be $N/2R(v/2)^2$. This is equivalent to take the coupling constant $g = 4R$ instead of $g = 2R$ in the notation of [17].

The four X_i are in the antisymmetric representation of $Sp(N)$ and corresponds to directions parallel to the orientifold, the fifth one is the longitudinal direction of the M(atrix)-model. The A_I with $I = 5,...,9$ give the transversal coordinates. The fermionic fields are S_a in the adjoint representation and $S_{\dot a}$ in the symmetric representation. We will perform a loop calculation using the background field method, As we will consider scattering in a direction transverse to the fixed plane, the background field B_I only has non-zero components B_I, $I = 5,\ldots,9$. B_I is chosen to satisfy the equations of motion, and is given by

$$B_9 = \frac{vt}{2}\sigma_3 \otimes 1_{N\times N} \quad B_8 = b\sigma_3 \otimes 1_{N\times N}, \tag{40}$$

where σ_3 is a Pauli matrix and $1_{N\times N}$ is the identity matrix of $N \times N$. This corresponds to a particle composed of N D-particles moving in a direction transverse to the orientifold with an impact parameter b and velocity $v/2$ relative to the fixed point. (The distance between the particle and its mirror image is thus $\sqrt{(2b)^2 + (vt)^2}$ and the relative velocity is v). We expand around the background using

$$Y_I = B_I + A_I, \tag{41}$$

and obtain terms quadratic in the fluctuations which determine the propagators and cubic and quartic terms which give vertices with three and four legs respectively. Massless propagators are associated with strings stretching between D-particles within the composite particle, and massive propagators are associated with strings stretching between a D-particle in the composite particle and another in the image. In terms of matrices, these are described by block diagonal and off-diagonal matrices respectively. The matrices $X_{\mu=5\ldots9}$ are in the adjoint representation which is $N(2N+1)$-dimensional. Of them, N^2+N become massive. On the other hand, matrices $X_{\mu=1\ldots4}$ are in the two index antisymmetric representation having $2N^2-N-1$ fields out of which N^2-N aquire a mass. In total we get the following massive fields:

- $8N^2$ bosons of mass $m^2=(2r)^2$.
- N^2+N bosons of mass $m^2=(2r)^2+2v$.
- N^2+N bosons of mass $m^2=(2r)^2-2v$.
- N^2+N complex ghosts of mass $m^2=(2r)^2$.
- $4N^2$ fermions of mass $m^2=(2r)^2+v$.
- $4N^2$ fermions of mass $m^2=(2r)^2-v$.

where $r^2=(b^2+(vt/2)^2)$ is the distance between the particle and the orientifold. Integrating out massive fields at one loop we obtain the potential

$$V=N\frac{v^2}{2(2r)^3}-\frac{v^4}{(2r)^7}\left(\frac{15}{32}N^2+\frac{5}{16}N\right). \tag{42}$$

We note that the N^2 contribution is one half of the one-loop potential between two D-particles in uncompactified space[4], while for $N=1$ we obtain the result of [32].

To obtain the two-loop correction we must evaluate vacuum diagrams using the cubic and quadratic vertices. The resulting expression for the phase shift is expanded in powers of the velocity and finally we calculate a corresponding potential. The result is:

$$\begin{aligned}(\text{bos.}+\text{gh.}+\text{ferm.}) &= \left(\frac{225}{256}\frac{v^6}{(2r)^{14}}+\ldots\right)N^3 \\ &+\left(\frac{45}{16}\frac{v^4}{(2r)^{10}}+\frac{675}{128}\frac{v^6}{(2r)^{14}}+\ldots\right)N^2 \\ &+\left(\frac{135}{32}\frac{v^4}{(2r)^{10}}-\frac{1575}{256}\frac{v^6}{(2r)^{14}}+\ldots\right)N.\end{aligned} \tag{43}$$

The first thing to note is that there is no v^2 term at any order of N showing that the metric is not changed at two-loops. This is in agreement with the non-renormalization theorem proved in [34], see also [35,36]. To compare the non-vanishing terms with the supergravity result it is necessary to include the coupling constant $g = 4R$ as discussed previously. We must also return to Minkowsky space by putting $v \to iv$. Powers of RM^3 are also restored using dimensional analysis with the result

$$L = \frac{N}{R}\frac{v^2}{8} - \frac{N}{RM^3}\frac{v^2}{2(2r)^3} + \frac{15}{32}\frac{N^2}{R^3M^9}\frac{v^4}{(2r)^7}$$
$$+\frac{225}{64}\frac{N^3}{R^5M^{18}}\frac{v^6}{(2r)^{14}} - \frac{45}{4}\frac{N^2}{R^3M^{12}}\frac{v^4}{(2r)^{10}} + \dots \tag{44}$$

which includes zero, one and two loop contributions and for each power of the velocity, only the leading order in N is written.

In a supergravity calculation, the background brings in powers of the orientifold charge which is proportional to $1/M^3$. Hence the lowest order $1/M$ must represent the interaction between the particle and its image. This interaction potential follows from the results of [16,17] and is given by

$$V = -\frac{15}{16}\frac{N^2}{R^3M^9}\frac{v^4}{r^7} - \frac{225}{32}\frac{N^3}{R^5M^{18}}\frac{v^6}{r^{14}} + \mathcal{O}(\frac{v^8}{r^{21}}). \tag{45}$$

This is exactly twice what is included in expression (44) in correspondence with the fact that in (44) only the action of the particle is considered and not that of the image. On the other hand the agreement is not surprising since it can be checked diagram by diagram because both calculations are very similar. Of more interest in our case is the subleading order in $1/M$. The v^2 terms are

$$\frac{1}{2}\frac{N}{R}\left(\frac{v}{2}\right)^2\left(1 - \frac{1}{2(Mr)^3}\right), \tag{46}$$

which is the one loop result that was obtained in [27,32].

The v^4 term is

$$V = -\frac{15}{32}\frac{N^2}{R^3M^9}\frac{v^4}{(2r)^7}(1 - \frac{3}{(Mr)^3}), \tag{47}$$

in precise agreement with the supergravity result of section 2.

One can also follow a different procedure to perform the M(atrix)-model calculation that gives more physical insight and provides a further check of the result. Let us, as above, consider the scattering of N D-particles (that are very close to each other) out of the orientifold fixed-plane. If we separate out one of the D-particles and calculate the force due to the other $N-1$ particles it is easy to see where the different powers of N come from. The only thing we need to do is to associate each of the Lie algebra elements with the appropriate string configuration.

As we have seen there are three types of strings: strings that connect a D-particle with its mirror mage, strings that connect a D-particle with the mirror image of another D-particle and strings that connect D-particles within the composite particle among themselves, or mirror images among themselves. The third category are massless if the D-particles are very close to eachother. The first category is relevant only for the adjoint (i.e symmetric) representation. For diagrams with two cubic vertices we can argue in the following way, noting from the algebra that there will always be precisely one massless string[30]. If the diagram includes one massless string and two of the second type we get a factor $N(N-1)^2$, if we in addition to the massless string have one of the first type and one of the second type we get $N(N-1)$ and if both are of the first type (in addition to the massless string) we get just N. Note that an extra power of N is included to account for the fact that anyone of the N particles can be singled out. In other words, there are N strings going from a D-particle to its own image and $N-1$ going from the D-particle to the image of some other D-particle. The argument for diagrams with one quartic vertex is almost identical. It is clear, then, that one can perform the calculation for the case $N=2$ and then carefully separating the different contributions, multiply each one by the corresponding factor of N. In this way the same result (44) is obtained.

4 Conclusions

In this talk I have reviewed yet another successful application of the perturbative M(atrix) model. From the supergravity point of view we are considering a three graviton effect, where we have one graviton each from the D-particle, the mirror image and the orientifold meeting in a vertex. It is interesting to note the agreement in view of the latest developments in the study of three graviton scattering. A natural extension of this work is to consider graviton scattering in the presence of a five brane. Two gravitons and one five brane allows more general kinematic configurations than the one studied in this talk.

At one loop[27] the result consists of a piece involving an interaction between the graviton and its mirror image of order N^2v^4/r^7 and an interaction directly with the five brane charge of the orientifold of order Nv^2/r^3. We have seen that there is no contribution to the v^2 at two loops but we have found a term of order N^2v^4/r^{10}, with supergravity and the M(atrix)-model giving the same coefficient. It is tempting to conjecture an agreement also at higher loops for the v^4 term. This would imply a vanishing of all N^kv^4 terms for $k > 2$ above one loop. This is needed also for our succesful two loop result to be protected. Furthermore, an extension of the calculation in section two provides the supergravity prediction for the N^2v^4 term at any loop.

Finally it would be interesting to have a more systematic understanding of when one can expect nonrenormalization theorems, and when one can trust the perturbative M(atrix)-model.

Acknowledgements

I would like to thank Martín Kruczenski and Pär Stjernberg for a fruitful collaboration.

References

1. T. Banks, W. Fischler, S. Shenker and L. Susskind, "M theory as a Matrix Model: A Conjecture," Phys. Rev. **D55** (1997) 5112; hep-th/9610043.
2. U.H. Danielsson, G. Ferretti and Bo Sundborg, "D-particle dynamics and bound states", Int. Journ. Mod. Phys. A, Vol. 11, No. 31 (1996) 5463, hep-th/9603081.
3. D. Kabat and P. Pouliot, "A Comment on Zero-brane Quantum Mechanics", Phys. Rev. Lett. **77** (1996), hep-th/9603127.
4. M. R. Douglas, D. Kabat, P. Pouliot and S. Shenker, "D-branes and Short Distances in String Theory", Nucl. Phys. **B485** (1997) 85, hep-th/9608024.
5. P.K. Townsend, Phys. Lett. **B350** (1995) 184, `hep-th/9501068`.
6. E. Witten, Nucl. Phys. **B443**, (1995) 85, `hep-th/9503124`.
7. L. Susskind, "Another Conjecture about Matrix Theory," hep-th/9704080
8. T. Banks, "Matrix Theory," hep-th/9710231.
9. D. Bigatti and L. Susskind, "Review of Matrix Theory," hep-th/9712072.
10. W. Taylor, "Lectures on D-Branes, Gauge Theory and Matrices," hep-th/9801182.
11. A. Bilal, "Matrix Theory: A Pedagogical Introduction," hep-th/9710136.
12. N. Seiberg, "Why is the Matrix Model Correct," Phys. Rev. Lett. **79** (1997) 3577, hep-th/9710009.
13. A. Sen, "D0 Branes on T^n and Matrix Theory," hep-th/9709220.
14. T.Banks, W.Fischler, I.R.Klebanov and L.Susskind, "Schwarzschild Black Holes from Matrix Theory", Phys.Rev.Lett. **80** (1998) 226-229,

hep-th/9709091.
15. I. R. Klebanov and L. Susskind, "Schwarzschild Black Holes in Various Dimensions from Matrix Theory", Phys.Lett. **B416** (1998) 62-66, hep-th/9709108.
16. K. Becker and M. Becker, "A Two Loop Test of Matrix Theory," Nucl.Phys. **B506** (1997) 48, hep-th/9705091.
17. K. Becker, M. Becker, J. Polchinski and A. Tseytlin, "Higher Order Graviton Scattering in Matrix Theory," Phys.Rev. **D56** (1997) 3174; hep-th/9706072.
18. M. Dine, R. Echols and J. Gray, "Renormalization of Higher Derivative Operators in the Matrix Model," hep-th/9805007.
19. G. Lifschytz, "Four Brane and Six Brane Interactions in M(atrix) Theory", hep-th/9612223.
O. Aharony and M. Berkooz, "Membrane Dynamics in M(atrix) Theory" Nucl.Phys. **B491** (1997):184-200, hep-th/9611215.
G. Lifschytz and S. Mathur "Supersymmetry and Membrane Interactions in M(atrix) Theory" Nucl.Phys. **B507** (1997):621-644, hep-th/9612087.
20. M. Berkooz and M. R. Douglas, "Five-Branes in M(atrix) Theory" Phys.Lett. B395: 196-202,1997 , hep-th/9610236.
21. Y. Okawa, T. Yoneya, "Multi-Body Interactions of D-Particles in Supergravity and Matrix Theory", hep-th/9806108 .
22. M. Dine and A. Rajaraman, "Multigraviton Scattering in the Matrix Model" hep-th/9710174.
23. M. Fabbrichesi, G. Ferretti and R. Iengo, "Supergravity and matrix theory do not disagree on multi-graviton scattering", JHEP 06:002,1998, hep-th/9806018. "About the Matrix model computation of three-graviton scattering", hep-th/9806166.
24. W. Taylor and M. Van Raamsdonk "Three graviton scattering in Matrix theory revisited", hep-th/9806066.
25. R. Echols and J. P. Gray, "Comment on Multigraviton Scattering in the Matrix Model", hep-th/9806109
26. J. McCarthy, L. Susskind and A. Wilkins, "Large N and the Dine-Rajaraman problem", hep-th/9806136.
27. O. J. Ganor, R. Gopakumar, S. Ramgoolam, "Higher Loop Effects in M(atrix) Orbifolds", Nucl.Phys. **B511** (1998) 243-263, hep-th/9705188.
28. U. H. Danielsson and G. Ferretti, "The Heterotic Life of the D-particle", Int.J.Mod.Phys. A12 (1997) 4581-4596, hep-th/9610082.
29. M. Dine, R. Echols and J. Gray, "Tree Level Supergravity and the Matrix Model",hep-th/9810021.
30. U.H. Danielsson, M. Kruczenski and P. Stjernberg, "Graviton Scattering

in an M Orientifold Background", hep-th/9807071.
31. E. Witten, "Five-branes and M-Theory on an Orbifold", Nucl.Phys. **B463** (1996) 383, hep-th/9512219.
32. N. Kim and S.-J. Rey "M(atrix) Theory on T^5 Orbifold and Five-Branes," hep-th/9705132.
33. A. Fayyazuddin and D.J. Smith, "A note on T^5/Z_2 compactification of the M-theory matrix model" Phys. Lett. **B407** (1997) 8, hep-th/9703208.
34. D.-E. Diaconescu and R. Entin, "A Non-Renormalization Theorem For The d=1, N=8 Vector Multiplet" Phys. Rev. **D56** (1997), 8045; hep-th/9706059.
35. S. Paban, S. Sethi and M. Stern, "Constraints From Extended Supersymmetry in Quantum Mechanics", hep-th/9805018
36. S. Paban, S. Sethi and M. Stern, "Supersymmetry and Higher Derivative Terms in the Effective Action of Yang-Mills Theories", hep-th/9806028

HIGHER SPIN N = 8 SUPERGRAVITY IN AdS$_4$

E. SEZGIN

Center for Theoretical Physics, Texas A&M University, College Station, TX 77843, USA

P. SUNDELL

Department of Engineering Sciences, Physics and Mathematics Karlstad University, S-651 88 Karlstad, Sweden

We review the basic structure of the higher spin extension of $D = 4, N = 8$ AdS supergravity. The theory is obtained by gauging the higher spin superalgebra $shs^E(8|4)$ by a procedure pioneered by Vasiliev. The algebra $shs^E(8|4)$ is a subalgebra of the enveloping algebra of $OSp(8|4)$. The physical states of the theory are in one to one correspondence with the symmetric product of two $OSp(8|4)$ singletons. This singleton theory, which may be viewed in a certain limit as the supermembrane theory on $AdS_4 \times S^7$, is expected to describe the dynamics of the higher spin theory. Thus, the higher spin $N = 8$ supergravity on AdS_4 is conjectured to describe the field theory limit of M-theory on $AdS_4 \times S^7$.

1 Introduction

A consistent interacting field theory of fields with spin higher than two has long interested the high energy physicists. Many attempts to construct such theories encountered severe difficulties in the past and led to no-go theorems. However, no-go theorems are usually based on a number of assumptions and such theorems may as well disappear when some of those assumptions are relaxed. Indeed, this is just what happened in the search for higher spin field theories as well. Firstly, and perhaps not surprisingly, the number of higher spin fields participating in the full theory need to be infinite, as opposed to one higher spin field at a time considered previously, and secondly, and that is the surprising part, the spacetime in which the theory is formulated must admit AdS space as a solution [1,2] rather than Minkowskian space assumed in previous attempts.

The occurrence of infinitely many fields is very suggestive of an underlying extended object theory. String theory in AdS_4 has been considered briefly in [3] but the results appear to be inconclusive. In a separate line of development, the eleven dimensional supermembrane theory appeared in 1987 [4]. Soon after, it was suggested [5] that the AdS_4 supersingleton may play a role in its description. This is the ultra short representation of the AdS_4 supergroup $OSp(4|8)$, consisting of 8 bosonic and 8 fermionic states [6,7] which cannot be described in terms of local fields in the bulk of AdS_4. Soon after that, a connection between the $OSp(8|4)$ supersingleton field theory, assumed to arise from the $AdS_4 \times S^7$ compactification of the supermembrane, and a higher spin supergravity theory in AdS_4 bulk was conjectured in [8, 9, 10]. This conjecture was especially aided by the remarkable property of the singletons, namely the fact that the symmetric product of two supersingletons yields an infinite tower of massless higher spin states [11]. The form of this result for $OSp(8|4)$ was spelled out in [10], and it was argued that all of these states should arise in the quantum supermembrane theory [8,9,10].

These developments provided further motivation for the analysis of the supermembrane on various manifolds of the form $AdS_4 \times M_7$, where M_7 is a positive curvature Einstein manifold, and various issues arising in the expansion of the supermembrane action around these vacua [12,13,14]. These were the early developments on the subject of bulk AdS and boundary CFT connection in the context of the $D = 11$ supermembrane its simplest form. The AdS/CFT connection has acquired a deeper and wider significance in the light the developments of the last year which were triggered by the work of Maldacena [15].

Turning to the story of higher spin fields, interestingly enough, at the same period when as the supermembrane/singletons/higher-spin-fields connections were being considered, Fradkin and Vasiliev [1,2] were in the course of developing a higher spin gauge theory in its own right (see [1] for references to earlier work). These authors succeeded in constructing interacting field theories for higher spin fields. As mentioned above, the previous difficulties in constructing higher spin theories were bypassed by formulating the theory in AdS_4 and by considering an infinite tower of gauge fields controlled by various higher spin algebras based on certain infinite dimensional extensions of AdS_4 superalgebras. An intriguing property of the higher spin field theories, which again is very suggestive of an underlying extended object, is the fact that the interactions are non-local. In particular, the AdS_4 radius can not be taken to infinity since its positive powers occur in the higher spin interactions and therefore one can not take a naive Poincaré limit.

In a series of papers Vasiliev [17–20] pursued the program of constructing the AdS higher spin gauge theory and simplified the construction considerably. In [18] the spin 0 and 1/2 fields were introduced to the system at the level of the equations of motion within the framework of free differential algebras. The equations of motion were furthermore cast into an elegant geometrical form in [19, 20] by extending the higher spin algebra to include new auxiliary spinorial variables. For $N = 1$, $R \wedge R$ type actions have been considered [1,2], but one drawback of these actions is that the spin $s \leq 1$ sector of the theory does not fit in a natural and geometrical way into the part of the action that describes the fields with spin $s \geq 3/2$.

Interestingly enough, applying the formalism of Vasiliev to a suitable higher spin algebra that contains the maximally extended super AdS algebra $OSp(8|4)$, the resulting spectrum of gauge fields and spin $s \leq \frac{1}{2}$ fields coincide with the massless states resulting from the symmetric product of two $OSp(8|4$ supersingletons [9]. In a recent paper [21], we examined the Vasiliev theory of higher spin fields (which is applicable to a wide class of higher spin superalgebras) and determine the precise manner in which the $N = 8$ de Wit-Nicolai gauged supergravity [22] which is the gauged version of the Cremmer-Julia $N = 8$ supergravity in four dimensions [23] can be described within this framework. We showed how this embedding works in the higher spin AdS supergravity based on the higher spin superalgebra known as $shs^E(8|4)$[24,25].

The most natural next step in this program is to compare the interactions in the higher study with those of gauged $N = 8$ supergravity, and to search for the $E_7/SU(8)$ structure [23] in the full theory. Even more tantalizing is the prospects of generating the full content of the higher spin theory from the $OSp(8|4)$ singleton field theory which lives in the boundary of AdS_4.

The purpose of this report is far more modest, namely to give a concise review of the work presented in [21], and to emphasize the elegant geometrical picture that underlies the higher spin field equations. In Sec. 2, we discuss the higher spin algebra, its gauging and the constraints which encode the dynamics. In Sec. 3, we discuss the linearization of the higher spin equations around the AdS_4 vacuum, the spectrum of physical states, their equations of motion and the comparison with the linearized field equations of the gauged $N = 8, D = 4$ supergravity. The uniqueness of the constraints is discussed in Sec. 4. Further comments on the results and on open problems are collected in Sec. 5.

2 The Higher Spin Gauge Theory

2.1 The Higher Spin Algebra

The higher spin algebra $shs^E(8|4)$ arises most naturally as a truncation of an extended higher algebra $\widehat{shs}^E(8|4)$. The latter is the algebra of polynomials in the bosonic $SO(3,2)$ Majorana spinors $Y_{\underline{\alpha}}$ and $Z_{\underline{\alpha}}$ ($\underline{\alpha} = 1, ..., 4$) and the real, fermionic $SO(8)$ vector θ^i which obey the following associative but noncommutative $\star$ product rule [20]

$$P(Z,Y)\star Q(Z,Y) \;=\; \int d^4U\, d^4V\, P(Z+U,Y+U)\, Q(Z-V,Y+V)e^{-iU^{\underline{\alpha}}V_{\underline{\alpha}}}\,, \tag{1}$$

where the integral is normalized such that $1\star P = P$. This formula in particular implies

$$\begin{aligned} Z_{\underline{\alpha}}\star Z_{\underline{\beta}} &= Z_{\underline{\alpha}}Z_{\underline{\beta}} - iC_{\underline{\alpha\beta}}\,, & Z_{\underline{\alpha}}\star Y_{\underline{\beta}} &= Z_{\underline{\alpha}}Y_{\underline{\beta}} + iC_{\underline{\alpha\beta}}\,,\\ Y_{\underline{\alpha}}\star Z_{\underline{\beta}} &= Y_{\underline{\alpha}}Z_{\underline{\beta}} - iC_{\underline{\alpha\beta}}\,, & Y_{\underline{\alpha}}\star Y_{\underline{\beta}} &= Y_{\underline{\alpha}}Y_{\underline{\beta}} + iC_{\underline{\alpha\beta}}\,, \end{aligned} \tag{2}$$

where $C_{\underline{\alpha\beta}}$ is the anti-symmetric charge conjugation matrix. The product rule Eq. (1) is isomorphic to the normal ordered product of a set of harmonic oscillators formed out of the spinor variables. The $\star$ product between θ's is that of the $SO(8)$ Clifford algebra, so that for example $\theta^i\star\theta^j = \delta^{ij} + \theta^i\theta^j$.

We next define the map

$$\tau P(Z;Y,\theta) \;:=\; P(-iZ;iY,i\theta)\,, \tag{3}$$

which acts as a graded anti-homomorphism of the $\star$ algebra:

$$\tau(P\star Q) \;=\; (-1)^{PQ}\tau(Q)\star\tau(P)\,. \tag{4}$$

The extended higher spin algebra $\widehat{shs}^E(8|4)$ is by definition the space of Grassmann *even* polynomials $P(Z;Y,\theta)$ obeying [25]

$$\widehat{shs}^E(8|4) \;:=\; \left\{P(Z;Y,\theta)\,\middle|\,\tau P = -P\,,\; P^\dagger = -P\right\}\,, \tag{5}$$

with Lie bracket

$$[P\,,Q]_\star \;:=\; P\star Q - Q\star P\,. \tag{6}$$

The higher spin algebra $shs^E(8|4)$ is defined by [25]

$$shs^E(8|4) \;:=\; \widehat{shs}^E(8|4)|_{Z=0}\,. \tag{7}$$

The $OSp(8|4)$ subalgebra generators are

$$OSP(8|4) \;=\; \{\, Y_{\underline{\alpha}} Y_{\underline{\beta}} \,,\; Y_{\underline{\alpha}} \theta^i \,,\; \theta^i \theta^j \,\} \;. \tag{8}$$

The algebra $shs^E(8|4)$ splits under $OSp(8|4)$ into levels labeled by $\ell = 0, 1, \ldots$ and the ℓ'th level generators are given by a homogeneous polynomials of degree $4\ell + 2$. The 0'th level is $OSp(8|4)$. Actually, $shs^E(8|4)$ is the space of odd polynomials in $OSp(8|4)$ generators which are fully symmetric in $SO(3,2)$ spinor indices and fully anti-symmetric in $SO(8)$ vector indices.

2.2 Gauging $\widehat{shs}^E(8|4)$

The gauging of $\widehat{shs}^E(8|4)$ proceeds by introducing a master gauge field $\hat{\omega}$ which is an $\widehat{shs}^E(8|4)$ valued 1-form on the space $\mathcal{M}$ which is the product of spacetime with the non-commutative Z space with coordinate $z^M = (x^\mu, Z^{\underline{\alpha}})$, where x^μ are the spacetime coordinates. Introducing the total exterior derivative

$$\hat{d} \;:=\; dz^M \partial_M \;=\; dx^\mu \partial_\mu + dZ^{\underline{\alpha}} \partial_{\underline{\alpha}} \;, \tag{9}$$

we define the $\widehat{shs}^E(8|4)$ valued connection 1-form[a]

$$\begin{aligned} \hat{\omega} \;&=\; dz^M \hat{\omega}_M(x, Z; Y, \theta) \;, \\ \tau\hat{\omega} \;&=\; -\hat{\omega} \;, \qquad \hat{\omega}^\dagger \;=\; -\hat{\omega} \;. \end{aligned} \tag{10}$$

The $\widehat{shs}^E(8|4)$ valued curvature 2-form $\hat{\mathcal{R}}$ is

$$\hat{\mathcal{R}} \;=\; \hat{d}\hat{\omega} - \hat{\omega} \star \hat{\omega} \;, \tag{11}$$

which obeys the Bianchi identity

$$\hat{D}\hat{\mathcal{R}} \;:=\; \hat{d}\hat{\mathcal{R}} - [\hat{\omega}\,,\,\hat{\mathcal{R}}]_\star \;=\; 0 \;. \tag{12}$$

The $\widehat{shs}^E(8|4)$ gauge transformations are

$$\delta_{\hat{\varepsilon}} \hat{\omega} \;=\; \hat{D}\hat{\varepsilon} \;, \qquad \delta_{\hat{\varepsilon}} \hat{\mathcal{R}} \;=\; [\,\hat{\varepsilon}\,,\,\hat{\mathcal{R}}\,]_\star \;, \tag{13}$$

---

[a] $\hat{\omega}_\alpha$ is related to the field S_α introduced by Vasiliev in [19] by $S_\alpha = z_\alpha + 2i\hat{\omega}_\alpha$.

where $\hat{\varepsilon}$ is an $\widehat{shs}^E(8|4)$ valued parameter.

The gauging of $\widehat{shs}^E(8|4)$ differs from ordinary Yang-Mills theory in two significant ways. Firstly, $\hat{\mathcal{R}}$ is a non-local $Z^{\underline{\alpha}}$ function of $\hat{\omega}$ due to the $\star$ operation in the quadratic term $\hat{\omega}\star\hat{\omega}$. Imposing constraints on $\hat{\mathcal{R}}$, this non-locality in Z-space manifests itself as non-locality in spacetime in the form of higher derivative dependence on the fields. Secondly, viewing the connection $\hat{\omega}$ as a map from $\mathcal{M}$ to the internal space coordinatized by (Y,θ), the values assumed by $\hat{\omega}$ at a given point in $\mathcal{M}$ depend on the Z-coordinates of the point.

We define the $shs^E(8|4)$ valued gauge field ω and its curvature $\mathcal{R}$ by

$$\omega \;:=\; i^*\hat{\omega} \;=\; dx^\mu \omega_\mu(x;Y,\theta)\;, \tag{14}$$

$$\mathcal{R} \;:=\; i^*\hat{\mathcal{R}} \;=\; \tfrac{1}{2}dx^\mu\wedge dx^\nu \mathcal{R}_{\mu\nu}(x;Y,\theta)\;, \tag{15}$$

where $i: shs^E(8|4)\hookrightarrow \widehat{shs}^E(8|4)$ is the embedding map. From the definition Eq. (5) of $shs^E(8|4)$ it then follows that ω has the expansion

$$\begin{aligned}\omega \;=\; \frac{1}{2i}\sum_{k=0}^{\infty}\Bigg(&\sum_{m+n=4k}\left(\tfrac{1}{2!}\omega_{ij}(m,n)\theta^{ij}+\tfrac{1}{6!}\omega_{i_1\cdots i_6}(m,n)\theta^{i_1\cdots i_6}\right)\\ &+\sum_{m+n=4k+1}\left(\omega_i(m,n)\theta^i+\tfrac{1}{5!}\omega_{i_1\cdots i_5}(m,n)\theta^{i_1\cdots i_5}\right)\\ &+\sum_{m+n=4k+2}\left(\omega(m,n)+\tfrac{1}{4!}\omega_{i_1\cdots i_4}(m,n)\theta^{i_1\cdots i_4}+\tfrac{1}{8!}\omega_{i_1\cdots i_8}(m,n)\theta^{i_1\cdots i_8}\right)\\ &+\sum_{m+n=4k+3}\left(\tfrac{1}{3!}\omega_{ijk}(m,n)\theta^{ijk}+\tfrac{1}{7!}\omega_{i_1\cdots i_7}(m,n)\theta^{i_1\cdots i_7}\right)\Bigg)\;,\end{aligned} \tag{16}$$

where we have used the notation

$$P(m,n) := \tfrac{1}{m!n!}P_{\alpha_1\cdots\alpha_m\dot{\alpha}_1\cdots\dot{\alpha}_n}y^{\alpha_1}\cdots y^{\alpha_m}\bar{y}^{\dot{\alpha}_1}\cdots\bar{y}^{\dot{\alpha}_n}\;. \tag{17}$$

Thus bosonic gauge fields are always in the 1, 28 and $35_+ + 35_-$ representations of $SO(8)$, while the fermionic fields are always in 8 and 56 representations.

To realize $shs^E(8|4)$ invariant dynamics irreducibly on a set of fields in spacetime, the $\widehat{shs}^E(8|4)$ valued curvature $\hat{\mathcal{R}}$ must be constrained. There exists a set of constraints which achieves this while maintaining spacetime diffeomorphism invariance. The solution of these constraints involve a 0-form master field $\hat{\phi}$ subject to certain conditions. The field $\hat{\phi}$ essentially consists of the non-vanishing components of the spacetime $shs^E(8|4)$ curvatures and the spin $s = 0, \frac{1}{2}$ fields listed in Table 1. The required properties of $\hat{\phi}$ turn out to be

$$\tau\hat{\phi} = \bar{\pi}\hat{\phi} , \qquad \hat{\phi}^\dagger = \pi(\hat{\phi}) \star \Gamma , \tag{18}$$

where

$$\Gamma = \theta^1\theta^2 \cdots \theta^8 \tag{19}$$

is the $SO(8)$ chirality matrix, and π and $\bar{\pi}$ are homomorphisms of the $\star$ algebra defined by

$$\begin{aligned} \pi P(z,\bar{z};y,\bar{y};\theta) &:= P(-z,\bar{z};-y,\bar{y};\theta) , \\ \bar{\pi} P(z,\bar{z};y,\bar{y};\theta) &:= P(z,-\bar{z};y,-\bar{y};\theta) . \end{aligned} \tag{20}$$

In Eq. (18) we have introduced an $SO(3,1)$ covariant notation where y_α and z_α are complex, two-component Weyl spinors with hermitian conjugates $\bar{y}_{\dot{\alpha}} := (y_\alpha)^\dagger$ and $\bar{z}_{\dot{\alpha}} := (z_\alpha)^\dagger$ and $Y_{\underline{\alpha}} := (y_\alpha, \bar{y}^{\dot{\alpha}})$ and $Z_{\underline{\alpha}} := (z_\alpha, -\bar{z}^{\dot{\alpha}})$. This amounts to formulating the $shs^E(8|4)$ invariant theory in an $SO(3,1)$ basis.

At $Z = 0$ we find that the field

$$\phi(x;Y,\theta) := i^*\hat{\phi} \tag{21}$$

has the expansion

$$\phi = \sum_{m>n\geq 0} \left[\phi(m,n;\theta) + (-1)^n \phi(m,n;\theta)^\dagger \star \Gamma\right] + \sum_{m\geq 0} \phi(m,m;\theta) , \tag{22}$$

where we have used the notation Eq. (17) and where the complex fields $\phi(m,n;\theta)$ have the θ-expansion

$$\phi(m,n;\theta) = \begin{cases} \phi(m,n)+\frac{1}{4!}\phi_{i_1 \cdot i_4}(m,n)\theta^{i_1 \cdot i_4}+\frac{1}{8!}\phi_{i_1 \cdot i_8}(m,n)\theta^{i_1 \cdot i_8} & m-n = 0 \bmod 4 \\ \frac{1}{3!}\phi_{ijk}(m,n)\theta^{ijk} + \frac{1}{7!}\phi_{i_1\cdots i_7}(m,n)\theta^{i_1\cdots i_7} & m-n = 1 \bmod 4 \\ \frac{1}{2!}\phi_{ij}(m,n)\theta^{ij} + \frac{1}{6!}\phi_{i_1\cdots i_6}(m,n)\theta^{i_1\cdots i_6} & m-n = 2 \bmod 4 \\ \phi_i(m,n)\theta^i + \frac{1}{5!}\phi_{i_1\cdots i_5}(m,n)\theta^{i_1\cdots i_5} & m-n = 3 \bmod 4 \end{cases} \tag{23}$$

for $m > n$, and

$$\begin{aligned} \phi(m,m;\theta) &= \phi(m,m) + \tfrac{1}{4!}\phi_{ijkl}(m,m)\theta^i\theta^j\theta^k\theta^l + (-1)^m\phi(m,m)^\dagger\Gamma \ , \\ \phi^{ijkl}(m,m)^\dagger &= \tfrac{(-1)^m}{4!}\epsilon_{ijklnpqr}\phi^{npqr}(m,m) \ . \end{aligned} \tag{24}$$

for $m = n$. Hence the bosonic fields are in the 1, 28 and $35_+ + 35_-$ representations of $SO(8)$ and the fermions are in the 8 and 56 representations. In particular we recognize that the scalar fields $\phi_{ijkl}(0,0)$ are the $35_+ + 35_-$ real scalars of the the $N = 8$ supergravity multiplet [23] (see Table 1), and that these obey the $SU(8)$ invariant reality condition which is crucial for the consistency of the $E_7/SU(8)$ coset construction. We also find that $\phi(0,0)$ is the complex scalar of the level 1 supermultiplet in Table 1 and that $\phi(1,0;\theta)$ contains the fermions of level 0 and 1 in Table 1. As we shall see in Sec. 3, the remaining spin $s \geq 1$ components of ϕ are auxiliary fields that are given by the derivatives of the spin $s \leq \frac{1}{2}$ fields and certain non-vanishing spacetime curvatures that are known as Weyl tensors.

2.3 The Master Curvature Constraint

A set of curvature constraints that realize the dynamics of the higher spin gauge theory irreducibly are

$$\hat{\mathcal{R}}_{\mu\nu} = 0 \ , \qquad \hat{\mathcal{R}}_{\mu\alpha} = 0 \ , \qquad \hat{\mathcal{R}}_{\alpha\dot{\beta}} = 0 \ , \tag{25}$$

and the 'twisted' reality condition

$$\epsilon^{\dot{\alpha}\dot{\beta}}\hat{\mathcal{R}}_{\dot{\alpha}\dot{\beta}} = \epsilon^{\alpha\beta}\hat{\mathcal{R}}_{\alpha\beta} \star K \ , \tag{26}$$

where the operator K is defined by

$$K \ := \ \Gamma \exp(iY^{\underline{\alpha}} Z_{\underline{\alpha}}) \ . \tag{27}$$

The condition Eq. (26) implies that $\hat{\mathcal{R}}_{\alpha\beta}$ can be expressed in terms of the field $\hat{\phi}$ introduced in Eq. (18) as

$$\hat{\mathcal{R}}_{\alpha\beta} \ = \ -\tfrac{i}{2}\epsilon_{\alpha\beta}\hat{\phi} \star \kappa\Gamma \ , \tag{28}$$

where

$$\kappa \ := \ \exp(iy^\alpha z_\alpha) \ . \tag{29}$$

This operator is an inner Kleinian operator[19] with the basic property

$$\kappa \star P(z, \bar{z}; y, \bar{y}; \theta) \ = \ \kappa \, P(y, \bar{z}; z, \bar{y}; \theta) \ , \tag{30}$$

from which it follows that

$$\kappa \star \kappa \ = \ 1 \ , \qquad \kappa \star P \ = \ \pi(P) \star \kappa \ . \tag{31}$$

The operator $K = \kappa\bar{\kappa}\Gamma$ and from Eq. (31) it follows that

$$K \star K \ = \ 1 \ , \qquad K \star P \ = \ \tau^2(P) \star K \ . \tag{32}$$

We shall discuss the consequences of the twisted reality condition Eq. (26) shortly. First, we need to establish the integrability and $\widehat{shs}^E(8|4)$ covariance of the constraints Eq. (25) and Eq. (28). To begin with we note these constraints can be written concisely as

$$\hat{\mathcal{R}} \ = \ \tfrac{i}{4}(dz^\alpha \wedge dz_\alpha \hat{\phi} \star \kappa\Gamma + d\bar{z}^{\dot{\alpha}} \wedge d\bar{z}_{\dot{\alpha}} \hat{\phi} \star \bar{\kappa}) \ , \tag{33}$$

Inserting the constraint Eq. (33) into the Bianchi identity Eq. (12) one finds the following constraint on $\hat{\phi}$

$$\hat{D}\hat{\phi} \ := \ \hat{d}\hat{\phi} - \hat{\omega} \star \hat{\phi} + \hat{\phi} \star \bar{\pi}\hat{\omega} \ = \ 0 \ . \tag{34}$$

Since $\hat{D}^2\hat{\phi} = 0$ there are no further conditions and the full set of constraints given by Eq. (33) and Eq. (34) is integrable. The constraint Eq. (33) is invariant under the $\widehat{shs}^E(8|4)$ transformation Eq. (13) provided that

$$\delta_{\hat{\varepsilon}}\hat{\phi} = \hat{\varepsilon} \star \hat{\phi} - \hat{\phi} \star \bar{\pi}\hat{\varepsilon} \ . \tag{35}$$

These gauge transformations close, that is

$$[\delta_{\hat{\varepsilon}_1}, \delta_{\hat{\varepsilon}_2}]\hat{\phi} = -\delta_{[\hat{\varepsilon}_1, \hat{\varepsilon}_2]_\star}\hat{\phi} \ . \tag{36}$$

The gauge transformations Eq. (13) and Eq. (35) also preserve the constraint Eq. (34), since

$$\delta_{\hat{\varepsilon}}\hat{D}\hat{\phi} = \hat{\varepsilon} \star \hat{D}\hat{\phi} - \hat{D}\hat{\phi} \star \bar{\pi}\hat{\varepsilon} \ . \tag{37}$$

In summary, the full set of integrable constraints which define the $N = 8$ higher spin gauge theory based on $shs^E(8|4)$ is:

$$\hat{D}_\alpha\hat{\phi} = 0 \ , \tag{38}$$

$$\hat{\mathcal{R}}_{\alpha\dot{\beta}} = 0 \ , \tag{39}$$

$$\hat{\mathcal{R}}_{\alpha\beta} = -\tfrac{i}{2}\epsilon_{\alpha\beta}\hat{\phi} \star \kappa\Gamma \ , \tag{40}$$

$$\hat{\mathcal{R}}_{\alpha\mu} = 0 \ , \tag{41}$$

$$\hat{\mathcal{R}}_{\mu\nu} = 0 \ , \tag{42}$$

$$\hat{D}_\mu\hat{\phi} = 0 \ , \tag{43}$$

We will discuss uniqueness in Sec. (5). While these equations fully capture the higher spin theory based on $shs^E(8|4)$, they contain many auxiliary fields. Their elimination and the derivation of the full equations of motion in terms of a set of dynamical fields is a highly non-trivial matter which has not been solved yet. The structure of the equations become clear, however, by linearizing them around a vacuum solution. One then finds that Eq. (38) determines the Z dependence of $\hat{\phi}$ in terms of the initial condition ϕ. Then Eqs. (39)-(40) determine the Z dependence of $\hat{\omega}_\alpha$ in terms of ϕ up to a pure gauge solution $\hat{D}_\alpha\hat{\xi}$ which can be used to impose the gauge

$$\hat{\omega}_\alpha|_{Z=0} \;=\; 0\;. \tag{44}$$

This amounts to that there are no spacetime degrees of freedom associated with the initial conditions for $\hat{\omega}_\alpha$. From Eq. (41) one then solves for the Z dependence of $\hat{\omega}_\mu$ in terms of ϕ and the initial condition ω. One is then left with Eqs. (42)-(43), which one can show are equivalent to

$$\begin{aligned} \mathcal{R} \;&:=\; i^*\hat{\mathcal{R}} \;=\; \tfrac{1}{2}dx^\mu\wedge dx^\nu\mathcal{R}_{\mu\nu}(x;Y,\theta) \;=\; 0\;,\\ D\phi \;&:=\; i^*\hat{D}\hat{\phi} \;=\; dx^\mu D_\mu\phi(x;Y,\theta) \;=\; 0\;, \end{aligned} \tag{45}$$

where $\mathcal{R}$ is the curvature two-form defined in Eq. (15). Observe that Eq. (45) is the pull-back by i^* of the full set of constraints Eq. (33) and Eq. (34) to spacetime. If the Z space had been an ordinary commuting space, then this pull-back operation would have implied the vanishing of the spacetime curvatures defined as

$$\begin{aligned} R \;&:=\; d\omega-\omega\star\omega\;,\\ d_\omega\phi \;&:=\; d\phi-\omega\star\phi+\phi\star\bar{\pi}\omega\;. \end{aligned} \tag{46}$$

The non-locality of the Z space, however, implies that $\mathcal{R}$ and $D\phi$ depend on all the coefficients of the Taylor expansions of $\hat{\omega}_\mu$ and $\hat{\phi}$ around $Z=0$, which in turn are determined in terms of ω and ϕ as explained above. Hence $\mathcal{R}$ and $D\phi$ are infinite expansions of the form $\mathcal{R}=R+\ldots$ and $D\phi=d_\omega\phi+\ldots$ in powers of the 0-form ϕ. Therefore Eq. (45) gives a set of non-trivial constraints on the spacetime curvatures R and $d_\omega\phi$.

The constraints Eq. (45) are invariant under the residual, gauge transformations [b] of Eq. (44), which are

$$\delta_\varepsilon\omega \;=\; i^*\delta_{i_*\varepsilon}\hat{\omega} \;=\; d\varepsilon - i^*[\hat{\omega}\,,\,i_*\varepsilon]_\star\;, \tag{47}$$

$$\delta_\varepsilon\phi \;:=\; i^*\delta_{i_*\varepsilon}\hat{\phi} \;=\; i^*(i_*\varepsilon\star\hat{\phi}-\hat{\phi}\star\bar{\pi}i_*\varepsilon)\;, \tag{48}$$

---

[b]Eq. (44) implies that $\hat{\omega}_\alpha \equiv z^\beta\hat{\omega}_{\alpha,\beta}(x,Z;Y,\theta)+\bar{z}^{\dot{\beta}}\hat{\omega}_{\alpha,\dot{\beta}}(x,Z;Y,\theta)$, so a residual gauge transformation parameter $\hat{\varepsilon}(x,Z;Y,\theta)$ obeys $\delta_{\hat{\varepsilon}}\hat{\omega}_\alpha|_{Z=0}=(\partial_\alpha+2i\omega_\alpha{}^\beta\partial_\beta+\cdots)\hat{\varepsilon}|_{Z=0}=0$. The $shs^E(8|4)$ valued and therefore Z-independent parameter $\hat{\varepsilon}=i_*\epsilon$ clearly obeys this condition.

where $\varepsilon(x;Y,\theta)$ is an $shs^E(8|4)$ valued parameter. From the closure of the $\delta_{\hat\varepsilon}$ transformations on $\hat\omega$ and $\hat\phi$, and using the property $[i_*\varepsilon_1\,,\,i_*\varepsilon_2]_\star = i_*[\varepsilon_1\,,\,\varepsilon_2]_\star$, it follows that the residual gauge transformations Eqs. (47)-(48) close.

Let us now return to the twisted reality condition Eq. (26). To illustrate the role of the operator K in Eq. (26), let us suppose that we have a solution to Eqs. (38)-(43) such that $\hat\phi^{(0)} = 0$. We can then set $\hat\omega_\alpha^{(0)} = 0$ by means of an $\widehat{shs}^E(8|4)$ gauge transformation. Then Eq. (41) implies that $\partial_\alpha\hat\omega^{(0)} = 0$ which is solved by $\hat\omega_\mu^{(0)} = \Omega_\mu^{(0)}(Y,\theta)$. Next, we observe from Eq. (38) that a small fluctuation field $\phi^{(1)}$ obeys $\partial_\alpha\hat\phi^{(1)} = 0$, which is solved by $\hat\phi^{(1)} = \phi(Y,\theta)$. Then Eq. (43) reads

$$d\phi - \Omega^{(0)} \star \phi + \phi \star \bar\pi\Omega^{(0)} \;=\; 0\,. \tag{49}$$

Since $\Omega^{(0)}$ is flat one might expect that this equation only allows rigid solutions, but a more careful analysis (see Sec. (3.2) in the case when $\Omega^{(0)}$ is the AdS vacuum) shows that these equations do not constrain the chiral components $\phi(m,0;\theta)$. Moreover, it follows from

$$\phi(y,\bar y;\theta)\star\bar\kappa|_{Z=0} \;=\; \phi(y,0;\theta) \tag{50}$$

that only the chiral components of ϕ contribute to the linearization of $\mathcal{R}$. Eq. (45) thus expresses the linearized spacetime curvatures $R^{(1)}$ in terms of the chiral components of ϕ (see Eqs. (56)-(58) in the case when $\Omega^{(0)}$ is the AdS vacuum)

The crucial role of the K-twisted reality condition Eq. (26) can now be understood as follows. If we impose an ordinary reality condition without the K-twist, namely if we impose the condition $\epsilon^{\dot\alpha\dot\beta}\hat{\mathcal{R}}_{\dot\alpha\dot\beta} = \epsilon^{\alpha\beta}\hat{\mathcal{R}}_{\alpha\beta}$, then $\hat{\mathcal{R}}_{\alpha\beta} = -\frac{i}{2}\epsilon_{\alpha\beta}\hat\varphi$ where $\hat\varphi^\dagger = \hat\varphi$ and the constraint Eq. (49) now becomes

$$d\varphi - \Omega^{(0)}\star\varphi + \varphi\star\Omega^{(0)} \;=\; 0\,, \tag{51}$$

where $\varphi = i^*\hat\varphi$. This constraint does not allow any local dynamics. This can be seen by observing that Eq. (51) is identical in form to the higher spin Killing equation for rigid gauge transformations that preserve the connection $\Omega^{(0)}$ (the only difference between φ and ϵ, which is insignificant in the present

$\ell \backslash s$	0	$\frac{1}{2}$	1	$\frac{3}{2}$	2	$\frac{5}{2}$	3	$\frac{7}{2}$	4	$\frac{9}{2}$	5	$\frac{11}{2}$	6	...
0	$35_+ + 35_-$	56	28	8	1									
1	$1+1$	8	28	56	$35_+ + 35_-$	56	28	8	1					
2					1	8	28	56	$35_+ + 35_-$	56	28	8	1	
3									1	8	28	56	$35_+ + 35_-$	...
4													1	...
⋮														

Table 1. The $SO(3,2) \times SO(8)$ content of the symmetric tensor product of two $N = 8$ supersingletons. Each entry refers to the $SO(8)$ content and the energy $E_0 = s + 1$, where s is the spin. The representations have been arranged into a tower of $OSp(8|4)$ supermultiplets labeled by a level number ℓ explained in Sec. (2.1). The vertical bar separates the spin $s \geq 1$ physical modes of the $shs^E(8|4)$ gauge field ω from the physical modes of the Weyl 0-form ϕ introduced in Sec. (2.2)

context, is that they have opposite τ-parity)[c].

We conclude this section by noting that the constraints Eq. (33) and Eq. (34) are invariant under the spacetime diffeomorphisms which are incorporated into the $\widehat{shs}^E(8|4)$ gauge group such that $\delta x^\mu = \xi^\mu$ is generated by the gauge transformation with parameter $\hat{\varepsilon}(\xi) = i_\xi \hat{\omega}$. The Z space diffeomorphisms, on the other hand, are broken down to local $SO(3,1)$ rotations by the reality condition Eq. (26). To preserve local $SO(3,2)$ invariance one would have to impose an untwisted reality condition which would not yield dynamics as argued under Eq. (51). The untwisted reality condition that leads to Eq. (51) is equivalent to the vanishing of $(\gamma^{\underline{a}})^{\underline{\alpha\beta}} \hat{\mathcal{R}}_{\underline{\alpha\beta}}$ where $(\gamma^{\underline{a}})_{\underline{\alpha\beta}}$ ($\underline{a} = 0,1,2,3,5$) are the $SO(3,2)$ γ-matrices.

[c]The flatness of $\Omega^{(0)}(Y,\theta)$ implies that $\Omega^{(0)} = g^{-1} \star dg$ where g is an $shs^E(8|4)$ group element. Eq. (51) is then solved formally by taking $\varphi = g^{-1} \star \tilde{P} \star g$ for an $shs^E(8|4)$ valued polynomial $\tilde{P}$ with constant coefficients. Applying this formal method to Eq. (49) gives $\phi = g^{-1} \star P \star \bar{\pi} g$ for a constant P. These results naively would imply that both Eq. (49) and Eq. (51) have only rigid solutions, whereas a more careful analysis shows that there is room for unconstrained chiral components in the solutions to Eq. (49) while Eq. (51) indeed only allows the rigid solutions.

3 The Anti-de-Sitter Vacuum and Spectrum

3.1 Expansion Around AdS Vacuum

The AdS vacuum solution is given by

$$\begin{aligned}\hat\phi^{(0)} &= 0\,,\\ \hat\omega_\alpha^{(0)} &= 0\,,\\ \hat\omega_\mu^{(0)} &= \Omega_\mu^{(0)} \;:=\; \tfrac{1}{4i}\left[\omega^{(0)}_{\mu\alpha\beta}y^\alpha y^\beta + \bar\omega^{(0)}_{\mu\dot\alpha\dot\beta}\bar y^{\dot\alpha}\bar y^{\dot\beta} + 2e^{(0)}_{\mu\alpha\dot\alpha}y^\alpha\bar y^{\dot\alpha}\right]\,, \end{aligned} \tag{52}$$

where $\omega^{(0)} = \frac{1}{4i}(\omega^{(0)}_{\alpha\beta}y^\alpha y^\beta + h.c)$ is the $SO(3,1)$ valued Lorentz connection of AdS spacetime of radius λ^{-1} and and with vierbein $e^{(0)\,a} = \lambda^{-1}(\bar\sigma^a)^{\dot\alpha\alpha}e^{(0)}_{\alpha\dot\alpha}$. The $SO(3,2)$ curvature vanishes, that is $d\Omega^{(0)} - \Omega^{(0)} \star \Omega^{(0} = 0$. In $SO(3,1)$ covariant component notation this means

$$\begin{aligned} d\omega^{(0)}_{\alpha\beta} &= \omega^{(0)}_{\alpha\gamma}\wedge\omega^{(0)\gamma}_{\beta} + e^{(0)}_{\alpha\dot\delta}\wedge e^{(0)\dot\delta}_{\beta}\,,\\ d\bar\omega^{(0)}_{\dot\alpha\dot\beta} &= \bar\omega^{(0)}_{\dot\alpha\dot\gamma}\wedge\bar\omega^{(0)\dot\gamma}_{\dot\beta} + e^{(0)}_{\delta\dot\alpha}\wedge e^{(0)\,\delta}{}_{\dot\beta}\,,\\ de^{(0)}_{\alpha\dot\beta} &= \omega^{(0)}_{\alpha\gamma}\wedge e^{(0)\,\gamma}{}_{\dot\beta} + \bar\omega^{(0)}_{\dot\beta\dot\delta}\wedge e^{(0)\dot\delta}_{\alpha}\,. \end{aligned} \tag{53}$$

We shall use a notation where

$$\nabla \;:=\; e^{(0)\,a}\nabla_a \tag{54}$$

is the $SO(3,1)$ covariant derivative with connection $\omega^{(0)}$. Its curvature, i.e. the Riemann tensor, is given by $r^{(0)}_{ab,cd} = -\lambda^2(\eta_{ac}\eta_{bd} - \eta_{ad}\eta_{bc})$. The AdS vacuum is invariant under rigid gauge transformations [d] with $shs^E(8|4)$ valued parameters $\epsilon^{(0)}$ which can be expressed in terms of the usual commuting $SO(3,2)$ Killing spinors η^r_α $(r=1,..,4)$ obeying

$$\nabla\eta^r_\alpha \;=\; e^{(0)}_{\alpha\dot\alpha}\bar\eta^{r\,\dot\alpha}\,. \tag{55}$$

---

[d] The Killing equation reads $d\varepsilon^{(0} - [\Omega^{(0)}\,,\,\varepsilon^{(0)}\,]_\star = 0$ and in the AdS case it consists of decoupled systems of equations for $\varepsilon^{(0)}(m,n;\theta)$ for each value of $m+n = 0,1,2,...$, and these equations admit only rigid solutions.

Defining $\eta^r := \eta^r_\alpha y^\alpha - \eta^r_{\dot\alpha}\bar y^{\dot\alpha}$, the rigid parameter $\varepsilon^{(0)}$ is then an arbitrary $shs^E(8|4)$ valued polynomial in η^r, $\bar y^r$ and θ^i with constant coefficients.

Solving for the Z dependence $\hat\phi$ and $\hat\omega$ from Eqs. (38)-(41), as explained in Sec. (2.2), in a perturbation expansion around the AdS vacuum solution Eq. (52) and then substituting these solutions into Eq. (45) one obtains its linearized form, which in $SO(3,1)$ covariant notation is given by [e]

$$R^{(1)}_{\alpha\beta,\gamma_1\cdots\gamma_{2s-2}}(\theta) = \phi_{\alpha\beta\gamma_1\cdots\gamma_{2s-2}}(\theta)\,,\ \ s = 1, \tfrac{3}{2}, 2, \dots, \tag{56}$$

$$R^{(1)}_{\dot\alpha\dot\beta,\dot\gamma_1\cdots\dot\gamma_{2s-2}}(\theta) = \phi_{\dot\alpha\dot\beta\dot\gamma_1\cdots\dot\gamma_{2s-2}}(\theta)\star\Gamma\,,\ \ s = 1, \tfrac{3}{2}, 2, \dots, \tag{57}$$

$$R^{(1)}_{\alpha\dot\beta,\gamma_1\cdots\gamma_k\dot\gamma_{k+1}\cdots\dot\gamma_{2s-2}}(\theta) = 0\,,\ \ s = \tfrac{3}{2}, 2, \tfrac{5}{2}, \dots,\ \ k = 0, \dots, 2s-3 \tag{58}$$

$$\nabla_{\alpha\dot\alpha}\phi_{\beta_1\cdots\beta_m\dot\beta_1\cdots\dot\beta_n}(\theta) = i\phi_{\alpha\beta_1\cdots\beta_m\dot\alpha\dot\beta_1\cdots\dot\beta_n}(\theta) - imn\epsilon_{\alpha\beta_1}\epsilon_{\dot\alpha\dot\beta_1}\phi_{\beta_2\cdots\beta_m\dot\beta_2\cdots\dot\beta_n}(\theta)$$
$$m, n = 0, 1, 2, \dots \tag{59}$$

where $R^{(1)} = d\omega - \{\Omega^{(0)}, \omega\}_\star$ is the linearization of the spacetime curvature R given in Eq. (46) and it reads

$$\begin{aligned} R^{(1)}_{\alpha_1\alpha_2,\beta_1\cdots\beta_m\dot\beta_1\cdots\dot\beta_n}(\theta) \ &:= \ 2\nabla\omega_{\alpha_1\alpha_2,\beta_1\cdots\beta_m\dot\beta_1\cdots\dot\beta_n}(\theta) \\ -m\,\epsilon_{\alpha_1\beta_1}\,\omega_{\alpha_2}{}^{\dot\gamma}{}_{,\beta_2\cdots\beta_m\dot\gamma\dot\beta_1\cdots\dot\beta_n}(\theta) &- n\,\omega_{\alpha_1\dot\beta_1,\alpha_2\beta_1\cdots\beta_m\dot\beta_2\cdots\dot\beta_n}(\theta)\,, \end{aligned} \tag{60}$$

which obeys the linearized Bianchi identity $dR^{(1)} - [\Omega^{(0)}, \omega]_\star = 0$, that is

$$\begin{aligned} &\nabla_{\dot\alpha}{}^{\gamma}R^{(1)}_{\gamma\alpha,\beta_1\cdots\beta_m\dot\beta_1\cdots\dot\beta_n} - \nabla_{\alpha}{}^{\dot\gamma}R^{(1)}_{\dot\gamma\dot\alpha,\beta_1\cdots\beta_m\dot\beta_1\cdots\dot\beta_n} \\ &-m\left[R^{(1)}_{\alpha\beta_1,\beta_2\cdots\beta_m\dot\alpha\dot\beta_1\cdots\dot\beta_n} - \epsilon_{\alpha\beta_1}R^{(1)\dot\gamma}_{\dot\alpha}{}_{,\beta_2\cdots\beta_m\dot\gamma\dot\beta_1\cdots\dot\beta_n}\right] \\ &+n\left[R^{(1)}_{\dot\alpha\dot\beta_1,\alpha\beta_1\cdots\beta_m\dot\beta_2\cdots\dot\beta_n} - \epsilon_{\dot\alpha\dot\beta_1}R^{(1)\gamma}_{\alpha}{}_{,\gamma\beta_1\cdots\beta_m\dot\beta_2\cdots\dot\beta_n}\right] \ = \ 0\,. \end{aligned} \tag{61}$$

---

[e] We use the following symmetrization convention: all $SO(3,1)$ spinor indices indices of the same type, such as $\alpha_1, \alpha_2, ...\alpha_m$, are to be understood to be symmetrized with unit strength.

We have converted the curved spacetime indices of the forms into flat indices using the AdS vierbein $e_\mu^{(0)\ a}$ and set

$$\begin{aligned} \omega_{\alpha\dot\beta} &:= (\sigma^a)_{\alpha\dot\beta}\,\omega_a\ , \qquad & R^{(1)}_{\alpha\beta} &:= \tfrac{1}{2}(\sigma^{ab})_{\alpha\beta}R^{(1)}_{ab}\ , \\ \nabla_{\alpha\dot\alpha} &:= (\sigma^a)_{\alpha\dot\alpha}\nabla_a\ , \qquad & \nabla\omega_{\alpha\beta} &:= \tfrac{1}{2}(\sigma^{ab})_{\alpha\beta}\,\nabla_a\omega_b\ . \end{aligned} \tag{62}$$

The linearized field equations Eqs. (56)-(61) are invariant under the linearization of the $shs^E(8|4)$ valued gauge transformations Eqs. (47)-(48), which assume the following form in $SO(3,1)$ covariant component language:

$$\begin{aligned} \delta\omega_{\alpha\dot\alpha,\beta_1\cdots\beta_m\dot\beta_1\cdots\dot\beta_n} &= \nabla_{\alpha\dot\alpha}\varepsilon_{\beta_1\cdots\beta_m\dot\beta_1\cdots\dot\beta_n} + m\,\epsilon_{\alpha\beta_1}\varepsilon_{\alpha\beta_2\cdots\beta_m\dot\alpha\dot\beta_1\cdots\dot\beta_n} \\ &\quad + n\,\epsilon_{\dot\alpha\dot\beta_1}\varepsilon_{\beta_1\cdots\beta_m\dot\beta_2\cdots\dot\beta_n}\ , \end{aligned} \tag{63}$$

$$\delta\phi = 0\ . \tag{64}$$

3.2 *The Spin $s \le 1$ Linearized Field Equations*

We begin by analyzing Eq. (59) from which we find that the non-chiral spin $s+l$ component $\phi(2s+l,l;\theta)$ $(l = 1,2,...)$ is expressed in terms of its chiral component $\phi(2s,0;\theta)$ as follows

$$\phi_{\alpha_1\cdots\alpha_{2s+l}\dot\alpha_1\cdots\dot\alpha_l}(\theta) = \nabla_{\alpha_1\dot\alpha_1}\nabla_{\alpha_2\dot\alpha_2}\cdots\nabla_{\alpha_l\dot\alpha_l}\phi_{\alpha_1\cdots\alpha_{2s}}(\theta)\ . \tag{65}$$

The chiral components of ϕ are identified with the chiral curvature components as in Eq. (56). These chiral curvature components are the Weyl tensors and they are the only spacetime curvature components that are non-vanishing on-shell, as follows from Eqs. (56)-(57). Inserting the expansions Eq. (16) and Eqs. (23)-(24) into Eq. (56) and Eq. (65) one verifies that the $SO(8)$ content of both sides of these equations indeed match. Thus, the spin $s \ge 1$ components of the master field ϕ are auxiliary in the sense that they can be solved for algebraically in terms of the gauge fields and their derivatives. The spin $s \le \frac{1}{2}$ components, on the other hand, are independent since they cannot be eliminated algebraically.

In obtaining Eq. (65) one uses only the fully symmetrized components

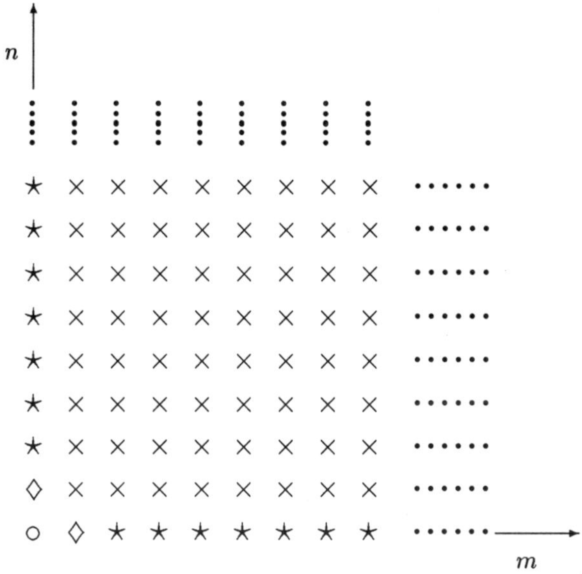

Figure 1. Each entry of the integer grid, $m, n = 0, 1, 2, \ldots$, represents a component $\phi(m,n;\theta)$ of ϕ : the O denotes the scalars, the two $\diamondsuit$'s denote the spin $\frac{1}{2}$ fields; a $\star$ denotes a chiral, auxiliary component $\phi(2s, 0, \theta)$ ($s = 1, \frac{3}{2}, \ldots$) which is equal to the spin s Weyl tensor given in Eq. (56); the $\times$'s denote auxiliary components which are given in terms of derivatives of the chiral components according to Eq. (65).

of Eq. (59). The remaining components are ϵ-traces. By combining Eqs. (56)-(57) with the Bianchi identity Eq. (61) for $s \geq \frac{3}{2}$ one can show by induction that the ϵ-traces of Eq. (59) for $m + n \geq 3$ are *identically* satisfied provided that its ϵ-traces for $m + n \leq 2$ are satisfied. These constitute the linearized equations of motion for the spin $s \leq 1$ fields and the Bianchi identity for the spin 1 fields, which we summarize as follows:

a) The complex scalar $\phi(0,0)$ and the $35_+ + 35_-$ scalars $\phi^{ijkl}(0,0)$ are dynamical fields which constitute $\phi(\theta) \equiv \phi(0,0;\theta)$ (see Eq. (24)) which obey the linearized field equation $(\nabla_\alpha{}^{\dot\alpha}\nabla_{\beta\dot\alpha} + 2\epsilon_{\alpha\beta})\phi(\theta) = 0$, that is

$$(\nabla^2 + 2)\phi(\theta) \;=\; 0\,. \tag{66}$$

b) The complex Weyl fermions ϕ_α^{ijk} and $\phi_\alpha^{i_1\cdots i_7}$ are dynamical fermionic fields which make up $\phi_\alpha(\theta)$ which obey the linearized field equation $\nabla_{\dot\alpha}{}^\alpha\phi_\alpha(\theta)=0$, that is

$$(\bar\sigma^a)^{\dot\alpha\alpha}\nabla_a\phi_\alpha(\theta) \;=\; 0\,. \qquad (67)$$

c) The spin 1 field strength $R^{(1)}_{\alpha\beta}(\theta)=\phi_{\alpha\beta}(\theta)$, obeys $\nabla_{\dot\alpha}{}^\beta R^{(1)}_{\beta\gamma}(\theta)=0$, which amounts to

$$\nabla^a R^{(1)}_{ab}(\theta) \;=\; 0\,, \qquad \epsilon^{abcd}\partial_b R^{(1)}_{cd}(\theta) \;=\; 0\,. \qquad (68)$$

It follows that $R^{(1)}(\theta)=d\omega(\theta)$, and in the Lorentz gauge $\nabla^a\omega_a(\theta)=0$ the resulting second order vector equation reads

$$(\nabla^2+3)\,\omega(\theta) \;=\; 0\,. \qquad (69)$$

Hence, taking into account the rank of the Dirac operator in Eq. (67) and the decoupling of the longitudinal mode in the spin 1 sector, one finds that there are two real physical component functions for each spin and $SO(8)$ label. These have mode function expansions with modes making up the massless representations of the AdS algebra, that is the representations $D(E_0=s+1,s)$ for $s=\frac{1}{2},1$ and $D(1,0)$ or $D(2,0)$ for $s=0$. We use a notation where $D(E_0,s)$ denotes a representation of $SO(3,2)$ in which E_0 labels the minimum energy eigenvalue of the energy operator M_{05} and s is the maximum eigenvalue of the spin operator M_{12} in the lowest energy sector.

3.3 Auxiliary and Dynamical Gauge Fields and Symmetries

We next turn to solving the vanishing curvatures in Eqs. (56)-(58) modulo the relationships among them due to the Bianchi identity Eq. (61). The end result, which is illustrated in Fig. (2), is that the *auxiliary gauge fields* $\omega(m,n;\theta)$ for $|m-n|\geq 2$ are given in terms of the *dynamical gauge fields* $\omega(m,n;\theta)$ for $|m-n|\leq 1$ as follows:

$$\omega_{\alpha\dot\alpha,\beta_1\cdots\beta_m\dot\beta_1\cdots\dot\beta_n}(\theta) \;=\; \tfrac{2}{m+1}\nabla\omega_{\dot\beta_1\dot\beta_2,\alpha\beta_1\cdots\beta_m\dot\alpha\dot\beta_3\cdots\dot\beta_n}(\theta)$$

$$+\epsilon_{\dot\alpha\dot\beta_1}\frac{2n}{n+1}\left[\frac{n-1}{m+n+2}\nabla\omega_{\dot\beta_2}{}^{\dot\gamma}{}_{,\alpha\beta_1\cdots\beta_m\dot\gamma\dot\beta_3\cdots\dot\beta_n}(\theta)\right.$$

$$+\frac{n+1}{m+n+2}\nabla\omega_{(\alpha}{}^{\gamma}{}_{,\beta_1\cdots\beta_m)\gamma\dot\beta_2\cdots\dot\beta_n}(\theta)$$

$$-\frac{m}{(m+1)(m+2)}\epsilon_{\alpha\beta_1}\nabla\omega^{\gamma\delta}{}_{,\gamma\delta\beta_2\cdots\beta_m\dot\beta_2\cdots\dot\beta_n}(\theta)\Big]$$

$$+m\,\epsilon_{\alpha\beta_1}\xi_{\beta_2\cdots\beta_m\dot\alpha\dot\beta_1\cdots\dot\beta_n}(\theta)\ , \qquad n-2\geq m\geq 0\ . \tag{70}$$

The components $\xi(m-1,n+1,\theta)$ $(m>0)$ are not determined by the constraint. Thus the auxiliary gauge fields $\omega(m,n;\theta)$ are given by $|m-n|-a$ derivatives of the dynamical gauge fields where $a=1$ for bosonic fields and $a=2$ for fermionic fields.

We refer to the dynamical gauge fields $\omega(s-1,s-1;\theta)$ $(s=2,3,...)$ as the generalized vierbeins and the dynamical gauge fields $\omega(s-\frac{3}{2},s-\frac{1}{2};\theta)$ $(s=\frac{3}{2},\frac{5}{2},...)$ as the generalized gravitini. The auxiliary gauge fields $\omega(s,s-2;\theta)$ $(s=2,3,...)$ are referred to as the generalized Lorentz connections.

The local gauge symmetries Eq. (63) are divided into auxiliary gauge symmetries with parameters $\varepsilon(m,n;\theta)$ for $|m-n|\geq 3$ and dynamical gauge symmetries with parameters $\varepsilon(m,n;\theta)$ for $|m-n|\leq 2$. The dynamical gauge symmetries in the bosonic spin s sector are the generalized reparametrizations with parameters $\varepsilon(s-1,s-1;\theta)$ and the generalized local Lorentz transformations with parameters $\varepsilon(s-2,s;\theta)$, and in the fermionic spin s sector they are the generalized local supersymmetries with parameters $\varepsilon(s-\frac{3}{2},s-\frac{1}{2};\theta)$ and the fermionic, Lorentz-like transformation with parameters $\varepsilon(s-\frac{5}{2},s+\frac{1}{2};\theta)$. From Eq. (63) it follows that

$$\delta_{\varepsilon_{aux}}\omega_{dyn} \;=\; 0\ , \tag{71}$$

where ε_{aux} is a parameter of an auxiliary gauge transformation and $\omega_{dyn}:=\sum_{|m-n|\leq 1}\omega(m,n;\theta)$. The auxiliary gauge transformation with parameter $\varepsilon(m-1,n+1;\theta)$ induces a shift in the undetermined component $\xi(m-1,n+1,\theta)$ in the auxiliary gauge field $\omega(m,n;\theta)$ given by Eq. (70). This Stueckelberg type gauge symmetry can be completely fixed by imposing the gauge condition

$$\xi(m,n;\theta) \;=\; 0\ . \tag{72}$$

The definition Eq. (71) also implies that the dynamical gauge symmetries form a subalgebra of $shs^E(8|4)$ which closes on the dynamical fields, since if $[\varepsilon^1_{dyn},\varepsilon^2_{dyn}]_\star=\varepsilon_{dyn}+\varepsilon_{aux}$ then from Eq. (71)

$$[\delta_{\varepsilon^1_{dyn}}, \delta_{\varepsilon^2_{dyn}}]\omega_{dyn} = \left(\delta_{\varepsilon_{dyn}} + \delta_{\varepsilon_{aux}}\right)\omega_{dyn} = \delta_{\varepsilon_{dyn}}\omega_{dyn} . \tag{73}$$

3.4 The Spin $s > 1$ Linearized Field Equations

The components of the curvature constraints Eqs. (56)-(58) that are not used up in solving for the auxiliary gauge fields are the field equations for the spin $s > 1$ dynamical gauge fields. They read

$$\begin{aligned} e^{(0)\dot\beta}_{\alpha_1} \wedge R^{(1)}_{\alpha_2\cdots\alpha_m\dot\beta\dot\alpha_1\cdots\dot\alpha_n}(\theta) &= 0 , \\ m = n \qquad &\text{for spin } s = m+1 = 2,3,\ldots \\ m = n+1 \ &\text{for spin } s = m + \tfrac{1}{2} = \tfrac{3}{2}, \tfrac{5}{2}, \ldots \end{aligned} \tag{74}$$

In the bosonic sector these are first order equations for the generalized Lorentz connections, which by virtue of Eq. (70) turn into second order field equations for the generalized vierbeins. In the fermionic sector Eq. (74) are first order Dirac like field equations for the generalized gravitini.

By making use of the dynamical gauge symmetries one can fix the generalized Lorentz type gauges

$$\begin{aligned} \nabla^{\alpha\dot\alpha}\omega_{\alpha\dot\alpha,\beta_1\cdots\beta_{s-1}\dot\beta_1\cdots\dot\beta_{s-1}}(\theta) &= 0 , \\ \omega_{\dot\alpha}{}^{\beta}{}_{,\beta\beta_1\cdots\beta_{s-2}\dot\beta_1\cdots\dot\beta_{s-1}}(\theta) &= 0 , \qquad s = 2,3,\ldots \end{aligned} \tag{75}$$

in the bosonic sector, and

$$\begin{aligned} \nabla^{\alpha\dot\alpha}\omega_{\alpha\dot\alpha,\beta_1\cdots\beta_{s-3/2}\dot\beta_1\cdots\dot\beta_{s-1/2}}(\theta) &= 0 , \qquad s = \tfrac{3}{2}, \tfrac{5}{2}, \ldots \\ \omega_{\alpha}{}^{\dot\alpha}{}_{,\beta_1\cdots\beta_{s-3/2}\dot\alpha\dot\beta_1\cdots\dot\beta_{s-3/2}}(\theta) &= 0 , \qquad s = \tfrac{3}{2}, \tfrac{5}{2}, \ldots \\ \omega^{\alpha}{}_{\dot\beta_1,\alpha\beta_1\cdots\beta_{s-5/2}\dot\beta_2\cdots\dot\beta_{s+1/2}}(\theta) &= 0 , \qquad s = \tfrac{5}{2}, \tfrac{7}{2}, \ldots \end{aligned} \tag{76}$$

in the fermionic sector. These conditions eliminate all $SO(3,1)$ irreducible components in the generalized vierbeins and gravitini except the spin s components $\omega_{\alpha_1\dot\alpha_1,\alpha_2\cdots\alpha_s\dot\alpha_2\cdots\dot\alpha_s}(\theta)$ and $\omega_{\alpha_1\dot\alpha_1,\alpha_2\cdots\alpha_{s-1/2}\dot\alpha_2\cdots\dot\alpha_{s+1/2}}(\theta)$, respectively, for which one obtains the field equations

$$[\nabla^2 + 3 - (s-1)^2]\,\omega_{\alpha_1\dot\alpha_1,\alpha_2\cdots\alpha_s\dot\alpha_2\cdots\dot\alpha_s}(\theta) \;=\; 0\,, \tag{77}$$

for spin $s = 2, 3, \ldots$, and

$$\nabla_\alpha{}^{\dot\gamma}\omega_{\beta_1\dot\gamma,\beta_2\cdots\beta_{s-1/2}\dot\beta_1\cdots\dot\beta_{s-1/2}}(\theta) - (s-\tfrac{1}{2})\omega_{\alpha\dot\beta_1,\beta_1\cdots\beta_{s-1/2}\dot\beta_2\cdots\dot\beta_{s-1/2}}(\theta) \;=\; 0\,, \tag{78}$$

for spin $s = \frac{3}{2}, \frac{5}{2}, \ldots$ These equations describe massless representations $D(E_0 = s+1, s)$ of the AdS algebra, as explained at the end of Sec. (3.2).

Any one of these representations are characterized by two real transverse functions. To see this we first note that in the bosonic sector the gauge condition Eq. (75) has residual gauge transformations given by generalized reparametrizations obeying

$$[\nabla^2 + 1 - s^2]\varepsilon_{\beta_1\cdots\beta_{s-1}\dot\beta_1\cdots\dot\beta_{s-1}}(\theta) \;=\; 0\,, \qquad \nabla^{\alpha\dot\alpha}\varepsilon_{\alpha\beta_1\cdots\beta_{s-2}\dot\alpha\dot\beta_1\cdots\dot\beta_{s-2}}(\theta) \;=\; 0\,. \tag{79}$$

These gauge transformations obey the gauge fixed equations of motion Eq. (77) and therefore decouple the longitudinal states from the physical spectrum. Hence the number of real transverse bosonic spin s functions is given by the number of components of the spin s irrep ($(s+1)^2$), minus the number of gauge conditions in Eq. (75) that are linear in derivatives (s^2), minus the number of residual gauge symmetries, which is equal to the number of degrees of freedom in $\varepsilon(s-1, s-1)$ (s^2) minus the number of constraints in Eq. (79) that are linear in derivatives ($(s-1)^2$), which indeed sum up to $(s+1)^2 - s^2 - [s^2 - (s-1)^2] = 2$. Similarly, in the fermionic sector one has residual generalized local supersymmetries obeying

$$\nabla_\alpha{}^{\dot\gamma}\varepsilon_{\beta_1\cdots\beta_{s-3/2}\dot\gamma\dot\beta_1\cdots\dot\beta_{s-3/2}}(\theta) - (s+\tfrac{1}{2})\,\varepsilon_{\alpha\beta_1\cdots\beta_{s-3/2}\dot\beta_1\cdots\dot\beta_{s-3/2}}(\theta) \;=\; 0\,. \tag{80}$$

Hence, taking into the account the fact that the Dirac operator has half the maximal rank, one finds that for any occurring $SO(8)$ representation there are indeed $(s+\frac{1}{2})(s+\frac{3}{2}) - (s-\frac{1}{2})(s+\frac{1}{2}) - [(s-\frac{1}{2})(s+\frac{1}{2}) - (s-\frac{3}{2})(s-\frac{1}{2})] = 2$ real transverse fermionic spin s functions.
Using rigid supersymmetries with parameter $\varepsilon^{(0)} = \frac{1}{2i}\varepsilon_{\alpha i} y^\alpha \theta^i + h.c$ constructed from the Killing spinors in Eq. (55) by setting $\varepsilon_{\alpha i} = \varepsilon_{ri}\eta^r_\alpha$, we find that the physical spectrum obtained in Eqs. (66)-(68) and Eqs. (77)-(78) can be arranged into the tower of supermultiplets given in Table 1. The lowest

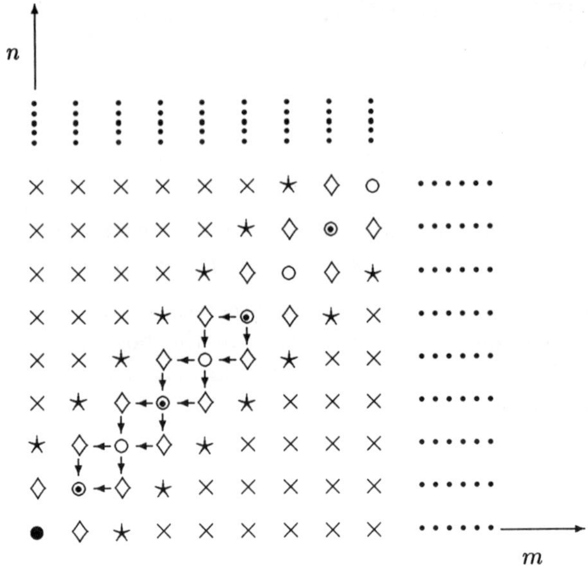

Figure 2. Each entry of the integer grid, $m, n = 0, 1, 2, \ldots$, represents a component $\omega(m, n; \theta)$ of the $shs^E(8|4)$ valued connection 1-form ω : the $\bullet$ denote the spin $s = 1$ gauge fields; a O or a ⊙ denotes a generalized vierbein; a ◇ denotes a generalized gravitino; a $\star$ denotes a generalized Lorentz connection and the $\times$'s denote the remaining auxiliary gauge fields. The $SO(8)$ content is suppressed in this Figure, but it can be read off from Table 1. The auxiliary gauge fields along the northwest-southeast diagonal denoted $\times$'s and $\star$'s are given in terms of the dynamical gauge fields in the middle of that diagonal denoted by O's and the ◇'s. The arrows indicate the action of supersymmetry on the members of the level $\ell = 2$ multiplet.

level, $\ell = 0$, is the CPT self-conjugate short supergravity multiplet, while the higher levels each consists of two short irreducible multiplets that are transformed into each other under CPT. Let us discuss the structure of the level $\ell = 0$ multiplet in more detail.

3.5 *Embedding of the $N = 8$ Gauged Supergravity*

From Eqs. (66)-(68) and Eq. (74) it follows that the linearized equations of motion of the level $\ell = 0$ multiplet are given by

$$\text{spin } s = 2 : \qquad r^{(1)}_{ab} = 0 ,$$

$$\text{spin } s=\tfrac{3}{2}: \qquad (\sigma^{abc})_{\alpha}{}^{\dot\alpha}\left(\nabla_b\omega^{\,i}_{c,\dot\alpha}-\tfrac{1}{2}\lambda\,(\sigma_b)_{\dot\alpha}{}^{\beta}\omega^{\,i}_{c,\beta}\right) = 0\,,$$

$$\text{spin } s=1: \qquad \nabla^a\partial_{[a}\omega^{\,ij}_{b]} = 0\,,$$

$$\text{spin } s=\tfrac{1}{2}: \qquad \nabla_{\alpha}{}^{\dot\alpha}\phi^{ijk}_{\dot\alpha} = 0\,,$$

$$\text{spin } s=0: \qquad \left(\nabla^2+2\lambda^2\right)\phi^{ijkl} = 0\,, \tag{81}$$

where λ is the inverse AdS radius and $r^{(1)}_{ab}$ is the linearization of the Ricci tensor $r_{ab}=e_a{}^{\mu}e_c{}^{\nu}r_{\mu\nu,bc}$ where $r_{\mu\nu,ab}$ is the $SO(3,1)$ valued Riemann curvature. These equations follow from varying the quadratic part of the gauged $N=8$ supergravity action [22]

$$e^{-1}\mathcal{L}_2=\frac{1}{2}R-\frac{\kappa^2}{8g^2}F_{\mu\nu,IJ}F^{\mu\nu,IJ}-\frac{1}{96}\partial_\mu\phi_{ijkl}\partial^\mu\phi^{ijkl}+\frac{g^2}{24\kappa^2}\phi_{ijkl}\phi^{ijkl}+\frac{6g^2}{\kappa^2}$$
$$+\left(\frac{i}{2}\bar\psi^i_{L\mu}\gamma^{\mu\nu\rho}\nabla_\nu\psi_{L\rho i}+\frac{ig}{\sqrt{2}\kappa}\bar\psi^i_{R\mu}\gamma^{\mu\nu}\psi_{L\nu,i}-\frac{1}{2}\bar\chi^{ijk}_L\gamma^\mu\nabla_\mu\chi_{Lijk}+h.c.\right), \tag{82}$$

where $i,j,..=1,...,8$ are $SU(8)$ indices and $I,J,..=1,...,8$ are $SO(8)$ indices, ϕ_{ijkl} are the $35+35$ scalars obeying the $SU(8)$ reality condition Eq. (24) and the fermions are Weyl. The complex conjugation changes chirality, and consequently both chiralities occur for the gravitini as well as the spin $\frac{1}{2}$ fields. Thus, the theory is vector like. In writing Eq. (82) we have assigned (energy) dimension $\frac{1}{2}$ to all the fermions, dimension 0 to the scalars and dimension 1 to the vector fields. We find that Eq. (81) is in perfect agreement with Eq. (82) provided that we make the identifications

$$\omega_\mu{}^{\alpha\dot\alpha} \to \lambda\, e_\mu{}^a\,, \qquad \omega^{ij}_\mu \to g\,A^{IJ}_\mu\,,$$
$$\omega^i_{\mu\alpha} \to \lambda^{\frac{1}{2}}\,,\psi^i_{L\mu}\,, \qquad \lambda^{-ft12}\,\phi^{ijk}_\alpha \to \chi^{ijk}_L\,, \tag{83}$$

and identify the following important relation between the Newton's constant κ, the $SO(8)$ gauge coupling g and the inverse AdS radius:

$$\frac{g^2}{\kappa^2} = \tfrac{1}{2}\lambda^2\,. \tag{84}$$

The free parameters of the higher spin theory are therefore the gauge coupling g and the inverse AdS radius λ. The gauge coupling is introduced into the full set of higher spin equations Eqs. (38)-(43) by replacing $\hat\omega\to g\,\hat\omega$. These equations are consistent with the assignment of dimension 0 to the master fields $\hat{\omega}$ and $\hat{\phi}$ provided that one makes use of the dimensionful parameter λ to rescale the component fields. Consequently the interaction terms will have negative powers of λ[2,16] and therefore the flat limit in which $\lambda \to 0$ is singular.

4 On the Uniqueness of the Constraints

A natural question is whether there are curvature constraints other than Eq. (33) that realize the $shs^E(8|4)$ dynamics irreducibly. However, the uniqueness of the $N = 8$, $D = 4$ gauged supergravity suggests that there should not be much room for interaction ambiguity. Nonetheless, one might relax Eq. (33). A relaxation that preserves irreducibility amounts to introducing more auxiliary fields over and above those already present in $\omega_\mu(Y, \theta)$ and $\phi(Y, \theta)$. A relaxation that does not maintain irreducibility could provide room in the curvature for propagating $shs^E(8|4)$ multiplets other than the gauge multiplet. These two types of generalizations could be useful in finding an action or matter couplings.

To relax the condition on $\hat{\mathcal{R}}_{\mu M}$ while preserving the local symmetries one must first convert the curved spacetime indices on $\hat{\mathcal{R}}_{\nu M}$ into flat indices using a preferred invertible gauge field such as the vierbein and then find the relaxed $\widehat{shs}^E(8|4)$ valued constraint. Insisting on the Yang-Mills like form of the $\widehat{shs}^E(8|4)$ gauge transformations this does not seem possible.

Turning to the remaining constraints, setting $\hat{\mathcal{R}}_{\alpha\dot{\beta}} = (\sigma^a)_{\alpha\dot{\beta}}\hat{\phi}_a$, $(a = 0, 1, 2, 3)$ modifies Eq. (34) such that $\hat{D}_\alpha\hat{\phi}$ will be proportional to $(\sigma^a)_\alpha{}^{\dot{\beta}}(\hat{D}_{\dot{\beta}}\hat{\phi}_a) \star \bar{\kappa}$. There are further integrability conditions involving Z-dependence and the (untwisted) condition $\hat{D}_\mu\hat{\phi}^a = 0$. The full consequences of introducing the field $\hat{\phi}^a$ is not clear to us at present.

Finally we consider modifying the constraint Eq. (33) as

$$\hat{\mathcal{R}} \;=\; \tfrac{i}{4}\left[dz^\alpha \wedge dz_\alpha \mathcal{V}(\hat{\phi} \star \kappa\Gamma) + d\bar{z}^{\dot{\alpha}} \wedge d\bar{z}_{\dot{\alpha}} \bar{\mathcal{V}}(\hat{\phi} \star \kappa^\dagger)\right] , \tag{85}$$

where $\mathcal{V}$ is an arbitrary $\star$ function and $\hat{\phi}$ obeys Eq. (18). The Bianchi identities $D_{\dot{\alpha}}\mathcal{V}(\hat{\phi} \star \kappa\Gamma) = 0$ and $D_\alpha\bar{\mathcal{V}}(\hat{\phi} \star \bar{\kappa}) = 0$ are integrable conditions which imply the condition Eq. (34). The constraint Eq. (85) implies the non-linear reality condition

$$\epsilon^{\dot{\alpha}\dot{\beta}}\hat{\mathcal{R}}_{\dot{\alpha}\dot{\beta}} \;=\; \bar{\mathcal{V}}(\mathcal{V}^{-1}(\epsilon^{\alpha\beta}\hat{\mathcal{R}}_{\alpha\beta}) \star K) \,. \tag{86}$$

The constraint Eq. (85) evidently describes a system with the same degrees of freedom as the system described by Eq. (33). To analyze further the consequences of Eq. (85) let us consider the field redefinition

$$\hat{\phi}' = \mathcal{F}(\hat{\phi} \star \kappa\Gamma) \star \kappa\Gamma , \tag{87}$$

where $\mathcal{F}$ is a $\star$ function. Defining $\mathcal{F}_\pm(-X) := \pm\mathcal{F}(X)$ and observing that $\mathcal{F}_-(\hat{\phi}\star\kappa\Gamma) = \mathcal{F}_-(\hat{\phi}\star\bar{\kappa})\star\kappa\Gamma$ and $\mathcal{F}_+(\hat{\phi}\star\kappa\Gamma) = \mathcal{F}_+(\hat{\phi}\star\bar{\kappa})$, we find from Eq. (34) that

$$D_{\dot{\alpha}}\hat{\phi}' = 0 , \qquad D_\alpha\hat{\phi}' = \mathcal{F}_+ \star D_\alpha K . \tag{88}$$

Moreover, $\hat{\phi}'$ inherits a reality condition from $\hat{\phi}$ which in general is non-linear. For real $\mathcal{F}$ it simplifies to:

$$\hat{\phi}'^\dagger = \pi(\hat{\phi}'_- + \hat{\phi}'_+ \star K) \star \Gamma \qquad \text{for} \quad \mathcal{F} = \bar{\mathcal{F}} , \tag{89}$$

where $\hat{\phi}'_\pm := \mathcal{F}_\pm(\hat{\phi} \star \kappa\Gamma) \star \kappa\Gamma$. Thus, there is a set of apparently inequivalent (see below) constraints given by Eq. (85) where $\mathcal{V}$ is a complex $\star$ function defined modulo the equivalence relation

$$\mathcal{V} \sim \mathcal{V}' \quad \text{if} \quad \mathcal{V} = \mathcal{V}' \circ \mathcal{F} \quad \text{for} \quad \mathcal{F}_+ = 0 , \quad \mathcal{F} = \bar{\mathcal{F}} . \tag{90}$$

In other words the odd real part of $\mathcal{V}$ in Eq. (85) can be redefined away. Note that if one attempts to set $\mathcal{V}(X) = X$ by a field redefinition, then the 'interactions' from its even or imaginary parts resurface in the reality condition on $\hat{\phi}$ and the right hand side of the expression for $D_\alpha\hat{\phi}$. However, other types of field redefinitions reduce the amount of freedom in $\mathcal{V}$. This can be seen even at the linearized level in a simple example as follows.

Starting from $\mathcal{V}(X) = aX + \cdots$, where $a = |a|e^{i\theta}$, a real field redefinition $\hat{\phi} \to |a|^{-1}\hat{\phi}$ leads to $\mathcal{V}(X) = e^{i\theta}X$. However, at the linearized level the phase factor only enters the curvature constraints Eqs. (56)-(57) and can therefore be eliminated by redefining the initial condition ϕ. In the notation $\phi \equiv \phi_+ + \phi_- + \phi_0$, where $\phi_\pm(m, n; \theta) = 0$ for $\pm(m-n) > 0$ and $\phi_0(m, n; \theta) = 0$ for $m \neq n$, the phase factor can be absorbed into the complex fields in $\phi_\pm$ by redefining $\phi_\pm \to e^{\mp i\theta}\phi_\pm$. Notice that unlike Eq. (87) these redefinitions treat $\phi_\pm$ and ϕ_0 differently.

5 Discussion

Despite the fact that the higher spin AdS supergravity theory based on $shs^E(8|4)$ has no flat space limit, such a limit may be achieved by spontaneous breaking of the higher spin symmetries, followed by sending the AdS radius to infinity in a suitable way.

Another consequence of spontaneous symmetry breaking is that the higher spin theory may be truncated to the gauged or ungauged $N = 8, D = 4$ supergravity by sending the symmetry breaking parameter to infinity. This truncation is similar to the decoupling of gravity from matter by setting Newton's constant κ equal to zero, in which case one does not necessarily require that the coupled field equations allow consistent truncation to the matter sector for finite κ.

Spontaneous symmetry breaking may require scalars with a suitable potential and couplings to the higher spin gauge fields. In particular, symmetry breaking that allows truncation down to $N = 8, D = 4$ supergravity might be achieved by using the 2 scalars at the first level. However, a constant value for this field does not appear to solve the field equations without involving higher spin fields. Therefore a more plausible scenario is to couple the theory to massive higher spin multiplets containing fields which can trigger the symmetry breaking.

The $N = 8$ supersingleton propagating at the boundary of AdS spacetime, assumed to arise from the $AdS_4 \times S^7$ compactification of the supermembrane, serves as a spectrum generating representation for the massless higher spin theory propagating in the bulk of the AdS spacetime. As mentioned in the introduction, it has been conjectured that the singletons play a fundamental role in the description of the higher spin supergravity theory in the bulk. Since the massless spectra of the bulk and the boundary theories agree, the essential test of the bulk/boundary duality is whether it is possible to represent the $shs^E(8|4)$ symmetry algebra as charges of anomaly free currents in the $N = 8$ supersingleton theory. We expect that the nonlinearities of the bulk theory would then be reproduced by the interactions between composite singleton states. The higher spin gauge coupling g can be introduced in the boundary theory by rescaling the composite operators corresponding to the bulk gauge fields. The boundary theory depends on λ as well, and the relation Eq. (84) should be predicted by the boundary interactions. The boundary singleton theory also involves massive composite states forming infinite dimensional higher spin multiplets. Therefore the proposed bulk/boundary duality also yields higher spin bulk interactions of both massless and massive sectors which could trigger the spontaneous breaking of $shs^E(8|4)$. While this program by and large remains to be realized, the construction of higher spin currents in the boundary of AdS_5 has been recently investigated [28,29].

Another interesting issue is whether it is possible to accommodate the auxiliary spinor variable $Z_{\underline{\alpha}}$ in the boundary theory. The higher spin curvature constraint might conceivably arise from the κ-like symmetry a by a supermembrane worldvolume embedded into a target space extended by Z-like dimensions.

The generalization of the interacting higher spin field theory to higher dimensions is another open problem. The intimate relations between $M2$ and $M5$-branes makes it particularly interesting to study higher spin theory in AdS_7. The $(2,0)$ tensor multiplet $D = 6$ is the doubleton representation of the AdS_7 supergroup and its symmetric tensor product yields an infinite dimensional multiplet of massless higher spins in AdS_7[30]. The superdoubleton theory living at the boundary of AdS_7 would now arise from the $AdS_7 \times S^4$ compactification of the $M5$-brane, and we expect that it generates the dynamics of the higher spin extension of the maximal gauged supergravity in $D = 7$.

It is tantalizing to consider the possibility of a massless higher spin field theory directly in $D = 11$. Even though a standard AdS supergroup does not exist in $D = 11$, a generalized version of such a group and its singleton like representations have been constructed [31]. Whether an action, or equations of motion, for these representations can be written down in the $D = 10$ boundary of AdS_{11} remains to be seen. In this context, it is interesting to note that the representations of $SO(9)$ as the little group classifying the massless degrees of freedom of $D = 11$ supergravity has been studied [32] with interesting results that suggest the existence of higher spin massless fields in $D = 11$.

Acknowledgments

We wish to thank M. Cederwall, B. Nilsson, M. Duff and M. Vasiliev for useful discussions and Chalmers University of Technology for their kind hospitality. This research has been supported in part by NSF Grant PHY-9722090.

References

1. E.S. Fradkin and M.A. Vasiliev, *On the gravitational interaction of massless higher spin fields*, Phys. Lett. **B189** (1987) 89.
2. E.S. Fradkin and M.A. Vasiliev, *Cubic interaction in extended theories of massless higher-spin fields*, Nucl. Phys. **B291** (1987) 141.
3. E.S. Fradkin and V.Ya. Linetskii, *On the anti de Sitter quantum (super) strings*, Phys.Lett. **B261** 26,1991.
4. E. Bergshoeff, E. Sezgin and P.K. Townsend, *Supermembranes and eleven dimensional supergravity*, Phys. Lett. **87B** (19189) 75.
5. M.J. Duff, *Supermembranes: the first fifteen weeks*, Class. Quantum Grav. **5** (1988) 189.
6. M. Günaydin, *Oscillator-like unitary representations of noncompact groups and supergroups and extended supergravity theories*, in *Lecture Notes in Physics, Vol. 180, eds. E. Inönüand M. Serdaroğlu* (1983).
7. H. Nicolai and E. Sezgin, *Singleton representations of* $OSp(N|4)$, Phys. Lett. **143B** (1984) 389.
8. E. Bergshoeff, E. Sezgin and Y. Tanii, *A quantum consistent supermembrane theory*, Trieste preprint, IC/88/5. Jan 1988.
9. E. Bergshoeff, A. Salam, E. Sezgin and Y. Tanii, *Singletons, higher spin massless states and the supermembrane*, Phys. Lett. **205B** (1988) 237.
10. E. Bergshoeff, A. Salam, E. Sezgin and Y. Tanii, $N = 8$ *supersingleton field theory*, Nucl. Phys. **B305** (1988) 497.
11. M. Flato and C. Fronsdal, *One massless particle equals two Dirac singletons*, Lett. Math. Phys. **2** (1978) 421.
12. E. Bergshoeff, M.J. Duff, C.N. Pope and E. Sezgin *Supersymmetric supermembrane vacua and singletons*, Phys. Lett. **199B** (1987) 69.
13. E. Bergshoeff, M.J. Duff, C.N. Pope and E. Sezgin, *Compactifications of the eleven-dimensional supermembrane*, Phys. Lett. **B224** (1989) 71.
14. M.J. Duff, C.N. Pope and E. Sezgin, *A stable supermembrane vacuum with a discrete spectrum*, Phys. Lett. **225B** (1989) 319.
15. J.M. Maldacena, *The large N Limit of superconformal field theories and supergravity*, Adv. Theor. Math. Phys. **2** (1998) 231, hep-th/9711200.
16. E.S. Fradkin and M.A. Vasiliev, *Superalgebra of higher spins and auxiliary fields*, Int. J. Mod. Phys. A **3** (1988) 2983.
17. M.A. Vasiliev, *Free massless fields of arbitrary spin in the de Sitter space and initial data for a higher spin superalgebra*, Fortschr. Phys. **35** (1987) 741.
18. M.A. Vasiliev, *Consistent equations for interacting massless fields of all spins in the first order curvatures*, Ann. Phys. **190** (1989) 59.

19. M.A. Vasiliev, *Consistent equations for interacting gauge fields of all spins in* $3 + 1$ *dimensions*, Phys. Lett. **B243** (1990) 378.
20. M.A. Vasiliev, *More on equations of motion for interacting massless fields of all spins in* $3 + 1$ *dimensions*, Phys. Lett. **B285** (1992) 225.
21. E. Sezgin and P. Sundell, *Higher spin* $N = 8$ *supergravity*, JHEP/018A/1198, hep-th/9805125.
22. B. de Wit and H. Nicolai, $N = 8$ *supergravity with local* $SO(8) \times SU(8)$ *invariance*, Phys. Lett. **B108** (1982) 285.
23. E. Cremmer and B. Julia, *The SO(8) supergravity theory. 1. The Lagrangian*, Phys. Lett. **B80** (1978) 48.
24. S.E. Konstein and M.A. Vasiliev, *Massless representations and admissibility condition for higher spin superalgebras*, Nucl. Phys. **B312** (1989) 402.
25. S.E. Konstein and M.A. Vasiliev, *Extended higher-spin superalgebras and their massless representations*, Nucl. Phys. **B331** (1990) 475.
26. M.A. Vasiliev, *Equations of motion for* $D = 3$ *massless fields interacting through Chern-Simons higher spin gauge fields*, Mod. Phys. Lett. **A7** (1992) 3689.
27. S.F. Prokushkin and M.A. Vasiliev, *Higher spin gauge interactions for massive matter fields in 3D AdS spacetime*, hep-th/9806236.
28. D. Anselmi, *Theory of higher spin tensor currents and central charges*, hep-th/9808004.
29. S. Ferrara and A. Zaffaroni, *Bulk gauge fields in AdS supergravity and supersingletons*, in *Strings'98* on-line proceedings, hep-th/9807090.
30. M. Günaydin, P. van Nieuwenhuizen and N.P. Warner, *General construction of the unitary representations of the anti de Sitter super algebras and the spectrum of the* S^4 *compactification of eleven-dimensional supergravity*, Nucl. Phys. **B255** (1985) 63.
31. M. Günaydin, *Unitary supermultiplets of OSp(1/32,R) and M-theory*, Nucl. Phys. **B528** (1998) 432, hep-th/9803138.
32. T. Pengpan and P. Ramond, *M(ysterious) patterns in* $SO(9)$, hep-th/9808190.

M(ysterious) $SO(9)$ Triality

Pierre Ramond
Institute for Fundamental Theory, Department of Physics, University of Florida, Gainesville FL 32611, USA

We observe that the light-cone $SO(9)$ triplet of eleven-dimensional supergravity generalizes to four-fold infinite families with the quantum numbers of massless higher spin states. Their mathematical structure are related to the exceptional group F_4 for which triality is manifest. These patterns generalize to the agebras of the magic square

1 N= 1 Supergravity in Eleven Dimensions

It has been recently pointed out that 11-dimensional supergravity is the local limit of M-theory [1], which also contains all known string theories in ten dimensions, although its exact formualtion remains elusive [2].

Since M-theory lives in eleven dimensions, its massless and massive degrees of freedom must be expressible as multiplets of $SO(9)$ and $SO(10)$, the respective Lorentz little groups of eleven dimensions. Among those are the fields of the local supergravi ty theory which appear in the local limit. Even if they are not local, one expects the physical degrees of freedom of M-theory to be expressible as infinite towers of representations of these little groups.

There is a pervading lore [3] against interacting theories that contain massless states of higher spin. Well-founded in fact for local theories with a finite number of fields and orders of derivatives, this prejudice could be evaded for non-loca l theories. However, even with supersymmetry, building such a theory seems like a hopeless task, and the many published attempts have only met with partial success. A four-dimensional formulation [4] uses an infinite-dimensional superalgeb ra, with the interesting feature that it necessarily contains a cosmological constant. Since M-theory is most likely non-local, it may evade the no-go theorems, and could contain an infinite number of fields. It is therefore interesting to examine the $SO(9)$ properties of eleven-dimensional supergravity, whose massless states form the local limit of M-theory.

In the following, we point out that the supergravity triplet of $SO(9)$ representations is the tip of a mathematical iceberg. After presenting group-theoretical evidence that the supergravity representations are the first of an infinite family of massl ess states of higher spin, we describe a mathematical resolution [5] in terms of embeddings of $SO(9)$ in the exceptional group F_4. Since there are no coincidences in the study of these highly constrained theories, it is tempting to muse that these extra higher-spin massless states represent the degrees of freedom of M-theory, even though we have not been able to obtain any dynamical evidence for this conjecture.

2 Group Phenomenology of $SO(9)$

The representations of SO(9) are best described in Dynkin's language [6]. Its irreducible representations are labelled by four positive integers $[a_1\ a_2\ a_3\ a_4]$. Its four basic representations are:

- Vector, [1000], with 9 components, V_i,
- Adjoint, [0100], with 36 components, $B_{[ij]}$,
- Three-form, [0010], with 84 components, $B_{[ijk]}$,
- Spinor, [0001], with 16 components, ψ_α.

All representations with odd a_4 are spinorial. The irreps of $SO(9)$ are characterized by five generalized Dynkin indices

$$I_p \equiv \sum_{\rm rep} w^p \ , \quad p = 0, 2, 4, 6, 8 \ , \tag{1}$$

where w are the weights in the representation. Thus I_0 is the dimension of the irrep, and I_2 is related to the quadratic Casimir invariant by $C_2 = 36 I_2/I_0$.

$N = 1$ supergravity in eleven dimension is a local field theory that contains three different massless fields, two bosonic that describe gravity and a three-form field, and one Rarita-Schwinger spinor. Its physical degrees of freedom are classified in term s of the light-cone little group, $SO(9)$,

- Graviton as a symmetric second-rank tensor, [2000], $G_{(ij)}$,
- Third-rank antisymmetric tensor, [0010], $B_{[ijk]}$,
- Rarita-Schwinger spinor-vector, [1001], $\Psi_{\alpha i}$.

Their group-theoretical properties are summarized in the following table [7]

irrep	[1001]	[2000]	[0010]
I_0	128	44	84
I_2	256	88	168
I_4	640	232	408
I_6	1792	712	1080
I_8	5248	2440	3000

We note that these indices, except for I_8, match between the fermion and the two bosons. As is well known, equality of the bosonic and fermionic dimensions is an indication of supersymmetry. On the light-cone, the supersymmetry algebra reduces to

$$\{Q_\alpha, Q_\beta\} = \delta_{\alpha\beta} \ , \tag{2}$$

where the supersymmetric generators transform as the **16** spinor of $SO(9)$. They split into creation and annihilation operators under the decomposition

$$SO(9) \supset SO(6) \times SO(3) \ ; \qquad \mathbf{16} = (\overline{\mathbf{4}}, \mathbf{2}) + (\mathbf{4}, \mathbf{2}) \ , \tag{3}$$

and we obtain a Clifford algebra

$$\{\tilde{Q}, \tilde{Q}^\dagger\} = 1 \ , \tag{4}$$

where $\tilde{Q}$ transforms as $(\overline{\mathbf{4}}, \mathbf{2})$, and $\tilde{Q}^\dagger$ as $(\mathbf{4}, \mathbf{2})$. The states of the Sugra multiplet are then obtained by successive applications of the $\tilde{Q}^\dagger$ on the vacuum state, to yield 128 bosons and 128 fermions

$$\left\{1 \ , \tilde{Q}^\dagger \ , (\tilde{Q}^\dagger)^2 \ , \dots \ , (\tilde{Q}^\dagger)^7 \ , (\tilde{Q}^\dagger)^8\right\} | \, 0 > \ . \tag{5}$$

The equality between the number of bosons and fermions is manifest. All three irreps have the same quadratic Casimir invariant, since they have the same I_2/I_0 ratio.

Surprisingly, we found that some higher spin representations of $SO(9)$ also occur in triples with the same quadratic Casimir invariant, and show remarkable group-theoretical kinships with the supergravity triplet. The higher spin triplets appear in f our different types, $S-T-T$, $S-S-S$, $T-T-S$, and $T-S-T$, where S describes fermionic (odd a_4), and T bosonic (even a_4) degrees of freedom. The largest representation is listed first, and its dimension is equal to the sum of dimensions of th e other two irreps of the triplet. Thus only triplet of the $S-T-T$ type display supersymmetry-like properties, although only the first is supersymmetric. The whole pattern is summarized by the general form of the triples

$$[1{+}a_2{+}a_3, a_1, a_2, 1{+}a_3{+}a_4] \oplus [2{+}a_2{+}a_3{+}a_4, a_1, a_2, a_3] \oplus [a_2, a_1, 1{+}a_2{+}a_3, a_4] \ , \tag{6}$$

where a_i are non-negative integers. The simplest are the S-T-T triples, which begin with the already-mentioned supergravity triple, followed by

$$[2100]_b + [0110]_b + [1101]_f \ , \tag{7}$$

with group-theoretic numbers

irrep	[2100]	[0110]	[1101]
I_0	910	1650	2560
I_2	3640	6600	10240
I_4	19864	34920	54784
I_6	130840	217320	348160
I_8	977944	1498344	2466304

and described by fields of the form

$$h_{(ijk)l} + A_{(ij)(kl)m} + \Psi_{\alpha(ij)k} \; . \tag{8}$$

Their index structure indicates the appearance of higher spin fields. It is not possible to generate this triple by repeated use of the light-cone supersymmetry algebra acting on some field $\mid \lambda >$, with dimension equal to 20

$$\Big(1 \;, \tilde{Q}^\dagger \;, (\tilde{Q}^\dagger)^2 \;, \dots \;, (\tilde{Q}^\dagger)^7 \;, (\tilde{Q}^\dagger)^8\Big) \mid \lambda > \; . \tag{9}$$

This would imply that $\mid \lambda >$ appears twice in the triple, but the triple contains no duplicate representations of $SO(6) \times SO(3)$ that add up to dimension equal to 20. These fields appear in the Kronecker product of the supergravity triplet with the two-form [0100]. The states of this triple could be understood as being generated by something having the Lorentz properties of a 2- or 7-form in the light cone little group acting on the supergravity states. Other classes of $S - T - T$ triples are

$$[3010]_b + [1020]_b + [2011]_f \; , \tag{10}$$

$$[4000]_b + [0012]_b + [1003]_f \tag{11}$$

$$[4002]_b + [0030]_b + [3003]_f \; , \tag{12}$$

The simplest S-S-S triple is

$$[2003] \oplus [4001] \oplus [0021] \; . \tag{13}$$

The T-T-S family begins with

$$[1002] \;\oplus\; [3000] \;\oplus\; [0011] \; , \tag{14}$$

The T-S-T class starts with

$$[2002] \;\oplus\; [3001] \;\oplus\; [0020] \; . \tag{15}$$

2.1 Basic Operations

It is possible to understand these different triples in terms of four basic operations, which starting from the supergravity multiplet, generate all triples:

- Δ_1: Increase the Dynkin labels all three irreps within a triple by [0100].
- Δ_2: Increase the Dynkin labels of all three irreps within a triple by [1010].
- Δ_3: Increase the Dynkin labels of the first and third irreps by [0001], the second by [1000].
- Δ_4: Increase the Dynkin labels of the first and second irreps by [1001], the third by [0010]

The $\Delta_{1,2}$ operations may be simplest to understand as they can be generated by applying representations that appear either as the light-cone 2-form [0100], or in its twice-antisymmetrized product, since

$$([0100] \otimes [0100])_A = [0100] \oplus [1010] \ . \tag{16}$$

A light-cone 2-form may indicate a brane state, and these triples could then be understood as bound states of the supergravity fields with these branes. The third and fourth operations are more complicated as they treat the different members differently. However, starting from the supergravity multiplet, they generate all other triples, as shown in the diagram below, where the upward arrow denotes Δ_4, and the downward arrow denotes Δ_3:

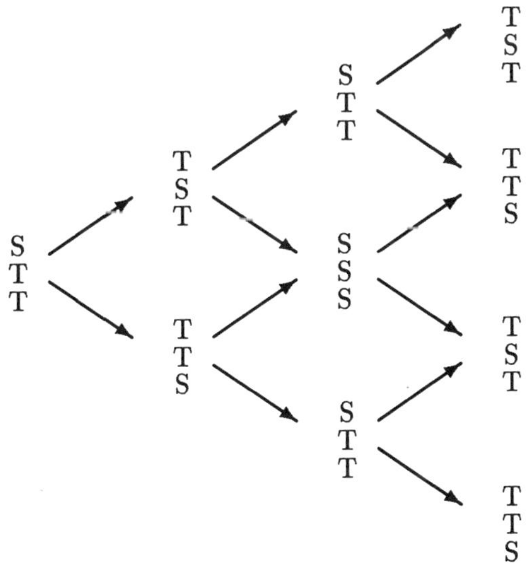

Clearly the supergravity multiplet sits at the beginning of a very intricate and beautiful complex of irreps of $SO(9)$. Limited by the two dimensions of the paper, we have not shown the effect of the $\Delta_{1,2}$ operations which act uniformly on any of the triples in the picture.

3 Mathematical Origin of the Triples

So far we have only offered numerical evidence for the remarkable structure of the $SO(9)$ representations. A recent paper [5] by B. Gross, B. Kostant, S. Sternberg and one of us (PR), unveils its mathematical origin. The following is a watered-d own version of its contents. It points to a construction of a more general character, but does not (yet) seem to shed light on its physical interpretation.

The triples stem from the triality of $SO(8)$, which is explicitly realized in F_4, and the three equivalent ways to embed $SO(9)$ into F_4. That very triality is already familiar to particle physicists: the three equivalent ways to embed $SU(2) \times U(1)$ in $SU(3)$, called I-spin, U-spin, and V-spin [9].

The general idea behind the mathematical construction goes as follows. Let F and B be two Lie algebras of equal rank such that $F \supset B$. The Weyl group of F, $\mathcal{W}(F)$, is bigger than that of B, $\mathcal{W}(B)$, with r-times as many operations. The fundamental Weyl chamber of F, is the sliver of weight space that contains all weights with positive or zero Dynkin labels; it is r times smaller than the fundamental Weyl chamber of the subgroup B.

Let λ be the highest weight of an irrep of F; it lies either inside or at the boundary of the Weyl chamber. We can choose r operations of $\mathcal{W}(F)$ not in $\mathcal{W}(B)$ which map the fundamental Weyl chamber of F into that of B. When applied to this highest weight, they produce r copies inside the chamber of B (unless the weight is at the chamber boundary). In order to make sure it is inside the chamber, we add to it the weight $\rho = [1, 1, 1, \ldots, 1]$, the Weyl vector (or half sum of positive roots). Then we are sure the Weyl group will act on this weight non-trivially. We now construct the r weights

$$\lambda_i \equiv w_i(\lambda + \rho_F) - \rho_B \ , \quad i = 1, 2 \ldots r \ , \tag{17}$$

where w_i are the operations $\mathcal{W}(F)$ not in $\mathcal{W}(B)$. They all lie on the fundamental Weyl chamber of B, and on its boundary, and therefore describe an r-plet of irreps of B.

Apply this reasoning to the case $F_4 \supset SO(9)$. The Weyl group of F_4 has dimension 1152, that of $SO(9)$, 384 so that $r = 3$. Starting from any representation of F_4 this construction generates a triplet of representations of $SO(9)$. There r emains to identify those three elements of the Weyl group, the reason for the relations among their invariants and the emergence of supersymmetry in the construction.

To understand the origin of the triality in F_4, the octonion language is convenient since the adjoint of F_4 is generated by antihermitian traceless 3×3 matrices over the octonions, supplemented by their automorphism group, G_2. Triality i s then related to the three inequivalent ways of picking out one of the matrix's off-diagonal elements, and this construction generalizes to the Lie algebras of the *magic square*[8].

Octonions, together with real numbers, R, complex numbers, C, quaternions, Q, are the four Hurwitz (division) algebras. 3×3 matrices with elements belonging to these algebras generate interesting mathematical structures.

- For real numbers, these matrices generate the Lie algebra $SO(3)$. Its maximal subgroup is $SO(2)$.
- For complex numbers, they generate the Lie algebra $A_2 \sim SU(3)$, and singling out one of the three off-diagonal elements picks out the subgroup $SU(2) \times U(1) \sim SO(3) \times SO(2)$.
- For quaternions, together with their automorphism group $A_1 \sim SU(2)$, they generate $C_3 \sim Sp(6)$. Two off-diagonal elements are treated equally by the subgroup $Sp(4) \times Sp(2) \sim SO(5) \times SO(3)$.
- With octonions, and their automorphism group G_2, they generate the exceptional group F_4. Its subgroup $B_4 \sim SO(9)$ naturally picks out one of the three off-diagonal elements.

Under $F_4 \supset SO(9)$, its adjoint breaks up as $\mathbf{52} = \mathbf{36} + \mathbf{16}$, where $\mathbf{36}$ is the adjoint of $SO(9)$ and $\mathbf{16}$ its spinor representation. Another way to look at this embedding is to say that it generates a 16-dimensional coset space acted on by the orthogonal group $SO(16)$. It yields the anomalous embedding $SO(16) \supset SO(9)$ according to which the spinor of $SO(9)$ fits in the vector of $SO(16)$.

4 The Magic Square

Starting from the four Hurwitz algebras, it is possible to construct the so-called composition algebras, which include all the exceptional groups, except G_2, the automorphism group of the octonions. That construction[8] relies on the triality of both $SO(8)$ and on the structure of 3×3 matrices. Start from 3×3 antihermitian traceless matrices with elements over the product of any two of the four Hurwitz algebras, with three off-diagonal elements, and two diagonal elements which are pure imaginary. T hey are acted on by the automorphism groups of the algebras, and each of their parameters generate a Lie algebra transformation, to produce one of the ten Lie algebras in the *magic square*:

	R	C	Q	O
R	$SO(3)$	$SU(3)$	$Sp(6)$	F_4
C	$SU(3)$	$SU(3) \times SU(3)$	$SU(6)$	E_6
Q	$Sp(6)$	$SU(6)$	$SO(12)$	E_7
O	F_4	E_6	E_7	E_8

In particular, the exceptional group F_4 is generated by 3×3 traceless antihermitian matrices over the octonions, together with G_2, the automorphism group of the octonion multiplication table. This produces the $3 \times 8 + 2 \times 7 + 14 = 52$ parameters of the algebra.

Each of the algebras appearing in the magic square have subalgebras of equal rank, and we can apply our mathematical construction to each. The results are summarized in table below, first for the exceptional groups,

Group	Subgroup	Coset Dimension	r
E_8	$SO(16)$	128	135
E_7	$SO(12) \times SO(3)$	64	63
E_6	$SO(10) \times SO(2)$	32	27
F_4	$SO(9)$	16	3

and then for the non-exceptional groups in the square

Group	Subgroup	Coset Dimension	r
$SO(12)$	$SO(8) \times SO(4)$	32	30
$SU(6)$	$SO(6) \times SO(3) \times SO(2)$	16	15
$Sp(6)$	$Sp(4) \times Sp(2)$	8	3
$SU(3) \times SU(3)$	$SO(3) \times SO(3) \times SO(2) \times SO(2)$	16	3
$SU(3)$	$SO(3) \times SO(2)$	4	3

The connection with supersymmetry occurs through the generation of representations of the subgroup in terms of polynomials in Clifford charges.

4.1 SU(3) Triples

By studying the simplest non-trivial example, of the embedding of $SU(2) \times U(1)$ into $SU(3)$, the emergence of supersymmetry in our general construction should become clear. We begin with some well known facts about $SU(3)$, the algebra generated by 3×3 antihermitian traceless matrices over the complex numbers. Its maximal subgroup is

$$SU(3) \supset SU(2) \times U(1) \sim SO(3) \times SO(2) \ . \tag{18}$$

There are three equivalent ways to imbed this subalgebra in $SU(3)$, corresponding, in particle physics language, to I-spin, U-spin and V-spin. These three embeddings can be understood in terms of the Weyl group. The Weyl group of $SU(3)$ is S_3, the permutation group on three objects. It is three times as big as the Weyl group of its maximal subgroup $SO(3) \times SO(2)$.

All the states spanned by this algebra live in a 2-dimensional lattice. In the Dynkin basis, the weights are labelled by two integers $[a_1, a_2]$. The action of the Weyl group is simplest acting on an orthonormal basis $\{e_i\}$ in which the roots have components which are half-integers between -2 and 2. The simple roots of $SU(3)$ are given in terms of the three vectors e_a which span a three-dimensional the Euclidean space

$$\alpha_1 = e_1 - e_2 \ ; \qquad \alpha_2 = e_2 - e_3 \ . \tag{19}$$

A weight is labelled in the orthonormal basis by three numbers, (b_1, b_2, b_3), such that $b_1 + b_2 + b_3 = 0$. It can also be expressed in the Dynkin basis as

$$w = a_1\omega_1 + a_2\omega_2 \ , \tag{20}$$

where the fundamental vectors $\omega_{1,2}$ are determined through

$$2\frac{(\alpha_i, \omega_j)}{(\alpha_i, \alpha_i)} = \delta_{ij} \ . \tag{21}$$

It follows that the same weight has the orthonormal basis components

$$b_1 = \frac{1}{3}(2a_1 + a_2) \ , \ b_2 = \frac{1}{3}(a_2 - a_1) \ , \ b_3 = -\frac{1}{3}(a_1 + 2a_2) \ . \tag{22}$$

The action of the Weyl group on the orthonormal components is just the permutation group on the three b_i's. Furthermore, the fundamental Weyl chamber is described by the inequalities $b_1 > b_2 > b_3$. We will use the Weyl vector, defined as half the sum of the positive roots:

$$\rho_{SU(3)} = (1, 0, -1) \sim [1, 1] \ , \tag{23}$$

where the square brackets in indicate the Dynkin basis and the curved brackets the orthonormal basis.

Start from an irrep of $SU(3)$, labelled by its maximum Dynkin weight $[a_1, a_2]$, or orthonormal components (b_1, b_2, b_3), and add the Weyl vector, so that the resulting weight is inside the chamber. Since the Weyl group of $SU(3)$ is three times as large as that of $SU(2)$, the fundamental Weyl chamber of $SU(2)$, is three times as large as that of $SU(3)$. Identify the three Weyl transformations (permutations) that produce a weight in the fundamental Weyl chamber of $SU(2)$, defined by $b'_1 \geq b'_2$. These yield three weights

$$(b_1+1, b_2, b_3-1) \qquad (b_1+1, b_3-1, b_2) \qquad (b_2, b_3-1, b_1+1)\ . \tag{24}$$

Next we subtract the $SU(2)$ Weyl vector, $\rho_{SU(2)} = (1/2, -1/2) \sim [1]$, and revert to the Dynkin basis. This construction yields the $SU(3)$-generated triples, with the $U(1) \sim SO(2)$ charge indicated as a subscript

$$SU(3): \qquad [a_1]_{\frac{1}{2}(a_1+1)} \oplus [a_1+a_2+1]_{\frac{1}{2}(a_1-a_2)} \oplus [a_2]_{-\frac{1}{2}(a_2+1)} \tag{25}$$

The three Weyl operations are best identified by their action on the fundamental irrep of $SU(3)$, u, d, s:

$$w_1 \quad : \quad (u,d,s) \to (u,d,s)\ , \tag{26}$$
$$w_2 \quad : \quad (u,d,s) \to (u,s,d)\ , \tag{27}$$
$$w_3 \quad : \quad (u,d,s) \to (s,u,d)\ . \tag{28}$$

Starting from the singlet of $SU(3)$, with $a_1 = a_2 = 0$, we get the simplest triple

$$[0]_{\frac{1}{2}} \oplus [1]_0 \oplus [0]_{-\frac{1}{2}}\ . \tag{29}$$

This triplet of representations forms a supersymmetric multiplet, generated by a charge that is a doublet under the $SU(2)$ and has $U(1)$ value of $-1/2$. To understand its origin, note that the group acting on the four-dimensional coset $SU(3)/SU(2) \times U(1)$ is $SO(4) \equiv SO(3) \times SO(3)$, defining an embedding of $SO(4) \supset SU(2) \times U(1)$. The states of this lowest triple correspond to the decomposition of the $SO(4)$ spinor.

An obvious interpretation is to view $SO(3) \times SO(2)$ as the compact subgroup as the light cone little group of either $SO(4,1) \times SO(2)$, or $SO(3) \times SO(3,1)$, which lead to theories in $d = 5$ and $d = 4$ dimensions, respectively.

Two fundamental operations generate the higher triples, in one to one correspondance with the rank of the mother algebra $SU(3)$. The operation that increases a_1 by two, produces on $[0]_{\frac{1}{2}}$ the infinite chain

$$[0]_{\frac{1}{2}} \oplus [2]_{\frac{3}{2}} \oplus [4]_{\frac{5}{2}} \oplus \cdots\ . \tag{30}$$

It is amusing that these states fit in the spin zero singleton (Rac) representation of $SO(3,2)$. The same operation on $[1]_0$ yields the Di representation. The $SO(2)$ charge is the energy and the $SO(3)$ representation determines the spin.

4.2 $Sp(6)$ Triples

We now apply the same construction to the next entry in the magic square. The Lie algebra $Sp(6) \equiv C_3$ is generated by the 3×3 traceless antihermitian matrices over quaternions, augmented by $SU(2)$, the automorphism group of the quaternions . By picking out one of its off-diagonal element, we obtain the embedding

$$Sp(6) \supset Sp(4) \times Sp(2) \sim SO(5) \times SO(3) , \tag{31}$$

with an eight-dimensional coset space acted on by $SO(8)$. $Sp(6)$ has three simple roots, given by

$$\alpha_1 = e_1 - e_2 , \quad \alpha_2 = e_2 - e_3 , \quad \alpha_3 = 2e_3 . \tag{32}$$

The same weight, written in Dynkin and orthonormal bases, has components $[a_1, a_2, a_3]$ and (b_1, b_2, b_3), respectively with the relations

$$b_1 = a_1 + a_2 + a_3 , \quad b_2 = a_2 + a_3 , \quad b_3 = a_3 , \tag{33}$$

derived from Equation (21), where now $i, j = 1, 2, 3$, after carefully accounting for the long root. Its Weyl vector is given by

$$\rho_{C_3} = (3, 2, 1) \sim [1, 1, 1] \tag{34}$$

The Weyl group of C_3 contains S_3, which permutes the three b_i's. As in the previous case, three of its operations relate the three equivalent ways to embed $SO(5) \times SO(3)$.

Add to the highest weight of an irrep of C_3, $[a_1, a_2, a_3]$ the Weyl vector, to put it inside the fundamental chamber of $Sp(6)$. There are three elements of the Weyl group which map this weight into the fundamental chamber of the subgroup $SO(5) \times SO(3)$, defined by $b_1 \geq b_2 \geq 0$, and $b_3 \geq 0$: the identity element, the parity P_{23}, which interchanges b_2 and b_3, and the cyclic element C_{231} which effects $(b_1, b_2, b_3) \to (b_2, b_3, b_1)$. After subtracting the Weyl vector of the subgroup $\rho = (2, 1; 1)$, where the last entry is for the $SO(3)$ subgroup, we obtain the three weights

$$(b_1 + 1, b_2 + 1; b_3) ; \qquad (b_1 + 1, b_3; b_2 + 1) ; \qquad (b_2, b_3; b_1 + 2) . \tag{35}$$

To rewrite these in the Dynkin basis, we use the expression for the simple roots of $SO(5)$ in the orthonormal basis

$$\alpha_1 = e_1 - e_2 , \qquad \alpha_2 = e_2 , \tag{36}$$

which yield the $Sp(6)$-generated triple

$$[a_1, a_2 + a_3 + 1; a_3] \oplus [a_1 + a_2 + 1, a_3; a_2 + a_3 + 1] \oplus [a_2, a_3; a_1 + a_2 + a_3 + 2] \tag{37}$$

The scalar irrep of $Sp(6)$ yields the simplest triplet

$$[0,1;0] \oplus [1,0;1] \oplus [0,0;2] , \tag{38}$$

or in terms of the dimensions of the representations of $SO(5) \times SO(3)$,

$$(\mathbf{5},\mathbf{1}) \oplus (\mathbf{4},\mathbf{2}) \oplus (\mathbf{1},\mathbf{3}) , \tag{39}$$

the representation content of super-Yang-Mills in ten dimensions. The orthogonal group $SO(8)$ acts on the eight-dimensional coset space, defining an embedding of the spinor of $SO(5) \times SO(3)$ into the vector irrep of $SO(8)$: $\mathbf{8}_V = (\mathbf{4},\mathbf{2})$. The triple is just the decomposition of the spinors of $SO(8)$: $\mathbf{8}_S = (\mathbf{4},\mathbf{2})$, $\mathbf{8}'_S = (\mathbf{5},\mathbf{1}) + (\mathbf{1},\mathbf{3})$. The subgroup can be interpreted as the light-cone little group of either $SO(6,1) \times SO(3)$, or $SO(5) \times SO(4,1)$, which imply Lorentz-invariant theories in either $d = 7$ or $d = 5$ space-time dimensions.

4.3 F_4 Triples

The F_4 simple roots are given by

$$\alpha_1 = e_2 - e_3 , \qquad \alpha_2 = e_3 - e_4 , \tag{40}$$

$$\alpha_3 = e_4 , \qquad \alpha_4 = \frac{1}{2}(e_1 - e_2 - e_3 - e_4) . \tag{41}$$

Through the use of Equation (21), we find the relation between the components of any weight in the Dynkin basis $[a_1, a_2, a_3, a_4]$ and the orthonormal basis (b_1, b_2, b_3, b_4)

$$b_1 = a_1 + 2a_2 + \frac{3}{2}a_3 + a_4 , \quad b_2 = a_1 + a_2 + \frac{1}{2}a_3 , \tag{42}$$

$$b_3 = a_2 + \frac{1}{2}a_3 , \quad b_4 = \frac{1}{2}a_3 . \tag{43}$$

The Weyl vector is given in the orthonormal and Dynkin bases by

$$\rho_{F_4} = \frac{1}{2}(11,5,3,1) \sim [1,1,1,1] . \tag{44}$$

To generate the triples, we start with an irrep of F_4, $[a_1, a_2, a_3, a_4]$, and add to it the Weyl vector to put it inside the fundamental chamber. Express it in the orthonormal basis. The three elements of the F_4 Weyl group which map weights inside the fundamental chamber of $SO(9)$, are the identity, a parity and an anticyclic permutation. These are most easily expressed by their action on the fundamental of F_4, which is a 3×3 hermitian traceless octonionic matrix. The parity interchanges two off-diagonal octonion elements, and the cyclic permutes the three off-diagonal octonion elements. The first is the original weight, the second and third are obtained in terms of permutations on the simple roots of D_4, for which we have

$$\alpha_1 = e_1 - e_2 \,, \qquad \alpha_2 = e_2 - e_3 \,, \tag{45}$$

$$\alpha_3 = e_3 - e_4 \,, \qquad \alpha_4 = e_3 + e_4 \,. \tag{46}$$

In this numbering, α_2 is at the center of the Dynkin diagram, and does not move under the permutations. The two permutations that produce the required weights are

$$S_{14} \; : \quad \alpha_1 \mapsto \alpha_4 \,, \; \alpha_4 \mapsto \alpha_1 \,, \; \alpha_3 \mapsto \alpha_3 \,, \tag{47}$$

$$C_{143} \; : \quad \alpha_1 \mapsto \alpha_4 \mapsto \alpha_3 \mapsto \alpha_1 \,. \tag{48}$$

Weights in the chamber of B_4 are determined by the inequalities $b_1 \geq b_2 \geq b_3 \geq b_4 \geq 0$. Next, to get in the closure of the Weyl chamber, we subtract the Weyl vector of $SO(9)$,

$$\rho_{B_4} = \frac{1}{2}(7, 5, 3, 1) \sim [1, 1, 1, 1] \,, \tag{49}$$

and convert in the Dynkin basis of $SO(9)$, using the B_4 relations

$$\alpha_1 = e_1 - e_2 \,, \qquad \alpha_2 = e_2 - e_3 \,, \tag{50}$$

$$\alpha_3 = e_3 - e_4 \,, \qquad \alpha_4 = e_4 \,, \tag{51}$$

from which

$$a_1 = b_1 - b_2 \,, \qquad a_2 = b_2 - b_3 \,, \tag{52}$$

$$a_3 = b_3 - b_4 \,, \qquad a_4 = 2b_4 \,. \tag{53}$$

This leads back to the F_4 generated triples in the Dynkin basis

$$\begin{aligned} &[1 + a_2 + a_3, a_1, a_2, 1 + a_3 + a_4] \oplus [2 + a_2 + a_3 + a_4, a_1, a_2, a_3] \\ &\oplus [a_2, a_1, 1 + a_2 + a_3, a_4] \,. \end{aligned} \tag{54}$$

This formula corresponds to that previously derived empirically. For the simplest case, the triple is the supergravity multiplet

$$[1,0,0,1] \oplus [2,0,0,0] \oplus [0,0;1,0] . \tag{55}$$

The group $SO(16)$ acts on the sixteen-dimensional coset, providing the embedding of the spinor of $SO(9)$ into the vector of $SO(16)$. Then the lowest order triple is the decomposition of the two spinors of $SO(16)$ into $SO(9)$, which explains the super symmetry.

4.4 $E_{6,7,8}$-*Multiples*

As indicated by the decompositions of the E-like exceptional groups in the magic square, we can generate higher order Clifford algebras.

- The embedding $E_6 \supset SO(10) \times SO(2)$ produces a 32-dimensional coset space. The lowest multiplet contains 27 terms which are decomposition of the two spinor irreps of $SO(32)$ in terms of $SO(10) \times SO(2)$. These are generated by a C lifford algebra with 2^{16} elements, half fermions, half bosons. The $SO(32)$ Clifford is also generated by the magic square embedding $SO(12) \supset SO(8) \times SO(4)$.
- With $E_7 \supset SO(12) \times SO(3)$, a 64-dimensional coset space is generated. The two spinor irreps of $SO(64)$ break into 63 irreps of the subgroup, with 2^{32} states generated by a Clifford algebra, again half of them fermions.
- Finally, $E_8 \supset SO(16)$ produces 2^{64} states describing the two spinors of $SO(128)$ in terms of 135 irreps of $SO(16)$!

In all these cases, the lowest multiplet that is generated contains states of spin higher than two, which means that any theory based on these structures probably have no local limit, except perhaps by compactification.

Finally we note that the magic square embedding $SU(6) \supset SO(6) \times SO(3) \times SO(2)$ yields in the lowest multiplet the states generated by the $SO(16)$-Clifford, corresponding to the dimensional reduction of the sugra multiplet, and it could be v iewed as coming from $d = 8-, 5-, 4-$dimensional theories or even on $AdS_4 \times S_6$. For the reader interested in proofs concerning the properties of these multiplets, we refer the reader to [5], where a novel formula between representations of Li e algebras of the same rank is derived.

In conclusion, we have described a very rich mathematical structure, which contains as special cases, the supermultiplets of eleven-dimensional supergravity and ten-dimensional super-Yang-Mills. While there is at present no clear physical interpretation, it is tempting to believe that it may shed some light on the eventual structure of M-theory.

5 Acknowledgements

I thank Professors L. Brink and R. Marnelius for allowing me to present the above at this workshop. I also benefited from the hospitality of the Aspen Center for Physics and illuminating conversations with Profs L. Brink, R. Moody and S. Sternberg. This work is supported in part by the United States Department of Energy under grant DE-FG02-97ER41029.

1. For a modern review see, A. Sen, hep/th-9802051
2. T. Banks, W. Fishler, S. Shenker and L. Susskind, Phys Rev D55, 5112(1997).
3. L. Durand III, Phys Rev 128, 434(1962); S. Weinberg and E. Witten, Physics Letters 96B, 59(1980).
4. M. A. Vasiliev, Ann. Phys. 190, 59(1989); Phys Lett B257, 111(1991); hep-th/9611024. E. Sezgin and P. Sundell, CTP TAMU-19/98, hep-th/9805125.
5. B. Gross, B. Kostant, P. Ramond and S. Sternberg, Proc Nat'l Acad Sci 95, 8441(1998). Unbeknown to us, some of these results were already known: see Lerche, C. Vafa, and N.P. Warner *Nucl. Phys.* **B 324**, 427(1989).
6. R. Slansky, Physics Reports 79, 1(1981).
7. See, W. G. McKay, J. Patera, Tables of Dimensions, Indices, and Branching Rules for Representations of Simple Lie Algebras, Marcel Dekker (New York and Basel), 1981.
8. B. Gross, B. Kostant, S. Sternberg, private communication; P. Ramond, Introduction to Exceptional Groups and Algebras, CALT-68-577, December, 1976, unpublished.
9. S. Meshkov, C. Levinson, and H. Lipkin, Phys. Rev. Lett.10, 361(1963)

COMMENTS ON STRING-THEORY SCALES AND THE DESERT HYPOTHESIS

C.P. BACHAS
Laboratoire de Physique Théorique de l' Ecole Normale Supérieure
24 rue Lhomond, 75231 Paris, France
E-mail: bachas@physique.ens.fr

I discuss the compactification and fundamental-string scales, both in the weakly-coupled heterotic theory and in type I string theory. A large two-dimensional space transverse to our Brane World in type I theory, can mimic in some ways a supersymmetric renormalizable QFT describing the large energy range between the electroweak and unification scales. Gauge coupling unification is, however, still an open problem in this Brane World picture.

1 The SQFT Hypothesis

String/M-theory [1,2] is a higher-dimensional theory with a single dimensionful parameter, which can be taken to be the fundamental string tension or the eleven-dimensional Planck scale. The theory has on the other hand a large number of 'dynamical parameters' characterizing its many distinct semiclassical vacua, such as compactification radii or sizes of defects localized in the compact space. Understanding how the Standard Model and Einstein gravity arise at low energies in one of those vacuum states is a central outstanding problem of String/M-theory.

The usual hypothesis is that the string, compactification and Planck scales lie all close to one another, and that the physics at lower energies is well described by some effective four-dimensional renormalizable supersymmetric quantum field theory (SQFT), which must include the Minimal Supersymmetric Standard Model (MSSM) and some hidden sectors. I will refer to this picture of the world as the 'SQFT hypothesis'. In this picture the breaking of the residual supersymmetry and the generation of the electroweak scale are believed to be triggered by non-perturbative gaugino condensation – a story that is ofcourse incomplete because of the notorious problems of vacuum stability and of the cosmological constant.

The minimal version of the SQFT hypothesis is obtained when there are no light fields charged under $SU(3)_c \times SU(2)_{ew} \times U(1)_Y$, besides those of the MSSM. This is the 'energy desert' scenario, a slight misnommer since the 'desert' may be populated by various fields which couple with gravitational strength to ordinary matter. This minimal scenario is supported, as is well known, by two pieces of strong, though indirect evidence: (a) the measured low-energy gauge couplings[3] *do meet* when extrapolated to higher energies with the MSSM β-functions, and (b) the energy $M_U \sim 2 \times 10^{16} GeV$ at which they meet is in the same ball-parc as the string scale. A detailed analysis within the weakly-coupled heterotic string leads, in fact, to a discrepancy of roughly one order of magnitude between the theoretical point of unification, and the one that fits the low-energy data [4]. This is a small discrepancy on a logarithmic scale, and it could be fixed by small modifications of the minimal scenario [5,6]. Besides being simple and rather natural, the minimal SQFT hypothesis makes thus two quantitative predictions which fit the low-energy data to better than one part in ten.

2 The Weakly-Coupled Heterotic Theory

The SQFT hypothesis is particularly compelling in the context of the weakly-coupled heterotic string [7]. Both the graviton and the vector bosons live in this case in the ten-dimensional bulk, and their leading interactions are given by the same order in string perturbation theory (i.e. the sphere diagram). This leads to the universal relation between the four-dimensional Planck mass (M_P) and the bare Yang-Mills couplings [8],

$$M_P^2 \sim M_h^2 / g_{\rm YM}^2 \ , \tag{1}$$

independently of the details of compactification. If we assume that $g_{\rm YM} \sim o(1)$, then the heterotic string scale (M_h) is necessarily tied to the Planck scale. Furthermore, the standard Kaluza-Klein formula for the four-dimensional gauge couplings is

$$1/g_{\rm YM}^2 \sim (RM_h)^6 / g_h^2 \ , \tag{2}$$

with R the typical radius of the six-dimensional compact space and g_h the dimensionless string coupling. Pushing the Kaluza-Klein scale ($M_{\rm KK} \sim R^{-1}$) much below M_h requires therefore a hierarchically-strong string coupling. Of course all radii need not be equal, but at least in orbifold compactifications T-duality allows us to take them all larger or equal to the string length, and then the above argument forbids any single radius from becoming large.

There is actually a loophole in the above reasonning. If some compact dimensions are much larger than the heterotic string length, loop corrections to the inverse squared gauge couplings will generically grow like a power of radius.[9] [a] It is thus logically conceivable that even though the observed low-energy couplings are $o(1)$, their tree-level values are *hierarchically smaller*[12,13].

[a] In special models, such as orbifolds without N=2 sectors, these large threshold corrections can be made to vanish, at least at one-loop. The evolution of gauge couplings with energy is thus unaffected by the openning of large extra dimensions [10]. However, since g_h must

Since it is the tree-level couplings that enter in the relation (1), the heterotic string scale could thus be significantly lower than the four-dimensional Planck mass (a rough estimate using one-loop thresholds gives $M_h \sim 10^9 GeV$ for two radii in the TeV range).

The main motivation for contemplating such possibilities in the past was the search for string models with low-energy supersymmetry broken spontaneously at tree level. The existing heterotic models employ a string variant [14,15] of the Scherk-Schwarz mechanism [16], which breaks supersymmetry in a way reminiscent of finite-temperature effects. The scale of (primordial) breaking is proportional to an inverse radius, so that lowering it to the electroweak scale requires the openning of extra dimensions at the TeV. This was shown [17] to be a generic feature in orbifold models. [b] Insisting on tree-level breaking is, on the other hand, only a technical requirement – there is no reason why breaking in nature should not be a non-perturbative effect. Furthermore, Scherk-Schwarz compactifications have not so far lead to any new insights on the problems of vacuum selection and stability. Thus, there seems to be little theoretical motivation for abandonning the SQFT hypothesis, and its successful unification predictions, in heterotic string theory.

3 Brane World and Open String Theory

Things are different in the theory of (unoriented) open and closed strings, in which gauge and gravitational interactions have different origins. While the graviton (a closed-string state) lives in the ten-dimensional bulk, open-string vector bosons can be localized on defects [20,21,22] now recognized as D(irichlet)-branes [23]. Furthermore while closed strings interact to leading order via the sphere diagram, open strings must be attached to a boundary and thus interact via a higher-genus surface, the disk or the real projective plane. The four-dimensional Planck mass and Yang-Mills couplings therefore read

$$1/g_{\rm YM}^2 \sim (R_{||}M_I)^{6-n}/g_I\ , \quad M_P^2 \sim R_\perp^n R_{||}^{6-n} M_I^8/g_I^2, \tag{3}$$

where $R_\perp$ is the typical radius of the n compact dimensions transverse to the brane, $R_{||}$ the typical radius of the remaining (6-n) compact longitudinal dimensions, M_I the type-I string scale and g_I the string coupling constant. As

in these models be hierarchically strong, the semiclassical string vacuum cannot be trusted. Power corrections to gauge couplings have been also discussed more recently in an effort to speed up the unification process.[11]

[b]For more general compactifications, the limit of supersymmetry restoration is also known to be a singular limit [18,19], even though there is no precise relation between the scale of symmetry breaking and some Kaluza Klein threshold.

a result (a) there is no universal relation between M_P, $g_{\rm YM}$ and M_I anymore, and (b) tree-level gauge couplings corresponding to different sets of branes have radius-dependent ratios and need not unify.

A couple of comments before going on. We are here discussing a theory of unoriented strings, because orientifolds [24] are required in order to cancel the tension and the RR charge of D-branes which fill the four non-compact space-time dimensions of our real world. Furthermore, using T-dualities we can ensure that both $R_\perp$ and $R_{||}$ are greater than or equal to the string scale. This may take us either to Ia or to Ib theory (also called I or I', respectively) – I will refer to the theory simply as type I. Finally, it should be stressed that D-branes are the only known defects which can localize non-abelian gauge interactions in string perturbation theory. Orbifold fixed points can only 'trap' matter fields and, in the type II theory, abelian vector bosons from twisted RR sectors of the spectrum. [c] These do not of course suffice to describe the interactions of the Standard Model.

In view of relations (3) type I string theory is more flexible, but also less predictive, than its heterotic counterpart. The string scale M_I is now a free parameter, even if one insists that both the string and the gauge coupling constants be kept fixed and of $o(1)$. This added flexibility can be used to remove the order-of magnitude discrepancy between the unification and string scales [25]. A much more drastic proposal [26,27,28] is to lower the string scale down to the TeV. Indeed, if all the Standard Model gauge fields were trapped on a brane, the relative weakness of gravity could be attributed to the spreading of gravitational flux in some large transverse space.[27] Keeping for instance g_I, $g_{\rm YM}$ and $R_{||}M_I$ of order one, leads to the relation

$$R_\perp \sim M_P^2/M_I^{2+n}. \tag{4}$$

A TeV string scale would then require from n=2 millimetric to n=6 fermi-size transverse dimensions.

It is at first surprising that such extreme possibilities are not already ruled out by experiment. The existing accelerator data cannot, however, rule out Kaluza-Klein and string excitations of the Standard Model particles, as long as they are more massive than a few TeV [29]. Furthermore, modifications of the gravitational force in the submillimeter range are not excluded by mesoscopic gravity experiments [30,31]. Cosmological and astrophysical considerations can put more stringent constraints in some cases, [27] but they are at best indirect. Other potential problems, such as proton decay and fermion masses, do not seem a priori fatal for the picture [32,33,34], even though they will make realistic model building a very delicate exercise indeed.

[c] I thank A. Sen for raising this point.

4 'Desert' in Energy or Transverse Space?

From the discussion of the previous section it is clear that in type I string theory the string scale could be anywhere between M_P and a TeV. The extreme choice $M_I \sim TeV$ has two relative advantages: (a) it is a natural starting point for addressing the question of the gauge hierarchy,[27] which now becomes the question 'why are some transverse radii so large?' (b) it brings string phenomena within the reach of future acceleretor experiments. This extreme choice is in a certain sense antipodal to the minimal SQFT hypothesis: although the MSSM is a stable renormalizable field theory, we are shrinking its range of validity to one order of magnitude at most! Nevertheless, the two scenaria share many common features, when the number of large dimensions transverse to our Brane (with capital B from now on) is exactly two.[35,36]

The argument [36] in a nutshell goes as follows: the parameters of the effective Brane action, which includes the interactions of the Standard Model, correspond to the values of bulk fields – such as the dilaton, massless scalar twist fields etc – evaluated at the position of our Brane. Now bulk fields couple in general to sources (these are the D-branes and orientifolds which are localized in the transverse space), and hence vary as functions of the transverse coordinates. Furthermore the sources have strength $\sim 1/g_I$, so that bulk field variations are of higher order in the string coupling constant. They are perceived by an observer on the Brane as radiative corrections to his parameters, similar to those produced by some unknown ultraviolet physics. If the transverse space is (effectively) two-dimensional, the field variations grow logarithmically. They can therefore become large for large $R_\perp$, but are reliably resummed by the dimensionnally-reduced classical supergravity equations. This situation is analogous to that of a renormalizable supersymmetric QFT, valid in an energy range of $M_I R_\perp$ orders of magnitude: its low-energy parameters receive large logarithmic corrections, which can be reliably resummed by the renormalization group equations.

The logarithmic sensitivity of Brane parameters on $R_\perp$ and hence also on M_P, may be our best hope for understanding the generation of the gauge hierarchy in thid picture. Let me conclude here instead with some comments on whether gauge-coupling unification could be eventually understood through the logarithmic variations of bulk fields in transverse space.[35] This may a priori seem impossible for the following reason: the three factor groups of the Standard Model cannot live on distant branes, because quarks and leptons represented by open strings stretching between them would be unacceptably heavy. But if the factor groups live at the same point in transverse space, one may worry that their gauge couplings receive universal corrections and dont split apart. This would indeed be the case if gauge couplings were determined solely by the value of the string dilaton. Fortunately there exist other bulk fields (such as twisted NS-NS scalars) with non-universal couplings to the various Yang-Mills actions on the branes.[37,38,39] The logarithmic variations of these fields could split the three Standard Model gauge couplings. It is however unclear why this splitting should be in the right proportion.

Acknowledgments

I thank the organizers for the invitation to speak and for their kind hospitality. This research was partially supported by the EEC grant TMR-ERBFMRXCT96-0090.

References

1. M.B. Green, J.H. Schwarz and E. Witten, *Superstring Theory* (Cambridge U. Press, 1987).
2. J. Polchinski, *String Theory* (Cambridge U. Press, 1998).
3. For a recent review of gauge coupling unification see for instance P. Langacker, hep-ph/9411247.
4. V. Kaplunovsky, *Nucl. Phys.* B **307**, 145 (1988).
5. As for instance in C. Bachas, C. Fabre and T. Yanagida, *Phys. Lett.* B **370**, 49 (1996).
6. For a review see K. Dienes, *Phys. Rept.* **287** 447 (1997) and references therein.
7. See V. Kaplunovsky, *Phys. Rev. Lett.* **55**, 1036 (1985) for an early discussion of this point.
8. P. Ginsparg, *Phys. Lett.* B **197**, 139 (1987).
9. L. Dixon, V. Kaplunovsky and J. Louis, *Nucl. Phys.* B **329**, 27 (1990).
10. I. Antoniadis, *Phys. Lett.* B **246**, 377 (1990).
11. K. R. Dienes, E. Dudas and T. Gherghetta, *Phys. Lett.* B **436**, 55 (1998); *Nucl. Phys.* B **537**, 47 (1999).
12. C. Bachas, unpublished (1995).
13. K. Benakli, hep-ph/9809582.
14. R. Rohm, *Nucl. Phys.* B **237**, 553 (1984).
15. S. Ferrara, C. Kounnas and M. Porrati, *Nucl. Phys.* B **304**, 500 (1988).
16. J. Scherk and J.H. Schwarz, *Phys. Lett.* B **82**, 60 (1979); *Nucl. Phys.* B **153**, 61 (1979).
17. I. Antoniadis, C. Bachas, D. Lewellen and T. Tomaras, *Phys. Lett.* B **207**, 441 (1988).
18. M. Dine and N. Seiberg, *Nucl. Phys.* B **301**, 357 (1988).

19. T. Banks and L.J. Dixon, *Nucl. Phys.* B **307**, 93 (1988).
20. J. Dai, R.G. Leigh and J. Polchinski, *Mod. Phys. Lett.* **A4** 2073 (1989).
21. G. Pradisi and A. Sagnotti, *Phys. Lett.* B **216**, 59 (1989).
22. P. Horava, *Phys. Lett.* B **231**, 251 (1989).
23. J. Polchinski, PRL **75**, 4724 (1995).
24. A. Sagnotti, in *Non-perturbative Quantum Field Theory*, eds. G. Mack *et al* (Pergamon Press, Oxford, 1988).
25. E. Witten, *Nucl. Phys.* B **460**, 541 (1996)
26. J.D. Lykken, *Phys. Rev.* D **54**, 3693 (1996).
27. N. Arkani-Hamed, S. Dimopoulos and G. Dvali, *Phys. Lett.* B **429**, 263 (1998); *Phys. Rev.* D **59**, 086004 (1999).
28. I. Antoniadis, N. Arkani-Hamed, S. Dimopoulos and G. Dvali, *Phys. Lett.* B **436**, 257 (1998).
29. See for instance I. Antoniadis, K. Benakli and M. Quirós, *Phys. Lett.* B **331**, 313 (1994).
30. J.E. Moody and F. Wilczek, *Phys. Rev.* D **30**, 130 (1984).
31. A. De Rujula, *Phys. Lett.* B **180**, 213 (1986).
32. G. Shiu and S.-H. H. Tye, *Phys. Rev.* D **58**, 106007 (1998).
33. N. Arkani-Hamed and S. Dimopoulos, hep-ph/9811353.
34. N. Arkani-Hamed, S. Dimopoulos, G. Dvali and J. March-Russell, hep-ph/9811448.
35. C. Bachas, *JHEP* **9811**, 023 (1998).
36. I. Antoniadis and C. Bachas, *Phys. Lett.* B **450**, 83 (1999).
37. G. Aldazabal, A. Font, L. E. Ibanez and G. Violero, *Nucl. Phys.* B **536**, 29 (1998).
38. L. E. Ibanez, R. Rabadan and A. M. Uranga, *Nucl. Phys.* B **542**, 112 (1999).
39. I. Antoniadis, C. Bachas and E. Dudas, hep-th/9906039.

The Power of Extra Dimensions[a]

KEITH R. DIENES[1], EMILIAN DUDAS[1,2], AND TONY GHERGHETTA[1]

[1] *CERN Theory Division, CH-1211 Geneva 23, Switzerland*
[2] *LPTHE, Univ. Paris-Sud, F-91405 Orsay Cedex, France*

In this talk, we discuss some of the recent developments concerning the possibility of large extra spacetime dimensions. We focus primarily on our recent proposal for lowering the scale of grand unification, but we also discuss how these ideas relate to the proposals of Witten and Lykken regarding lowering the string scale, and of Arkani-Hamed, Dimopoulos, and Dvali regarding lowering the Planck scale. We also give an overview of some of our recent work concerning how large radii might be generated dynamically, and how small neutrino masses might also be explained in this context. We then conclude by giving a brief guide to the current literature.

1 Introduction and overview

The possibility of large extra spacetime dimensions has recently received considerable attention[1−48]. This is clearly an exciting prospect. One of the earliest proponents of TeV-scale extra dimensions was Antoniadis[1], who attempted to use such extra dimensions to explain supersymmetry breaking. Later, Witten[2] pointed out that extra large dimensions could lower the string scale below its usual Planck-scale value, and subsequently Lykken[3] proposed that Witten's idea could be extended to lower the string scale all the way to the TeV-range. However, at the time this proposal did not seem feasible for two reasons. First, since string theory generically predicts a unification of gauge couplings and gravitation couplings, it was unclear how gauge coupling unification could be achieved at reduced energy scales below the usual GUT scale $M_{\text{GUT}} \approx 10^{16}$ GeV. Second, it was not clear how gravity — which is extraordinarily weak at low energies — could become comparable in strength to the gauge couplings at energy scales below the usual Planck scale $M_{\text{Planck}} \approx 10^{18-19}$ GeV. In other words, it was not clear how such "TeV-scale strings" could explain the weakness of gravity at low energies as well as the measured values of the low-energy couplings. It therefore did not seem feasible to lower the string scale below the usual GUT or Planck scales.

[a]Invited talk given by KRD at the 22nd Annual Johns Hopkins Workshop, "Novelties in String Theory", held in Göteborg, Sweden, 20–22 August 1998.

The turning point came in March 1998 with two independent realizations. First, it was shown in Ref.[4] that extra large spacetime dimensions also have the power to lower the fundamental Planck scale at which the gravitational force becomes comparable in strength to the fundamental gauge forces. Indeed, it was shown[4] that the fundamental Planck scale could be lowered to the TeV-range, apparently without violating experimental constraints! As pointed out in Ref.[4], this could provide a new way of thinking about the hierarchy problem without requiring supersymmetry. Second, it was shown in Ref.[5] that extra large spacetime dimensions also have the power to lower the fundamental GUT scale (*i.e.*, the scale at which the gauge couplings unify). Once again, it turns out that this new fundamental GUT scale can be lowered all the way to the TeV-range. These were then the two missing pieces to the puzzle. Together, these pieces formed a coherent picture of TeV-scale gravity, grand unification, and string theory. Joining these ideas, it became apparent that large extra spacetime dimensions have the power to lower the fundamental GUT, Planck, and string scales in a unified and consistent manner, perhaps all the way to the TeV-range.

Since March 1998, there has been a small but increasing number of people working in the field. This growth is clearly evident in Fig. 1. If this pattern persists, we will gain an ever-deepening understanding of the possibilities of extra dimensions and their potential for affecting physics beyond the Standard Model.

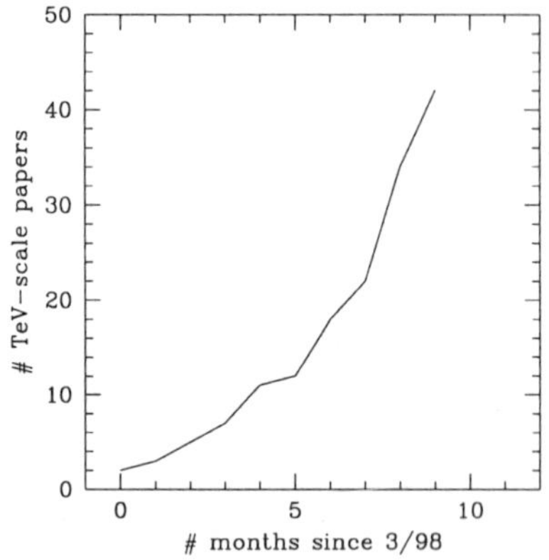

Figure 1: The total number of papers on TeV-scale GUTs, gravity, and strings, plotted as a function of time. This data for this plot was compiled from the Reference list given at the end of this article. Amusingly, this curve exhibits power-law behavior, one of the hallmarks of higher dimensions!

The extra dimensions that are required in order to lower the GUT scale[5,10] are bounded by experimental non-observation to be smaller than 10^{-18} meters, *i.e.*, $R^{-1} \gtrsim 700$ GeV. These extra dimensions are felt by all of the fundamental forces, including the gauge forces as well as the gravitational force; we shall refer to these types of extra dimensions as "universal". The value of the corresponding (lowered) GUT scale then depends on the chosen radius. With R^{-1} taken in the TeV-range, the GUT scale is only an order of magnitude higher in energy, and is therefore also in the TeV-range. However, this scenario turns out to have considerable flexibility[5,10], and it is not necessary for the TeV-limit to be taken.

The extra dimensions that are required in order to lower the Planck scale[4,6] are somewhat different, however. Because their purpose is simply to lower the (gravitational) Planck scale, they can be imagined to be dimensions that only the gravitational force can feel — the ordinary gauge forces are presumed to be immune to their effects. Thus, in this scenario[4,6], the gauge forces are imagined to be confined to a "brane" with respect to these extra dimensions. We shall refer to such extra dimensions as being of "gravity-only" type. It turns out that the experimental bounds on such "gravity-only" extra dimensions are much weaker, and require only that $R^{-1} \gtrsim 10^{-4}$ eV, or $R \leq$ millimeter. Moreover, it is found that two extra dimensions of this type and size can lower the Planck scale all the way to the TeV-range. As pointed out in Ref. [4], this might then provide a new non-supersymmetric solution to the gauge hierarchy problem. However, nothing in this scenario requires that the Planck scale be pushed all the way to the TeV-scale (provided supersymmetry is re-introduced), and in fact lowered Planck scales can generally be achieved if the associated radii are smaller or if the number of such "gravity-only" extra dimensions is larger.

Finally, it turns out that both of these scenarios can naturally be embedded into string theories with reduced string scales. As we stated above, the fact that the string scale can be lowered below its usual (weakly coupled heterotic) value was first pointed out in the context of Type I string theories and M-theory in Ref. [2], and later an extension to the TeV-range was advocated in Ref. [3] and more recently developed in Refs. [6,7,10]. However, once again nothing requires that the string scale be pushed all the way to the TeV-range, and indeed intermediate values are also possible. However, such reduced-scale string theories cannot be phenomenologically viable unless their low-energy gauge and gravitational couplings can be made consistent with observation. It is here that the GUT-lowering and Planck-lowering scenarios involving large extra dimensions come into play: the GUT-lowering scenarios[5,10] explain how a single unified coupling at an energy scale $E < 10^{16}$ GeV can reproduce the different low-energy gauge couplings as measured at the Z-scale, and likewise the Planck-lowering scenarios [4,6] explain how a relatively strong (order one) gravitational coupling (Newton constant) at energies $E < 10^{19}$ GeV can be consistent with the extremely weak gravitational coupling observed in our everyday world. Thus, large extra spacetime dimensions of various types and sizes together conspire to give pheonomenological consistency to reduced-scale string theories.

In this talk, we will start by concentrating on our proposal [5,10] for lowering the fundamental GUT scale. Although it is not necessary to extrapolate this idea to the TeV-range, we will focus on this case because this possibility is of greater experimental interest. We emphasize, however, that this extrapolation is not required on any theoretical grounds, and indeed this TeV-extrapolation may ultimately be ruled out upon detailed confrontation with low-energy experiment. The important point is that extra spacetime dimensions have the power to alter the fundamental high energy scales of physics. This shows that these scales are no longer immutable, which might have profound consequences for our understanding of the possibilities for physics beyond the Standard Model. After discussing our proposal for lowering the GUT scale, we will then consider how this proposal naturally joins with that of Ref. [4] for lowering the Planck scale within the context of string theory. We will then give a discussion of two outstanding questions: how might the necessary large radii be generated dynamically, and how might we explain the relatively small value of the neutrino masses? We will then close by giving a short guide to the TeV-scale literature.

2 Lowering the GUT scale

Let us begin by focusing on how the GUT scale might be lowered (perhaps all the way to the TeV-range). By definition, one of the unique issues faced in attempting to lower the GUT scale to the TeV-range is that the scale of gauge coupling unification must be substantially shifted from its usual MSSM value near $M_{\rm GUT} \approx 2 \times 10^{16}$ GeV. This in turn requires that the usual logarithmic running of the MSSM gauge couplings must somehow be altered. Remarkably, however, as demonstrated in Refs. [5,10], extra spacetime dimensions have precisely the effect we want: they modify the running of the gauge couplings in such a way that not only is the unification preserved, but in fact it occurs more rapidly. This then leads to a reduction in the unification scale. Thus, through extra dimensions, it becomes possible for the first time to contemplate grand unification occurring at an intermediate scale — indeed, as low as the TeV scale. In other words, we now see that it becomes possible to replace a four-dimensional GUT at 10^{16} GeV with a higher-dimensional GUT at the

TeV scale.

There is one aspect of this proposal that deserves special attention. By its very nature, such a radical change in the GUT scale requires that the running of the gauge couplings fully experience the effects of the extra dimensions. This in turn requires that the Standard Model fields actually propagate in the extra dimensions, and moreover that the effects of these extra dimensions not be arranged to cancel as the result of the effects of other symmetries. Thus, in this respect, the extra dimensions required in this proposal are fundamentally different from those of Ref. [1] (in which the effects of the extra dimensions on the gauge couplings are arranged to cancel as the result of $N = 4$ supersymmetry); from those of Ref. [4] (with respect to which the Standard Model particles are trapped on an effective brane); and also from those of Ref. [13] (in which the effects of the extra dimensions are restricted to certain open-string sectors). As far as we are aware, therefore, the effects of extra dimensions on the unification of the gauge couplings had not been investigated prior to Refs. [5,10], and as we shall see, this leads to special subtleties that we shall discuss below. Note that the effects of Kaluza-Klein states on the running of gauge couplings has been previously discussed in Ref. [49].

We begin by quickly recalling the usual four-dimensional gauge coupling unification scenario. In the usual MSSM calculation, one derives the one-loop running of the $SU(3) \times SU(2) \times U(1)$ gauge couplings by evaluating one-loop wavefunction renormalization diagrams, where all MSSM states can propagate in the loop. This then leads to the logarithmic running equation

$$\alpha_i^{-1}(\mu) \;=\; \alpha_i^{-1}(M_Z) \;-\; \frac{b_i}{2\pi} \ln \frac{\mu}{M_Z} \;. \qquad (1)$$

where the one-loop MSSM beta-function coefficients are $(b_1, b_2, b_3) = (33/5, 1, -3)$. Given the experimentally measured gauge couplings at the Z-scale, we can use Eq. (1) to extrapolate upwards in energy. This then leads to the conventional unification near $M_{\rm GUT} \approx 2 \times 10^{16}$ GeV.

How do we extend this calculation into higher dimensions? The basic idea is relatively simple. First, we imagine that there exist $\delta \equiv D-4$ extra spacetime dimensions, each compactified on a circle of radius R. Here $\delta = 1, 2, \ldots$ can take any integer value, and likewise $\mu_0 \equiv R^{-1}$, which sets the energy threshold for the extra dimensions, can range anywhere from several hundred GeV all the way up to the usual GUT scale $M_{\rm GUT} \approx 2 \times 10^{16}$ GeV. Corresponding to each MSSM field, there will be an infinite tower of Kaluza-Klein states whose masses are separated by μ_0. For various technical reasons discussed in Refs. [5,10], we actually must compactify on orbifolds rather than circles, and likewise there are some subtleties involved in arranging the Kaluza-Klein towers corresponding to the MSSM states. The upshot is that the states at the excited levels of the Kaluza-Klein towers must fall into $N = 2$ supermultiplets (even though the ground-state zero-mode MSSM states are only $N = 1$ supersymmetric). It also turns out that one has the freedom to choose whether or not to have Kaluza-Klein towers for the MSSM fermions. We shall therefore denote by the additional free parameter $\eta = 0, 1, 2, 3$ the number of MSSM chiral generations which we shall assume to have Kaluza-Klein towers.

Given this setup, it is then relatively straightforward to calculate the running of the gauge couplings: we simply re-evaluate the standard one-loop wave-function renormalization diagrams with the usual MSSM states as well as their corresponding Kaluza-Klein excitations in the loop. Of course, strictly speaking, such a theory is non-renormalizable due to the infinite towers of Kaluza-Klein states, but we are free to truncate these towers at some arbitrarily high excitation level without affecting our results. Therefore, for the purposes of calculating gauge coupling renormalization effects, we may consider this to be a fully renormalizable field theory. Evaluating the one-loop diagrams, we then find the general result [10]

$$\alpha_i^{-1}(\Lambda) = \alpha_i^{-1}(\mu) - \frac{b_i - \tilde{b}_i}{2\pi} \ln \frac{\Lambda}{\mu} - \frac{\tilde{b}_i}{4\pi} \int_{r\Lambda^{-2}}^{r\mu^{-2}} \frac{dt}{t} \left\{ \vartheta_3 \left(\frac{it}{\pi R^2} \right) \right\}^{\delta} \quad (2)$$

where the Jacobi theta-function $\vartheta_3(\tau) \equiv \sum_{n_i=-\infty}^{\infty} e^{\pi i \tau n^2}$ reflects the sum over the Kaluza-Klein states. Here the new beta-function coefficients $\tilde{b}_i$ corresponding to the excited Kaluza-Klein levels are given by

$$(\tilde{b}_1, \tilde{b}_2, \tilde{b}_3) = (3/5, -3, -6) + \eta\,(4, 4, 4)\,, \quad (3)$$

and the numerical coefficient r is given by $r \equiv \pi (X_\delta)^{-2/\delta}$ where $X_\delta \equiv 2\pi^{\delta/2}/\delta\Gamma(\delta/2)$. While the final term in Eq. (2) reflects the contributions from the Kaluza-Klein towers, the penultimate term reflects the fact that the zero-mode MSSM states and their Kaluza-Klein towers are not identical. [10] Note that all of the terms in Eq. (2) together form the one-loop result, and thus this expression is renormalization-group scheme-independent.

For many practical purposes, it is possible to *approximate* the result (2) in the following way: for $\mu \leq \mu_0$, we may replace Eq. (2) by Eq. (1), while for $\mu \geq \mu_0$, we may replace Eq. (2) by the power-law expression

$$\alpha_i^{-1}(\mu) = \alpha_i^{-1}(\mu_0) - \frac{b_i - \tilde{b}_i}{2\pi} \ln \frac{\mu}{\mu_0} - \frac{\tilde{b}_i X_\delta}{2\pi\delta} \left[\left(\frac{\mu}{\mu_0} \right)^{\delta} - 1 \right] . \quad (4)$$

However, for certain precision calculations (as will be discussed below), it may be necessary to use the full one-loop result given in Eq. (2).

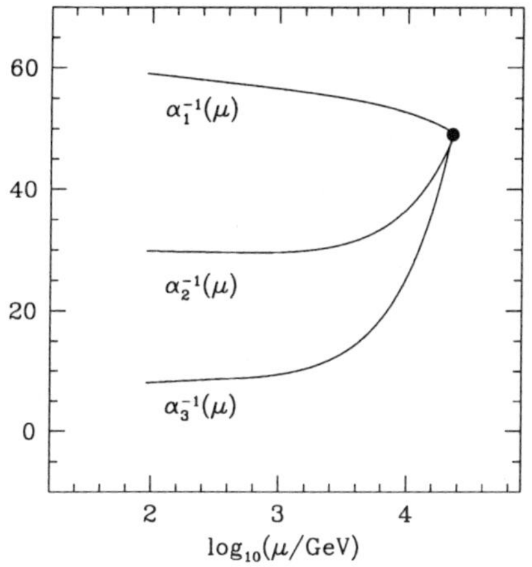

Figure 2: Unification of gauge couplings at the new unification scale $M'_{\text{GUT}} \approx 20$ TeV, assuming the appearance of a single extra spacetime dimension of radius $R^{-1} = 1$ TeV.

The remarkable feature of this higher-dimensional running is that gauge coupling unification is nevertheless preserved. As the most interesting case, let us consider $\mu_0 = 1$ TeV, $\delta = 1$, and $\eta = 0$. We then find the unification shown in Fig. 2. Thus, we see that extra dimensions are consistent with the emergence of a grand unified theory in the TeV-range — *i.e.*, the GUT scale has been lowered all the way to the TeV scale! However, this is not the only possibility, and it is shown in Refs. [5,10] that such a unification with a reduced unification scale occurs regardless of the values of $\mu_0 \equiv R^{-1}$, δ, or η.

This reduced-scale unification leads to many important quantitative questions. How predictive is this unification? How perturbative is it? How sensitive is it to unification scale effects? What about higher-loop corrections? These issues are discussed in some detail in Ref. [14]. The upshot is that this sort of unification scenario is predictive, perturbative, and not unreasonably sensitive to unification-scale effects. Since our original work, there have been further extensions of these basic ideas. These include the incorporation of two-loop effects [16], the incorporation of extra matter beyond the MSSM [35] (which has the potential to make the unification as precise as it is in the usual MSSM), the derivation of these RGE's from a Wilsonian perspective [38], and the extension of these ideas to the power-law running of the Yukawa couplings [5,10,19]. The higher-dimensional evolution of soft SUSY-breaking mass parameters has also been considered in Refs. [24,38,45].

3 Embedding this scenario into string theory

Let us now consider how this scenario might be embedded into string theory. It is here that we shall see the connection between the GUT-lowering scenario that we just described and the Planck-lowering scenario of Ref. [4].

Just as with any GUT theory, we ultimately seek an embedding into string theory. There are many reasons for this. First, string theory is automatically consistent with gauge coupling unification. Second, string theory provides a unification with gravity. Third, string theory is *finite* — this resolves issues of non-renormalizability. Moreover, string theory has many other compelling phenomenological features [50,51].

But now our fundamental GUT scale is very low! This in turn would require a very low string scale. Is such a thing possible?

Usually, the string scale is tied to the Planck scale:

$$M_{\text{string}} \sim g_{\text{string}} \, M_{\text{Planck}} \tag{5}$$

But, as was emphasized in Ref. [2], this holds only for *weakly coupled heterotic strings*! As the ten-dimensional coupling increases, this behavior changes due to non-perturbative effects, and often, the resulting theory is better described as a Type I (open) string theory. However, for open strings, the relation between these scales is different:

$$M_{\text{string}} \sim e^{\langle\phi\rangle/2} \, g_{\text{gauge}} \, M_{\text{Planck}} \tag{6}$$

where ϕ is the ten-dimensional *dilaton* field and g_{gauge} is the Type I gauge coupling. Thus, for Type I strings, it is possible to separate the string scale from the Planck scale by varying the VEV of the dilaton field.

Of course, if we vary the dilaton field, then this also changes the value of g_{gauge} in Eq. (6). Therefore, the true dependence on the dilaton is not obvious from Eq. (6). However, it is possible to eliminate the dilaton dependence completely from this equation, whereupon one finds that the separation between the string scale and the Planck scale ultimately depends on only the six-volume of compactification of the string theory from ten flat dimensions to four flat dimensions:

$$M_{\text{string}} \sim \sqrt{\frac{1}{\alpha_{\text{GUT}} M_{\text{Planck}}}} \, V^{-1/4} \, . \tag{7}$$

This equation is the Type I counterpart of the heterotic relation (5).

It is now clear how our GUT-lowering scenario can be embedded in such a reduced-scale string theory. Recall that in our scenario, the two input parameters are R (the radius of the extra "universal" extra dimensions felt by the gauge couplings as well as gravity), and $\delta \equiv D - 4$ (the total number of such extra dimensions). For any value of these parameters, we then predict M'_{GUT} (the new gauge coupling unification scale) and α'_{GUT} (the new unified gauge coupling). In order to embed this in string theory, let us choose to identify $M_{\text{string}} = M'_{\text{GUT}}$. From Eq. (7), we then see that this uniquely fixes the total six-volume V. Thus, if we write $V \sim R^\delta r^{6-\delta}$, we can actually *solve* for the radii of the remaining $6 - \delta$ extra dimensions. This is the remaining radius that would be needed to lower M_{string} to M'_{GUT} in a self-consistent way.

In order to see what radii would be necessary, let us consider the most interesting case of *one* extra dimension at $R \approx 0.5$ TeV. (This is may already be ruled out experimentally, but it will serve our illustrative purpose.) This in turn implies that $M'_{\text{GUT}} \approx 10$ TeV, which in turn implies that the remaining five extra dimensions must have radius

$$r \approx (10 \text{ MeV})^{-1} \approx 10 \text{ fermi} . \qquad (8)$$

This radius is therefore so large that it would seem, at first sight, to be excluded.

But now we must recall an important point: these remaining five extra dimensions were not part of our gauge coupling running — *i.e.*, they were not assumed to be felt by the Standard-Model particles. Thus, these extra dimensions are felt *only by gravity*, with the gauge forces presumed immune to their effects. As emphasized in Refs. [4,11], extra dimensions of this type and size are not excluded by any laboratory, astrophysical, or cosmological constraints. Indeed, this is nothing but the Planck-scale lowering scenario of Ref. [4], which utilizes extra gravity-only dimensions of (sub-)millimeter to lower the *Planck scale* to the TeV-range! Thus, we see that if we wish to embed the GUT-scale lowering scenario of Ref. [5] into a reduced-scale string theory, we are automatically led to the Planck-scale lowering scenario of Ref. [4].

This makes sense, of course, since string theory is a theory of both gauge and gravitational couplings. It is therefore to be expected that the gauge-lowering scenario of Ref. [5] requires the gravity-lowering scenario of Ref. [4], and vice versa, once they are embedded within the unified framework of string theory.

Thus, putting the pieces together, we are led to a unified embedding into string theory, as illustrated in Fig. 3 for the example we have just described. Above the string scale $M_{\text{string}} = 10$ TeV, the physics is described in terms of a full Type I string theory. (Actually, the string theory in question is a so-called Type I′ theory, but this is a technical detail we need not consider here.) Below 10 TeV, by contrast, the physics is described by an effective field theory. Let us consider what happens as we move upwards in energy, starting from the ex-

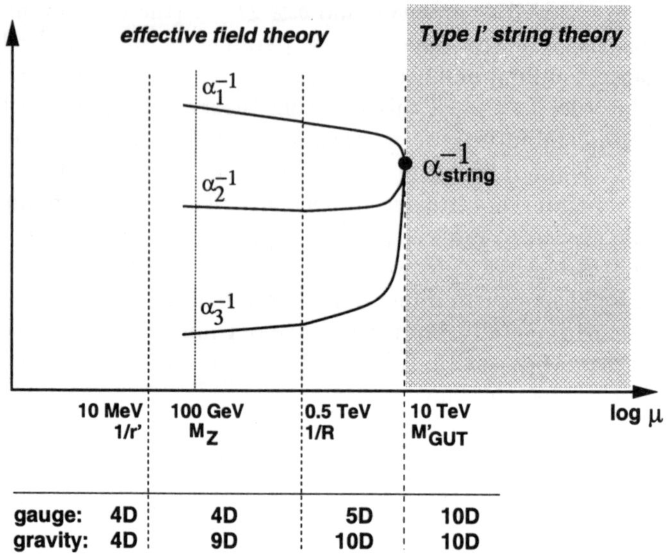

Figure 3: The unification of gauge and gravitational couplings within the framework of a Type I string theory, assuming a string scale at $M_{\rm string} = 10$ TeV.

treme left of the sketch in Fig. 3. Although we have indicated only the gauge couplings in Fig. 3, one can also imagine the inverse gravitational coupling (inverse Newton constant) G_N^{-1} superimposed on this plot. Since gravity is extremely weak at low energies, the value of G_N^{-1} starts out at exponentially high values (completely off the scale of the sketch). At energies below 10 MeV, both the gauge and gravitational sectors of the theory live in four dimensions, and therefore the inverse gravitational coupling scales like μ^{-2}, falling relatively slowly. At this pace, the gravitational coupling itself would ordinarily not become of order one until the Planck scale. But then we cross the first radius threshold at $\mu = 10$ MeV, where the gravitational coupling suddenly feels five extra spacetime dimensions. The evolution of the gravitational coupling suddenly becomes more dramatic, with the inverse gravitational coupling now falling like μ^{-7}. Meanwhile, the gauge couplings continue to evolve logarithmically, because the gauge sector of the theory is still four-dimensional. Next, we cross the final radius threshold at $\mu = 0.5$ TeV, with one additional "universal" extra dimension opening up. This causes the inverse gauge couplings to begin to run with a power-law scale-dependence (linear in this case), while the gravitational coupling now falls even more quickly, as μ^{-8}. Effectively, the Standard Model is now in five dimensions, while the gravitational world is fully ten-dimensional. Finally, the gauge and gravitational couplings all unify to form one string coupling $\alpha_{\rm string}^{-1}$ at 10 TeV. At this stage, the theory is fully ten-dimensional. Thus, we see that *all of fundamental physics* — the GUT scale, the Planck scale, and the string scale — have been successfully lowered to the TeV-range!

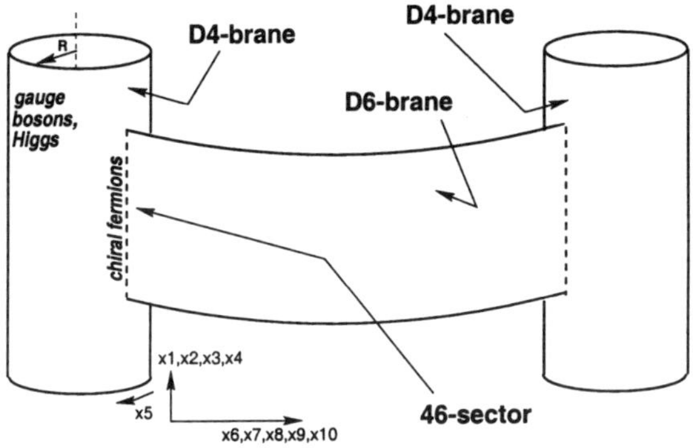

Figure 4: The new "brane" view of spacetime. A detailed discussion can be found in Ref. [10].

This picture also leads to a new "brane" view of spacetime, as shown in Fig. 4. At low energies, the (MS)SM lives on the four-dimensional intersection of different "branes". At higher energies, however, the gauge forces feel the extra dimension of radius R, and the Standard-Model world effectively becomes a cylinder. This extra dimension of radius R lowers the GUT scale, and ensures gauge coupling unification. (In the above figure, we have illustrated the $\eta = 0$ case wherein the chiral fermions do not feel this extra dimension.) By contrast, the remaining orthogonal directions (of sub-millimeter size) are felt only by gravitational interactions. This lowers the Planck scale, and explains the weakness of gravity. Finally, as we have seen, this entire picture combines to produce a lowered string scale. This lowering of the string scale ensures theoretical consistency, as well as calculability. Note that while Fig. 4 illustrates one possible (non-supersymmetric) realization in terms of four-branes and six-branes, there also exist (supersymmetric) realizations in terms of different sets of purely five-branes.

4 Generating a large compactification radius

Thus far, we have focused on the *consequences* of large extra compactified dimensions. But in string theories with spacetime SUSY, the compactification radius is a *modulus*, which means that the potential $V(R)$ is exactly flat to all orders in perturbation theory. Thus, there is no dynamical method for choosing radii, whereupon it is natural to assume that the radii are compactified at the string scale: $R/\sqrt{\alpha'} \sim \mathcal{O}(1)$. The above scenarios therefore face two important questions: (1) How are large compactification radii *generated*? (2) How are large compactification radii *stabilized*? In particular, we seem to require a *hierarchy* relative to the string scale:

$$\begin{array}{ll} \text{"universal" extra dimensions:} & R/\sqrt{\alpha'} \leq \mathcal{O}(20) \\ \text{"gravity-only" extra dimensions:} & R/\sqrt{\alpha'} \leq \mathcal{O}(10^6)\ . \end{array} \tag{9}$$

We shall now describe one dynamical mechanism [17] that may be able to accomplish this.[a] The basic idea is that the radius is a modulus because of spacetime supersymmetry. However, *thermal effects* necessarily break spacetime supersymmetry, because bosonic states feel Bose-Einstein statistics and fermionic states feel Fermi-Dirac statistics. Such thermal effects are expected in the early (hot) universe. Thus, *thermal effects* may be able to establish a non-zero potential for the radius.

In order to analyze this idea, one must calculate an open-string effective potential for the radius. To this end, let us consider a toy model consisting of the $SO(32)$ Type I string compactified to $D = 9$ (*i.e.*, with *one* compactification radius R). For $R \geq \sqrt{\alpha'}$, this has a natural interpretation as a "universal" extra dimension. For $R < \sqrt{\alpha'}$, by contrast, one must T-dualize, in which case this radius is interpreted as a gravity-only dimension (*i.e.*, a distance between branes). Given such a string model, we then calculate the one-loop effective potential $\Lambda(R, T)$ as a function of the compactification radius R and the temperature T. Calculating such an effective potential involves summing over the Matsubara frequencies for the bosonic and fermionic states in the model. Also, because we are essentially calculating a one-loop string amplitude in an open (unoriented) Type I string theory, we must add together the individual contributions from the four possible unoriented one-loop worldsheet topologies corresponding to the torus, Klein bottle, cylinder, and Möbius strip.

Before even doing the calculation, note that the torus contribution must always be invariant under the T-duality $(R/\sqrt{\alpha'}) \leftrightarrow (R/\sqrt{\alpha'})^{-1}$. This corresponds to an exchange between string momentum and winding modes, and implies that there is no distinction between large and small radii. This in turn forces the extremum of the potential to always lie at $R\sqrt{\alpha'} = 1$. Of course, this is precisely what we are seeking to avoid. The crucial point, however, is that for the *open-string* sectors (Klein-bottle, cylinder, and Möbius strip), the

[a]The work described in this section was performed in collaboration with A. Riotto.

winding modes are absent. This is because such winding modes do not survive the orientifold projection. Thus, unlike the torus contribution, we see that the open-string contributions *distinguish* between large and small radii, and can thereby set up a preferred flow, as desired! Specifically, the potential for the "universal" extra dimensions ($R > \sqrt{\alpha'}$) can be different from the potential for the "gravity-only" extra dimensions ($R < \sqrt{\alpha'}$), which might explain how these different types of extra dimensions can have different sizes.

What we find is shown in Fig. 5. First, looking at the torus contribution alone, we see that although the T-duality is manifest, thermal effects leave the self-dual point *unstable* rather than stable! This is contrary to the usual expectations. Moreover, looking at the cylinder and Möbius contributions, we see that the radius-dependence indeed "freezes out" for $R/\sqrt{\alpha'} \ll 1$, reflecting the loss of winding modes from the spectrum due to the orientifold projection that created the open string. This is what ultimately distinguishes between the universal and gravity-only extra dimensions. The Klein-bottle contribution vanishes for this toy model. Finally, looking at the total effective potential (lower right plot of Fig. 5), we see that thermal effects indeed push the compactification radius out to large values! (Although not shown in this plot, the potential also diverges towards $-\infty$ as $R \to 0$, reflecting the contribution from the torus amplitude. The functional form of this divergence as $R \to 0$ is, however, *different* from that as $R \to \infty$.) Thus, we see that these thermal potentials provide a *dynamical* method of generating large radii.

The behavior of the total thermal potential as a function of the temperature $a_T \equiv 2\pi T/M_{\text{string}}$ is shown in Fig. 6. Although the potential diverges at the Hagedorn temperature $a_T = 1/\sqrt{2}$, this potential weakens as the universe cools, ultimately becoming flat (reflecting the restoration of supersymmetry) as $T \to 0$. Thus, it is easy to imagine the following qualitative scenario. When the universe is originally very hot, there is a strong effective potential that drives the different radii to extremely large values. As the universe cools, the effective potential becomes weaker and flatter, and the "force" pushing the radius dissipates. Ultimately, as $T \to 0$, the radius is deposited out at large value. Thus, in this way, the large-radius *pre-conditions* are set for other dynamical (low-temperature) effects to come into play. Of course, there may also be other (non-thermal) mechanisms for generating a large compactification radius. One of the more intriguing possibilities involves so-called *brane inflation* [43].

Once a large radius is *generated*, many effects can potentially *stabilize* it. These include ordinary (low-T) supersymmetry-breaking potentials; a bulk cosmological constant (to halt expansion) [18]; the cosmological dynamics of the four original radii (which we have thus far neglected); possible inter-brane forces (*e.g.*, classical and Casimir forces) [18]; topological mechanisms (*e.g.*, trapped

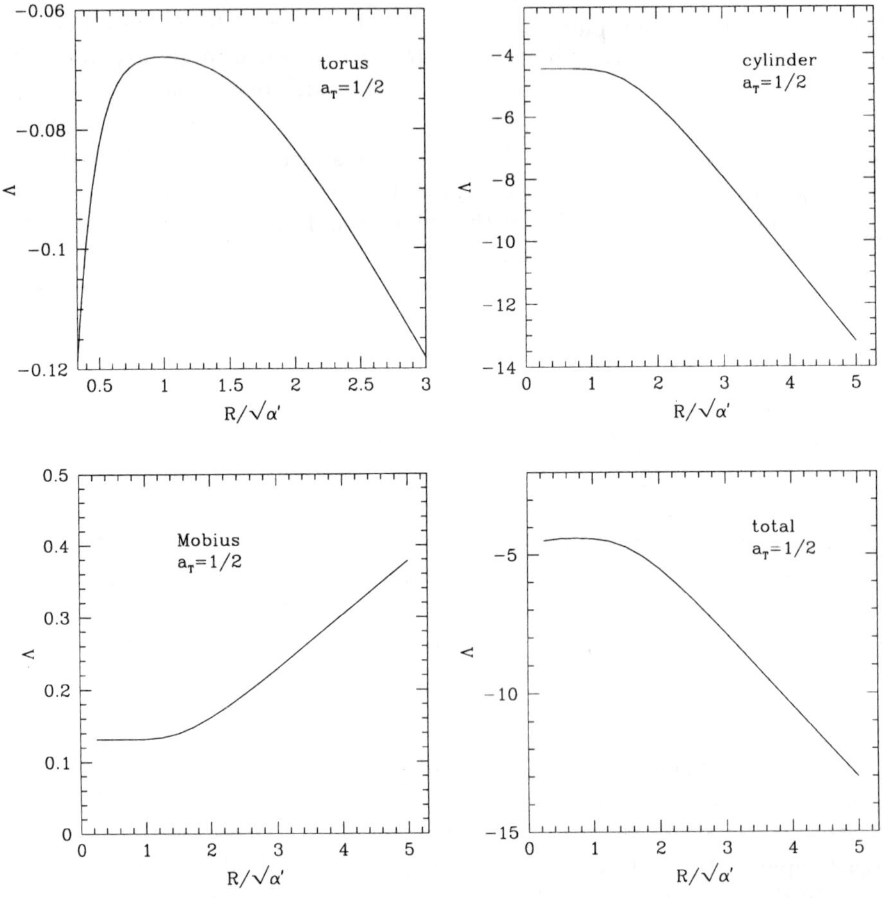

Figure 5: Individual contributions to the effective potential as a function of the radius of compactification, evaluated at the temperature $T/M_{\mathrm{string}} = 1/4\pi$, or $a_T \equiv 2\pi T/M_{\mathrm{string}} = 1/2$. The total effective potential (lower right) shows a clear tendency to push the radius out to large values. See Ref. [17] for further details.

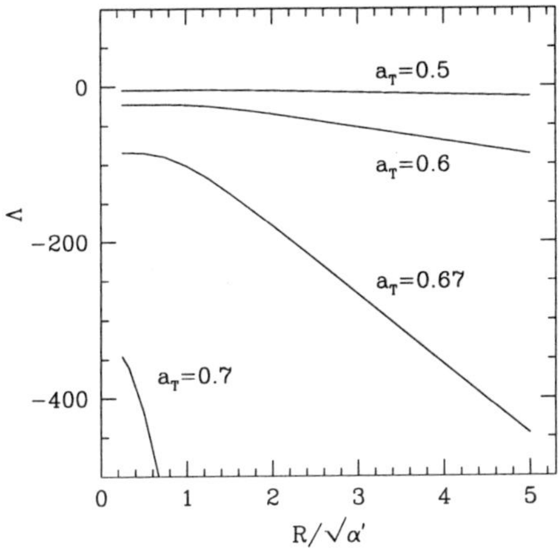

Figure 6: The effective potential as a function of the radius of compactification, for temperatures ranging from $a_T = 0.5$ to $a_T = 0.7$. In all cases, the radius is pushed out to large values. Note that the potential diverges to negative values for all radii at the Hagedorn temperature $a_T = 1/\sqrt{2} \approx 0.707$.

fluxes [12,18], or even string or brane winding modes [52]). Moreover, within a cosmological context, another simple possibility is *entropy conservation*[17]: the radii cannot expand without limit if the universe undergoes an adiabatic evolution. The latter idea is explored in Ref. [17], where an intriguing connection with the string Hagedorn phenomenon is found.

5 Light neutrino masses without heavy mass scales

As we have seen, the lesson from the recent developments has been that four-dimensional theories with heavy mass scales can be replaced by higher-dimensional theories without heavy mass scales, provided the extra spacetime dimensions are sufficiently large. Indeed, this situation may seem to suggest that there is no further need for high energy scales. However, low-energy neutrino data seem to provide independent evidence for yet another high mass scale. Specifically, if neutrinos have light but non-zero masses (as suggested by recent SuperKamiokande data), then these masses are most naturally explained in the context of $SO(10)$ unification via the well-known seesaw mechanism [53]. The seesaw mechanism relies on the existence of a new heavy mass scale M associated with a right-handed neutrino singlet field N. Indeed, in the simplest scenarios, light neutrino masses in the 10^{-2} eV range imply that M should be of the same order of magnitude as the usual four-dimensional GUT scale $\approx 10^{16}$ GeV. This therefore appears to provide a further need for a high fundamental GUT scale.

The question that emerges, then, is whether it is possible to generate light neutrinos *without* the introduction of heavy mass scales. Is there some intrinsically higher-dimensional mechanism which might permit naturally light neutrino masses to be generated?

To date, there have been two proposals [34,36] concerning how this might be accomplished. Both originate with the same observation: because the right-handed neutrino is a Standard-Model gauge singlet, it need not be restricted to a "brane" with respect to the full higher-dimensional space. It is therefore possible for this field to experience extra spacetime dimensions and thereby accrue an infinite tower of Kaluza-Klein excitations. This then leads to a number of higher-dimensional mechanisms [34,36] for suppressing the resulting neutrino mass without a heavy mass scale. In the following, we shall concentrate on one of the mechanisms advanced in Ref. [34], namely the intriguing possibility that in higher dimensions, neutrino oscillations need not imply the existence of neutrino masses at all! This would then eliminate the need for a high fundamental scale. Other mechanisms for explaining light but non-zero neutrino masses are also discussed in Refs. [34,36].

We begin by assuming that the right-handed neutrino feels extra dimensions, while the left-handed neutrino ν_L does not. Specifically, in higher dimensions (*e.g.*, in five dimensions, for concreteness), we consider a Dirac fermion Ψ, which in the Weyl basis can be decomposed into two two-component spinors: $\Psi = (\psi_1, \bar{\psi}_2)^T$. When the extra spacetime dimension is compactified on a $\mathbb{Z}_2$ orbifold, it is natural for one of the two-component Weyl spinors, *e.g.*, ψ_1, to be taken to be even under the $\mathbb{Z}_2$ action $y \to -y$, while the other spinor ψ_2 is taken to be odd. If the left-handed neutrino ν_L is restricted to a brane located at the orbifold fixed point $y = 0$, then ψ_2 vanishes at this point and so the most natural coupling is between ν_L and ψ_1. For generality, we will also include a possible "bare" Majorana mass M_0 for Ψ. This then results in a Lagrangian of the form

$$\begin{aligned} \mathcal{L} &= \int d^4x\, dy\, M_s \left\{ \bar{\psi}_1 i\bar{\sigma}^\mu \partial_\mu \psi_1 + \bar{\psi}_2 i\bar{\sigma}^\mu \partial_\mu \psi_2 + \tfrac{1}{2} M_0 \left(\psi_1 \psi_1 + \psi_2 \psi_2 + \text{h.c.} \right) \right\} \\ &\quad + \int d^4x \left\{ \bar{\nu}_L i\bar{\sigma}^\mu D_\mu \nu_L \;+\; \left(\hat{m} \nu_L \psi_1 |_{y=0} + \text{h.c.} \right) \right\} . \end{aligned} \tag{10}$$

Here y is the coordinate of the extra compactified spacetime dimension, and M_s is the mass scale of the higher-dimensional fundamental theory (*e.g.*, a reduced

Type I string scale). The first line represents the kinetic-energy term for the five-dimensional Ψ field as well as the bare Majorana mass term $\frac{1}{2}M_0\bar{\Psi}^c\Psi$. By contrast, the second line represents the kinetic energy of the four-dimensional two-component neutrino field ν_L as well as the Dirac Yukawa coupling between ν_L and ψ_1. Note that in five dimensions, a bare Dirac mass term for Ψ would not have been invariant under the action of the $\mathbb{Z}_2$ orbifold, since $\bar{\Psi}\Psi \sim \psi_1\psi_2+$ h.c.

Next, we compactify the Lagrangian (10) down to four dimensions by expanding the five-dimensional Ψ field in Kaluza-Klein modes. Imposing the orbifold relations $\psi_{1,2}(-y) = \pm\psi_{1,2}(y)$ implies that our Kaluza-Klein decomposition takes the form

$$\begin{aligned} \psi_1(x,y) &= \frac{1}{\sqrt{2\pi R}}\sum_{n=0}^{\infty}\psi_1^{(n)}(x)\cos(ny/R) \\ \psi_2(x,y) &= \frac{1}{\sqrt{2\pi R}}\sum_{n=1}^{\infty}\psi_2^{(n)}(x)\sin(ny/R)\ . \end{aligned} \qquad (11)$$

For convenience, we shall also define the linear combinations $N^{(n)} \equiv (\psi_1^{(n)} + \psi_2^{(n)})/\sqrt{2}$ and $M^{(n)} \equiv (\psi_1^{(n)} - \psi_2^{(n)})/\sqrt{2}$ for all $n > 0$. Inserting Eq. (11) into Eq. (10) and integrating over the compactified dimension then yields the four-dimensional Lagrangian

$$\begin{aligned} \mathcal{L} = \int d^4x \Bigg\{ & \bar{\nu}_L i\bar{\sigma}^\mu D_\mu \nu_L + \bar{\psi}_1^{(0)} i\bar{\sigma}^\mu \partial_\mu \psi_1^{(0)} \\ &+ \sum_{n=1}^{\infty}\left(\bar{N}^{(n)} i\bar{\sigma}^\mu \partial_\mu N^{(n)} + \bar{M}^{(n)} i\bar{\sigma}^\mu \partial_\mu M^{(n)}\right) \\ &+ \Bigg\{ \tfrac{1}{2} M_0\, \psi_1^{(0)}\psi_1^{(0)} \\ &+ \tfrac{1}{2}\sum_{n=1}^{\infty}\left[\left(M_0 + \frac{n}{R}\right) N^{(n)}N^{(n)} + \left(M_0 - \frac{n}{R}\right) M^{(n)}M^{(n)}\right] \\ &+ m\nu_L\psi_1^{(0)} + \sum_{n=1}^{\infty}\left(m\nu_L N^{(n)} + m\nu_L M^{(n)}\right) + \text{h.c.}\Bigg\}\Bigg\}\ . \end{aligned} \qquad (12)$$

Here the first and second lines give the four-dimensional kinetic-energy terms, while the third and fourth lines give the Kaluza-Klein and Majorana mass terms. The fifth line of Eq. (12) describes the coupling between the four-dimensional neutrino ν_L and the five-dimensional field Ψ. Note that in obtaining this Lagrangian, it is necessary to rescale the individual $\psi_1^{(0)}$, $N^{(n)}$, and $M^{(n)}$ Kaluza-Klein modes so that their four-dimensional kinetic-energy terms are canonically normalized. This then results in a suppression of the Dirac neutrino mass coupling $\hat{m}$ by the factor $(2\pi M_s R)^{1/2}$. In the third line, we have therefore simply defined the effective Dirac neutrino mass coupling $m \equiv \hat{m}/\sqrt{2\pi M_s R}$. Note that under some circumstances, this volume suppression factor may already be able to explain the lightness of the neutrino mass.[36] Specifically, even with $\hat{m} \sim \mathcal{O}(\text{TeV})$, we might obtain neutrino masses m of the correct size by properly choosing M_s and R.

Our interest here, however, will be in constructing a higher-dimensional seesaw mechanism.[34] This is possible because given the Lagrangian (12), the Standard-Model neutrino ν_L will mix with the entire tower of Kaluza-Klein states of the higher-dimensional Ψ field. Indeed, if we restrict our attention for the moment to the case of only one extra dimension for simplicity and define

$$\mathcal{N}^T \equiv (\nu_L, \psi_1^{(0)}, N^{(1)}, M^{(1)}, N^{(2)}, M^{(2)}, \ldots) , \tag{13}$$

we see that the mass terms in the Lagrangian (12) take the form $\frac{1}{2}(\mathcal{N}^T \mathcal{M} \mathcal{N} + \text{h.c.})$ where the mass matrix is symmetric and takes the form[a]

$$\mathcal{M} = \begin{pmatrix} 0 & m & m & m & m & m & \cdots \\ m & M_0 & 0 & 0 & 0 & 0 & \cdots \\ m & 0 & M_0 + 1/R & 0 & 0 & 0 & \cdots \\ m & 0 & 0 & M_0 - 1/R & 0 & 0 & \cdots \\ m & 0 & 0 & 0 & M_0 + 2/R & 0 & \cdots \\ m & 0 & 0 & 0 & 0 & M_0 - 2/R & \cdots \\ \vdots & \vdots & \vdots & \vdots & \vdots & \vdots & \ddots \end{pmatrix} . \tag{14}$$

Before proceeding further, let us discuss the possible values for the bare Majorana mass M_0. While in principle any values are allowed, depending on the structure of the full effective Lagrangian derived from the underlying string

[a]Note that the upper-left corner entry of this matrix is taken to be zero — *i.e.*, we have not assumed a possible primordial Majorana mass term for the left-handed neutrino. This is because we are assuming that lepton-number is conserved on the brane where the left-handed neutrino is located, and possibly broken only in the bulk via a Majorana mass M_0 for the higher-dimensional Ψ field. Of course, lepton-number breaking in the bulk can induce lepton-number violation on the brane in several different ways. One way would be through the coupling between Ψ and ν_L, but this is essentially taken into account in what follows when we fully diagonalize the resulting mass matrix. The second way is through an effective supergravity interaction. However, as we shall discuss shortly, we shall be mostly be interested here in a *topological* breaking of lepton-number in the bulk which occurs through the Scherk-Schwarz mechanism. In such cases, any resulting ν_L-mass operator must vanish as $R \to \infty$ and hence must be suppressed by additional powers of $M_s R \gg 1$.

model in question, there is a natural mechanism in string theory by which such a lepton-number violating bare Majorana mass M_0 can be generated. Specifically, as discussed in Ref. [34], it is possible to break lepton number globally (but not locally) via the Scherk-Schwarz mechanism [54]. In such a scenario, the possible values of M_0 are fixed by topological constraints to take only two values: $M_0 = 0$ (corresponding to no breaking), and $M_0 = (2R)^{-1}$. Let us here consider the non-trivial possibility $M_0 = (2R)^{-1}$. It is then possible to solve for the eigenvalues of Eq. (14), and we find [34] that these are given as the solutions of the transcendental equation

$$\lambda R \;=\; -\,\pi(mR)^2 \,\tan\left(\pi\lambda R\right) \;. \tag{15}$$

From this equation, we see that in addition to an infinite set of non-zero eigenvalues (corresponding to the masses of the infinite tower of Kaluza-Klein mass eigenstates), there also exists an exact *zero eigenvalue* $\lambda = 0$. The mass eigenstate corresponding to this eigenvalue is given exactly by

$$|\tilde{\nu}_L\rangle \;=\; \frac{1}{\sqrt{1+\pi^2 m^2 R^2}} \left\{ |\nu_L\rangle \;-\; mR \sum_{k=1}^{\infty} \frac{1}{k-1/2} \left[|N^{(k-1)}\rangle - |M^{(k)}\rangle \right] \right\} \tag{16}$$

where we have defined $N^{(0)} \equiv \psi_1^{(0)}$. This result is exact for all mR, but in most realistic scenarios (see Refs. [34,36]), we have $mR \ll 1$. Thus, we see that the above zero mass eigenstate is primarily composed of the neutrino gauge eigenstate ν_L. Although this neutrino mass eigenstate also contains a small, non-trivial admixture of Kaluza-Klein states, the dominant component of our massless eigenstate remains the gauge-eigenstate neutrino ν_L, as required phenomenologically. Nevertheless, this combined neutrino mass eigenstate is exactly massless in the limit that the full, infinite tower of Kaluza-Klein states participates in the mixing! In other words, the effects of the infinite tower of Kaluza-Klein states for the Ψ field has driven the neutrino mass exactly to zero! We stress that this remarkable result is valid *regardless* of the value of neutrino Yukawa coupling m or the radius scale R^{-1} of the Kaluza-Klein states.

One might still worry that a zero neutrino mass is unacceptable because of the recent evidence for neutrino oscillations. However, even though the neutrino mass is vanishing in this scenario, there continue to exist oscillations because of the non-trivial mixings between the neutrino and the infinite tower of Kaluza-Klein states. Specifically, upon diagonalizing the mass matrix (14) and calculating the resulting oscillation probabilities in the usual way [34], we find the result shown in Fig. 7. This figure thus provides an explicit verification that neutrino oscillations do indeed occur, even though the physical neutrino is exactly massless. Of course, in this section we have been discussing only the simple case of neutrino/anti-neutrino oscillations. However, this mechanism can easily be generalized to include the case of flavor oscillations as well.

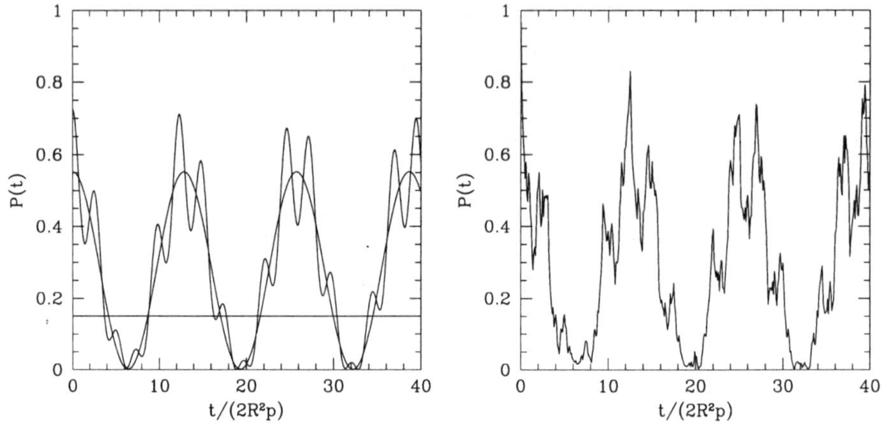

Figure 7: Higher-dimensional neutrino oscillations, even when the neutrino itself is massless. In each case, we plot the probability that the gauge neutrino eigenstate is *preserved* as a function of time. (a) Here we show this probability as more and more Kaluza-Klein states are included in the calculation. We have taken $mR = 0.4$ for concreteness. The flat line shows the contribution when only the massless neutrino is included (no oscillations); the cosine shows the probability when the first excited Kaluza-Klein states are also included; and the irregular curve shows the interference that results when the second excited Kaluza-Klein states are also included. Note that the initial probability $P(t = 0)$ approaches 1 as the full spectrum of Kaluza-Klein states is included, as required. (b) The final result: the total probability that the gauge neutrino ν_L is preserved as a function of time when all Kaluza-Klein states are included. The multi-component nature of the neutrino oscillation is reflected in the jagged shape of the oscillations. Note that the neutrino deficits are total even though the neutrino regenerations are only partial. See Ref. [34] for further details.

At first sight, it may seem strange that within the context of a higher-dimensional seesaw mechanism, we are able to have neutrino oscillations without neutrino masses. However, the crucial point is that once the infinite towers of Kaluza-Klein states are included, our mixing mass matrix (14) with $M_0 = (2R)^{-1}$ has the remarkable property that it yields a zero eigenvalue *without becoming diagonal*. This is not possible in the usual four-dimensional neutrino/anti-neutrino scenarios, but reflects the additional freedom that we have in higher dimensions to explain neutrino oscillations independently of neutrino masses. Indeed, the neutrino mass eigenstate oscillates into the entire tower of higher-dimensional Kaluza-Klein neutrinos, even though it has no mass of its own. The masses of the right-handed Kaluza-Klein states themselves are sufficient to generate the desired oscillations indirectly.

We conclude, then, that sizable neutrino oscillations can occur *regardless of the actual mass of the neutrino*, thanks to the indirect masses and mixings of the Kaluza-Klein states. Thus, if such a scenario can be realized within the context of a fully realistic string model, then the recent observations of neutrino oscillations can be re-interpreted not as providing evidence for neutrino masses, but rather as providing evidence for extra spacetime dimensions!

6 A brief guide to the TeV-scale literature

Since the original papers on this subject appeared in March 1998, a small body of TeV-scale literature has already developed[1–48]. Moreover, as Fig. 1 suggests, this body of literature can be expected to grow rapidly in the near future. Therefore, in this section, we shall give a brief but hopefully comprehensive guide to the literature. Note that papers are listed in the References in the strict chronological order corresponding to the receipt of their first versions on the `hep-th` or `hep-ph` electronic archives. Also note that we have included only those papers that directly deal with the possibility of substantially lowering the GUT, Planck, or string scales below their usual values through large-radius compactifications.

6.1 Original proposals

The original proposal for lowering the compactification scale to the TeV-range was made in Ref. [1] within the context of weakly coupled heterotic string theory. This idea has been pursued more recently in Refs. [9,24,45].

The original proposal for lowering the Planck scale was made in Ref. [4], and further pursued in Refs. [6,11].

The original proposal for lowering the GUT scale, as discussed in Sect. 2, was made in Ref. [5] and further pursued in Refs. [10,14].

The original proposal for lowering the string scale to the usual GUT scale was made in Ref. [2]. This proposal was subsequently generalized to lower the string scale to the TeV-range in Ref. [3]. The idea of embedding the proposal of Ref. [4] into string theory was discussed in Ref. [6], and Refs. [7,10] discussed a unified embedding of the combined proposals of Refs. [4,5] into string theory.

In various forms, the following issues are considered as different extensions of the above proposals.

6.2 Supersymmetry breaking, electroweak symmetry breaking, effective potentials

One possibility for supersymmetry breaking is to identify the scale of supersymmetry breaking with a compactification scale via the Scherk-Schwarz mechanism [54]. This was the original motivation behind the proposal in Ref. [1]. More recent papers [24,45] use this idea in order to derive an effective potential as a function of the compactification radius as a means of possibly explaining the dynamical origin of such a TeV-scale radius. This calculation is mathematically similar to the finite-temperature calculation of the one-loop effective potential in Ref. [17]. An analysis of the possibility of generating electroweak symmetry breaking is also discussed in Refs. [24,45,38], and the idea of using extra large dimensions to trigger electroweak symmetry breaking within the framework of a top-condensate model is discussed in Ref. [40]. Other issues pertaining to supersymmetry breaking (particularly whether this occurs in the bulk or on the brane) are discussed in Refs. [20,21,25,30]. Note that explicit realizations of this sort of partial supersymmetry breaking in Type I string theory can be found in Ref. [55].

6.3 Gauge coupling unification

As noted above, the original proposal for using extra spacetime dimensions to lower the GUT scale was made in Refs. [5,10,14]. As discussed in Sect. 2, this proposal utilizes the power-law running of gauge couplings that is expected in higher-dimensional theories, and which can be understood from a four-dimensional point of view as resulting from a summation over Kaluza-Klein states [49]. Since then, there have been a number of generalizations or extensions of this idea. Two-loop corrections were incorporated in Ref. [16]. Likewise, in Ref. [14], it is pointed out that the unification of gauge couplings achieved in this manner is not exact, and in Ref. [35] this proposal is therefore generalized to include extra matter states in such a way that the resulting unification is exactly as precise as in the usual MSSM. Discussions of higher-loop effects, perturbativity, and precision can be found in Refs. [14,35]. The relation of the higher-dimensional gauge coupling evolution equations to those obtained via a Wilson renormalization-group approach is discussed in Ref. [38]. Likewise, the possible unification of the gauge couplings in a non-supersymmetric context is discussed in Refs. [10,45].

There are also various alternative proposals for understanding gauge coupling unification. Ref. [7] points out that in Type I string theory, gauge coupling unification need not occur if the different Standard-Model group factors are realized on different branes. In this case, the question of gauge coupling unification becomes a question of fine-tuning the various brane compactification volumes. In Ref. [41], this observation is further generalized: even if the different gauge groups are realized on the same brane, gauge coupling need not occur if the brane is located sufficiently close to a singularity. An alternative possibility is advanced in Ref. [13], where it is pointed out that the usual logarithmic running of the gauge couplings can occasionally be preserved in Type I string theory due to a decoupling of the relevant Kaluza-Klein modes, even if the string scale is substantially lowered. Finally, yet another possibility [25] is that power-law running for the gauge couplings is not required if extra matter states with the correct quantum numbers are present in the spectrum so as to lower the scale of unification. This proposal is similar to a situation that can arise in the heterotic context [50,56], but does not suffice in this context to lower the string scale all the way to the TeV-range. The advantages and disadvantages of the different possibilities for gauge coupling unification are also discussed in Refs. [21,41].

6.4 Fermion mass hierarchy, flavor physics

Several different ideas have been advanced for addressing the fermion mass hierarchy problem. In Refs. [5,10], it is proposed that extra spacetime dimensions can induce power-law running for the Yukawa couplings in much the same way as they do for the gauge couplings. This was also previously pointed out in Ref. [57]. Therefore, if the different fermions can be arranged to feel different numbers of extra spacetime dimensions (as might arise in a Type I string model with a non-trivial brane configuration or as the result of a non-trivial vacuum shift), then this power-law running can "amplify" a small flavor-dependence at high scales into a full-fledged hierarchy at lower scales [5,10]. In Ref. [19], this proposal was further studied through a fixed-point analysis, following an earlier fixed-point proposal along these lines in Ref. [57]. The running of fermion Yukawa couplings was also studied in Ref. [38].

An alternative approach towards understanding general issues of flavor physics was proposed in Ref. [30]. In this paper, it is pointed out that approximate symmetries (of whatever kind) can be explained as being strongly broken on a distant brane; this breaking is then communicated to our brane via a bulk messenger field, and results in a breaking on our brane which is therefore suppressed by power-law or exponential factors arising from the distance between the branes. In a flavor context involving flavor symmetries, this might explain how to prevent dangerous flavor-changing neutral currents; the fermion mass hierarchy might also be interpreted as arising due to the varying distances between different branes. A complementary proposal for suppressing flavor-changing neutral currents via non-abelian flavor symmetries in the bulk was made in Ref. [33].

6.5 Proton decay

When the fundamental physical scale is reduced below the usual GUT, Planck, or string scales, the higher-order operators that can mediate proton decay (and more generally, nucleon decay) are no longer suppressed by heavy mass scales. They can therefore mediate proton decay at an unacceptably large rate. Many papers have proposed mechanisms to avoid this. In Refs. [5,10], it was proposed that a certain discrete $\mathbb{Z}_2$ orbifold symmetry could prevent a large class of such operators from arising. However, this is still not a full solution to the problem. In Ref. [7], it was pointed out that within the context of a certain set of reduced-scale Type I string models, the proton is exactly stable due to an unbroken $U(1)$ gauge symmetry corresponding to transverse center-of-mass motion of the branes. However, this results in an unacceptable massless gauge boson, and it is unclear how to break this $U(1)$ gauge symmetry. In Ref. [42], it was then proposed that the idea of discrete symmetries could be generalized, in a manner similar to that proposed in Ref. [58], such that all proton-destabilizing operators are eliminated. The success of this method relies on extending the MSSM via certain extra matter states in such a way that gauge coupling unification is preserved, as discussed in Ref. [35].

There are also other ideas pertaining to proton stability. In Ref. [25], for example, it was pointed out that reducing the string scale to an *intermediate* range (*e.g.*, $\approx 10^{10-12}$ GeV) would not require as full a suppression of the higher-dimension terms as is required in the full TeV-scale limit. Likewise, in Ref. [30], it was pointed out that the idea of approximate symmetries arising from breaking on distant branes can also be applied to explain proton stability.

6.6 Neutrino masses

To date, there have been two different sets of proposals [34,36] for explaining the small size of the neutrino masses. Ordinarily, the most popular idea for explaining such neutrino masses is through the so-called seesaw mechanism [53]. However, the seesaw mechanism relies on the existence of heavy mass scale $M \approx 10^{12-16}$ GeV associated with a right-handed neutrino; the neutrino mass then "seesaws" against this heavy mass scale, and becomes light. However, in the reduced-scale theories, such a heavy mass scale is not available, which would seem to suggest that obtaining light neutrinos would be problematic.

Refs. [34,36] both start from the assumption, as discussed in Sect. 5, that the analogue of the right-handed neutrino is actually a bulk field which experiences extra spacetime dimensions. In Ref. [34], several proposals are then made for suppressing the resulting left-handed neutrino mass. One proposal is that the neutrino Yukawa coupling will be suppressed due to the power-law running that this coupling experiences due to the Kaluza-Klein modes of the right-handed bulk neutrino. Another proposal is that the seesaw mechanism itself can be generalized to higher dimensions; instead of the simple 2×2 mixing between the left-handed and right-handed neutrinos, we actually have an infinite-dimensional mixing between the left-handed neutrino and an infinite Kaluza-Klein tower of right-handed neutrinos. This too can effectively suppress the left-handed neutrino mass. Within the context of such a higher-dimensional seesaw mechanism, further observations [34] include the surprising effects of adding a bare Majorana mass term for the right-handed neutrino, and the proposal of a new method for breaking lepton-number which involves smoothly shifting the position of the branes within Type I string theory. All of these mechanisms result in light *Majorana* mass terms for the left-handed neutrino. Finally, perhaps the most radical proposal in Ref. [34] is the possibility of having neutrino oscillations entirely without neutrino masses, as outlined in detail in Sect. 5.

In Ref. [36], it is pointed out that putting the right-handed neutrino in the bulk naturally suppresses the Dirac coupling between this right-handed neutrino (a bulk field) and the left-handed neutrino (a brane field) by a factor corresponding to the volume of the bulk. In some situations, this then can result in a sufficiently light *Dirac* neutrino mass. In order to generate Majorana neutrino masses, it is then proposed to break lepton-number on a distant brane, following the approach described in Ref. [30].

Since these two proposals [34,36] originally appeared, there have been two further analyses of these ideas. In Ref. [42], it was shown how to suppress dangerous higher-order Majorana neutrino mass operators which could otherwise destabilize the light neutrino. Once again, the idea is to employ a discrete symmetry along the lines of Ref. [58]. Finally, in Ref. [47], a short phenomenological analysis of the possible resulting lepton non-universality is made.

6.7 Field-theoretic implementations

As discussed in Refs. [2,3,6,7,10,20,41], the most natural implementation of these reduced Planck-scale and reduced GUT-scale ideas is within the context of a reduced-scale string theory. Nevertheless, for many purposes it is useful to have an effective field-theoretic description of brane physics. This subject is developed in Refs. [8,12]. Related earlier works can also be found in Ref. [59].

6.8 Radius generation and stabilization

The fundamental idea underpinning the possibility of lowering the Planck, GUT, and string scales is that there exist compactification radii whose sizes exceed the fundamental underlying string scale. In the case of "universal" extra dimensions, these radii need only be one or two orders of magnitude larger than the string scale, while in the case of "gravity-only" (off the brane) extra dimensions, these radii must be larger than the string scale by many orders of magnitude. Note that in the presence of supersymmetry, the compactification radius is generally a string modulus field, and hence any value is permitted, with none preferred. There are then two important questions that emerge in this context. First, how are such large radii generated? This is a dynamical question. Second, how are such large radii stabilized? This may or may not be a dynamical question.

The first question has been addressed in Ref. [17], where it was shown that thermal effects (such as might have existed in the early universe) necessarily break supersymmetry and thereby establish a set of potentials for the radii of the "universal" and "gravity-only" extra dimensions. As discussed in Sect. 4, these potentials can push these radii out to large but different values, as required. This calculation is analogous to similar calculations that were subsequently performed [24,45] in the case of "universal" extra dimensions participating in Scherk-Schwarz supersymmetry breaking.

As we remarked at the end of Sect. 4, once a large radius is *generated*, many effects can potentially *stabilize* it. These include ordinary (low-T) supersymmetry-breaking potentials; a bulk cosmological constant (to halt expansion) [18]; the cosmological dynamics of the four original radii (which we have thus far neglected); possible inter-brane forces (*e.g.*, classical and Casimir forces) [18]; topological mechanisms (*e.g.*, trapped fluxes [12,18], or even string or brane winding modes [52]). Moreover, within a cosmological context, another simple possibility is *entropy conservation* [17], since the radii cannot expand without limit if the universe undergoes an adiabatic evolution. All of these ideas are rather speculative, and it remains to be seen which, if any, can be incorporated into a full dynamical (or even cosmological) model.

6.9 Other "Standard-Model" issues

There are numerous other theoretical and phenomenological issues that are relevant for these reduced-scale theories, and which have been briefly discussed in various papers. These include the issues of anomalous $U(1)$ gauge symmetries (as discussed in Ref. [41]); strong CP / axion physics and the Peccei-Quinn mechanism (as discussed in Refs. [11,21,25]); the idea that for many purposes an intermediate ($\approx 10^{10-14}$ GeV) scale is preferable to the TeV-limit (as discussed in Refs.[21,25,41]); general constraints from rare decays and precision observables (as discussed in Ref.[11]); and the possibility of bulk technicolor (as discussed in Ref.[30]). More formal issues include the effects of extra spacetime dimensions on the general behavior of scattering cross sections (as discussed in Ref.[27]); the effects of extra spacetime dimensions on the form of the Kähler potential (as discussed in Ref.[48]); the issue of dilaton stabilization in Type I string theory (as discussed in Ref.[20]); and a careful investigation of the technical gauge hierarchy in Type I string theory (as discussed in Ref.[39]). General discussions of possible choices for compactification, unification, gravitational, and string scales are given in Refs.[2,3,7,20,21,25,41]. Earlier discussions, particularly in reference to large-radius compactifications in string theory and their phenomenological implications, can be found in Refs.[60,61]. We also note that there exists an alternative proposal[62] for solving the gauge hierarchy problem in higher-dimensional theories; however, this proposal does not utilize these extra dimensions to lower the fundamental scales of physics.

6.10 Collider signatures

One of the most appealing features of the TeV-scale proposals is the possibility that the physics of unification, quantum gravity, and even string theory might actually be seen in upcoming collider experiments. This has given rise to a number of papers dealing with explicit calculations of possible experimental observables at the Tevatron, the LHC, HERA, *etc.* In particular, Feynman rules appropriate for the interactions of Kaluza-Klein gravitons with ordinary Standard-Model matter fields have been derived in Refs.[26,28,31], and these have been used[26,27,28,31,32,37,44,46] to study the missing-energy signatures that result from Kaluza-Klein graviton production, the unique signatures that come from virtual Kaluza-Klein graviton exchange, and the rates and implications of Kaluza-Klein decays. By contrast, discussions of TeV-scale "universal" extra dimensions and the collider signatures of the Kaluza-Klein modes corresponding to Standard Model fields can be found in Refs.[1,10]. Together, these analyses have been used to determine experimental bounds on the respective sizes of "universal" and "gravity-only" extra dimensions, and to quantify the extra-dimension discovery potential of different future colliders.

6.11 Astrophysics and cosmology

Finally, it is of course self-evident that many of these ideas concerning extra "universal" and "gravity-only" extra dimensions and reduced fundamental energy scales will play an important role in astrophysics and cosmology. Just as in particle physics, there are two ways in which astrophysical and cosmological concerns can influence these reduced-scale proposals: they can be used to provide "experimental" (*i.e.*, observational) constraints on these proposals, and they can in turn stimulate the development of new theoretical scenarios or models for explaining cosmological evolution.

There already exist a number of papers dealing with these subjects. Astrophysical constraints (for example, coming from stellar evolution and supernovae) are discussed in Ref. [11]. In such situations, the primary worry is that energy loss (cooling) might be unacceptably accelerated via the production of Kaluza-Klein graviton modes. Various cosmological issues have also been examined. For example, dark matter is discussed in Refs. [11,21], and cosmic rays are discussed in Refs. [11,27]. Big-bang nucleosynthesis and baryogenesis are discussed in Refs. [11,22]. The properties of sub-millimeter black holes are discussed in Ref. [15], where it is shown that these black holes are smaller, colder, and longer-lived than would otherwise be expected. Higher-dimensional cosmological phase transitions are discussed in Ref. [17], where it is shown that the presence of extra spacetime dimensions lowers the associated critical temperatures and prevents such phase transitions from being first-order. Issues pertaining to the cosmological constant problem in higher dimensions are discussed in Refs. [7,12,18,27], and an earlier proposal that might be interesting in this regard is made in Ref. [63]. The properties of the string Hagedorn phenomenon as it affects string phenomenology and cosmology in the presence of extra large spacetime spacetime dimensions and reduced energy scales are also discussed in Ref. [17].

A final important issue is inflation. In Refs. [23,29], a traditional approach is taken wherein the inflaton is strictly a "brane field". It is then pointed out that the inflaton mass must be extremely light, which would appear to require supersymmetry or some other analogous mechanism to protect the inflaton mass. In Ref. [29], it is further pointed out that such a scenario might be more viable if the fundamental Planck scale is lowered to an intermediate value, rather than all the way to the TeV-range. Nevertheless, horizon and flatness problems continue to persist [29]. More recently, however, an entirely new approach to inflation was proposed [43]. In such a scenario, one imagines branes that are initially separated; the resulting positive vacuum energy corresponding to this brane configuration then triggers an exponential inflation in the three non-compact "brane" directions. In such a scenario, the inflaton essentially corresponds to the distance between the branes. The flatness of the inter-brane potential at such large distances guarantees that the inflation is exponential and persists for many e-foldings. Ultimately, as the branes come together, the inter-brane potential becomes steeper and the inflation naturally ends. Subsequent oscillations of the branes about their equilibrum collision point can then induce brane reheating, and density perturbations in this context are understood as arising from the transverse quantum deformations of the branes. However, there are still many unresolved issues in this approach. One such issue is whether sufficient brane reheating and density perturbations can be generated. More crucially, however, it is necessary to explain the origin of the initial conditions whereby the branes have a relatively large initial separation.

Note that a general discussion of cosmology in higher dimensions can be found in Ref. [64].

Acknowledgments

KRD wishes to thank the organizers of the 22nd Annual Johns Hopkins Workshop for the invitation to attend such a stimulating meeting, and for the opportunity to give a talk on some of the recent developments concerning large extra spacetime dimensions. He is also appreciative of the fact that no page limit was placed on these proceedings contributions.

References

Note: Refs. [1–48] are given in the strict chronological order in which their original versions were received on the electronic archives. Note that in this set of references, we have included only those papers that directly deal with the possibility of substantially lowering the GUT, Planck, or string scales below their usual values using extra spacetime dimensions. The remaining references correspond to other important related work.

1. I. Antoniadis, *Phys. Lett.* **B246** (1990) 377;
   I. Antoniadis, C. Muñoz, and M. Quirós, *Nucl. Phys.* **B397** (1993) 515;
   I. Antoniadis and K. Benakli, *Phys. Lett.* **B326** (1994) 69;
   I. Antoniadis, K. Benakli, and M. Quirós, *Phys. Lett.* **B331** (1994) 313.
2. E. Witten, *Nucl. Phys.* **B471** (1996) 135.
3. J.D. Lykken, *Phys. Rev.* **D54** (1996) 3693.
4. N. Arkani-Hamed, S. Dimopoulos, and G. Dvali, `hep-ph/9803315`.
5. K.R. Dienes, E. Dudas, and T. Gherghetta, `hep-ph/9803466`.
6. I. Antoniadis, N. Arkani-Hamed, S. Dimopoulos, and G. Dvali, `hep-ph/9804398`.
7. G. Shiu and S.-H.H. Tye, `hep-th/9805157`.
8. R. Sundrum, `hep-ph/9805471`.

9. A. Pomarol and M. Quirós, hep-ph/9806263.
10. K.R. Dienes, E. Dudas, and T. Gherghetta, hep-ph/9806292.
11. N. Arkani-Hamed, S. Dimopoulos, and G. Dvali, hep-ph/9807344.
12. R. Sundrum, hep-ph/9807348.
13. C. Bachas, hep-ph/9807415.
14. K.R. Dienes, E. Dudas, and T. Gherghetta, hep-ph/9807522.
15. P.C. Argyres, S. Dimopoulos, and J. March-Russell, hep-th/9808138.
16. D. Ghilencea and G.G. Ross, hep-ph/9809217.
17. K.R. Dienes, E. Dudas, T. Gherghetta, and A. Riotto, hep-ph/9809406.
18. N. Arkani-Hamed, S. Dimopoulos, and J. March-Russell, hep-th/9809124.
19. S.A. Abel and S.F. King, hep-ph/9809467.
20. Z. Kakushadze and S.-H.H. Tye, hep-th/9809147.
21. K. Benakli, hep-ph/9809582.
22. K. Benakli and S. Davidson, hep-ph/9810280.
23. D.H. Lyth, hep-ph/9810320.
24. I. Antoniadis, S. Dimopoulos, A. Pomarol, and M. Quirós, hep-ph/9810410.
25. C.P. Burgess, L.E. Ibáñez, and F. Quevedo, hep-ph/9810535.
26. G.F. Giudice, R. Rattazzi, and J.D. Wells, hep-ph/9811291.
27. S. Nussinov and R. Shrock, hep-ph/9811323.
28. E.A. Mirabelli, M. Perelstein, and M.E. Peskin, hep-ph/9811337.
29. N. Kaloper and A. Linde, hep-th/9811141.
30. N. Arkani-Hamed and S. Dimopoulos, hep-ph/9811353.
31. T. Han, J.D. Lykken, and R.-J. Zhang, hep-ph/9811350.
32. J.L. Hewett, hep-ph/9811356.
33. Z. Berezhiani and G. Dvali, hep-ph/9811378.
34. K.R. Dienes, E. Dudas, and T. Gherghetta, hep-ph/9811428.
35. Z. Kakushadze, hep-th/9811193.
36. N. Arkani-Hamed, S. Dimopoulos, G. Dvali, and J. March-Russell, hep-ph/9811448.
37. P. Mathews, S. Raychaudhuri, and K. Sridhar, hep-ph/9811501.
38. T. Kobayashi, J. Kubo, M. Mondragón, and G. Zoupanos, hep-ph/9812221.
39. I. Antoniadis and C. Bachas, hep-th/9812093.
40. B. Dobrescu, hep-ph/9812349.
41. L.E. Ibáñez, C. Muñoz, and S. Rigolin, hep-ph/9812397.
42. Z. Kakushadze, hep-th/9812163.
43. G. Dvali and S.-H.H. Tye, hep-ph/9812483.
44. P. Mathews, S. Raychaudhuri, and K. Sridhar, hep-ph/9812486.

45. A. Delgado, A. Pomarol, and M. Quirós, hep-ph/9812489.
46. T.G. Rizzo, hep-ph/9901209.
47. A.E. Faraggi and M. Pospelov, hep-ph/9901299.
48. T.E. Clark and S.T. Love, hep-th/9901103.

49. T.R. Taylor and G. Veneziano, *Phys. Lett.* **B212** (1988) 147.
50. For a review, see K.R. Dienes, *Phys. Reports* **287** (1997) 447 [hep-th/9602045].
51. See, *e.g.*: L.E. Ibáñez, hep-th/9112050; hep-th/9505098;
I. Antoniadis, hep-th/9307002;
M. Dine, hep-ph/9309319;
J.D. Lykken, hep-ph/9511456; hep-th/9607144;
F. Quevedo, hep-th/9603074; hep-ph/9707434.
52. R. Brandenberger and C. Vafa, *Nucl. Phys.* **B316** (1989) 391.
53. M. Gell-Mann, P. Ramond, and R. Slansky, in *Supergravity* (edited by P. van Nieuwenhuizen and D.Z. Freedman), North-Holland, Amsterdam (1979);
T. Yanagida, in *Proceedings of the Workshop on Unified Theories and Baryon Number in the Universe* (edited by A. Sawada and A. Sugamoto), K.E.K. preprint 79-18 (1979).
54. J. Scherk and J.H. Schwarz, *Phys. Lett.* **B82** (1979) 60.
55. I. Antoniadis, G. D'Appollonio, E. Dudas, and A. Sagnotti, hep-th/9812118.
56. K.R. Dienes and A.E. Faraggi, *Phys. Rev. Lett.* **75** (1995) 2646; *Nucl. Phys.* **B457** (1995) 409.
57. M. Lanzagorta and G.G. Ross, *Phys. Lett.* **B349** (1995) 319.
58. L.E. Ibáñez and G.G. Ross, *Nucl. Phys.* **B368** (1992) 3.
59. See, *e.g.*: V.A. Rubakov and M.E. Shaposhnikov, *Phys. Lett.* **B125** (1983) 136;
A. Lukas, B.A. Ovrut, K.S. Stelle, and D. Waldram, hep-th/9803235;
60. T. Banks and M. Dine, *Nucl. Phys.* **B479** (1996) 173; *ibid.* **B505** (1997) 445; hep-th/9609046.
61. E. Caceres, V.S. Kaplunovsky and I.M. Mandelberg, *Nucl. Phys.* **B493** (1997) 73.
62. H. Hatanaka, T. Inami, and C.S. Lim, hep-th/9805067.
63. R. Sundrum, hep-ph/9708329.
64. E.W. Kolb, Proceedings of the 1986 Erice School, p. 225.

SOME ASPECTS OF NONCOMMUTATIVE GEOMETRY AND PHYSICS

A. DIMAKIS
Department of Mathematics
University of the Aegean
GR-83200 Karlovasi, Samos, Greece
E-mail: dimakis@aegean.gr

F. MÜLLER-HOISSEN
Institut für Theoretische Physik
Bunsenstr. 9, D-37073 Göttingen, Germany
E-mail: hoissen@theorie.physik.uni-goettingen.de
and
Max-Planck-Institut für Strömungsforschung
Bunsenstr. 10, D-37073 Göttingen, Germany
E-mail: fmuelle@gwdg.de

An introduction is given to some selected aspects of noncommutative geometry. Simple examples in this context are provided by finite sets and lattices. As an application, it is explained how the nonlinear Toda lattice and a discrete time version of it can be understood as generalized σ-models based on noncommutative geometries. In particular, in this way one achieves a simple understanding of the complete integrability of the Toda lattice. Furthermore, generalized metric structures on finite sets and lattices are briefly discussed.

1 Introduction

Noncommutative (differential) geometry extends notions of classical differential geometry from differentiable manifolds to discrete spaces, like finite sets and fractals, and even 'noncommutative spaces' ('quantum spaces') which are given by noncommutative associative algebras (over $\mathbb{R}$ or $\mathbb{C}$). Such an algebra $\mathcal{A}$ replaces the commutative algebra of C^∞-functions on a smooth manifold $\mathcal{M}$.

A basic concept of classical differential geometry is the notion of a vector field. The latter is a derivation $C^\infty(\mathcal{M}) \to C^\infty(\mathcal{M})$. One may think of generalizing the notion of a vector field as a derivation $\mathcal{A} \to \mathcal{A}$ to associative algebras.[1] But there are interesting algebras, like the (commutative) algebra of functions on a finite set, on which there is no derivation at all except the trivial one (which maps all functions to zero). It seems that there is no adequate general definition of a vector field.

Perhaps of even more importance than vector fields are differential forms in classical differential geometry and theoretical physics. They extend the algebra $\Omega^0(\mathcal{M}) := C^\infty(\mathcal{M})$ to a graded associative algebra $\Omega(\mathcal{M}) = \bigoplus_{r\geq 0} \Omega^r(\mathcal{M})$, the algebra of differential forms. A central part of its structure is encoded in the action of the exterior derivative $\mathrm{d} : \Omega^r(\mathcal{M}) \to \Omega^{r+1}(\mathcal{M})$ which is a (graded) derivation. Given any associative algebra $\mathcal{A}$, one can always associate with it a differential algebra $\Omega(\mathcal{A})$ which should be regarded as an analogue of the algebra of classical differential forms on a manifold. Further geometric structures, like connections and tensors, can then be built on such a differential algebra in very much the same way as in classical differential geometry.

There is no universal way of associating a unique differential algebra with a given associative algebra $\mathcal{A}$. In general, there is no distinguished differential algebra and one has to understand what the significance is of the different choices. Even in the case of the algebra of smooth functions on a manifold, there is actually no longer a good argument to single out the ordinary calculus of differential forms. The latter is distinguished, however, via our classical conception how to measure volumes. The exploration of other differential calculi opened a door to a new world of geometry and applications in physics, as well as relations with other fields of mathematics. Some aspects have been briefly reviewed in a recent paper [2] to which we refer for further information.[a]

The formalism of noncommutative geometry is an extremely radical abstraction of ordinary differential geometry. It includes the latter as a special case, but allows a huge variety of different structures. In particular, it is possible to 'deform' ordinary differential geometry and models built on it while keeping basic concepts and recipes on which the models are based.

In the following we review some aspects of noncommutative geometry concentrating on a few easily accessible examples. Section 2 collects basic definitions of differential calculus on associative algebras. Section 3 recalls some facts about differential calculus on finite sets and a correspondence between first order differential calculi and digraphs.[3] Also a relation with an approach of Alain Connes [4] to noncommutative geometry is explained. Section 4 treats differential calculus on linear and hypercubic lattices. The corresponding differential calculi may be regarded as deformations of the ordinary differential calculus on $\mathbb{R}^n$. A more general class of such differential calculi is briefly discussed in section 5. The 'lattice differential calculus' underlies an important example of a discrete 'generalized σ-model'. As explained in section 6, such models generalize a class of completely integrable two-dimensional classical σ-models [5] by replacing the ordinary differential calculus by a noncommutative

[a]At present, this can be accessed online via *http://kaluza.physik.uni-konstanz.de/2MS.*

one.[6,7] They involve a generalized Hodge $\star$-operator. In classical (Riemannian) differential geometry, the Hodge operator is obtained from a metric which defines the distance between two points of the manifold. Distance functions on a rather general class of algebras have been introduced by Connes.[4] In particular, his definition applies to discrete sets. But the relation with a Hodge operator has still to be worked out. Section 7 is devoted to a discussion of some metric aspects. Section 8 contains some final remarks.

2 Differential calculi on associative algebras

Let $\mathcal{A}$ be an associative algebra over $\mathbb{R}$ or $\mathbb{C}$ with unit $\mathbb{I}$. A *differential calculus* on $\mathcal{A}$ is a $\mathbb{Z}$-graded associative algebra (over $\mathbb{R}$, respectively $\mathbb{C}$)[b]

$$\Omega(\mathcal{A}) = \bigoplus_{r\geq 0} \Omega^r(\mathcal{A}) \tag{1}$$

where the spaces $\Omega^r(\mathcal{A})$ are $\mathcal{A}$-bimodules[c] and $\Omega^0(\mathcal{A}) = \mathcal{A}$. There is a ($\mathbb{R}$-respectively $\mathbb{C}$-) linear map

$$\mathrm{d}\ :\quad \Omega^r(\mathcal{A}) \to \Omega^{r+1}(\mathcal{A}) \tag{2}$$

with the following properties,

$$\mathrm{d}^2 = 0 \tag{3}$$

$$\mathrm{d}(w\,w') = (\mathrm{d}w)\,w' + (-1)^r\,w\,\mathrm{d}w' \tag{4}$$

where $w \in \Omega^r(\mathcal{A})$ and $w' \in \Omega(\mathcal{A})$. The last relation is known as the (generalized) *Leibniz rule*. One also requires $\mathbb{I}\,w = w\,\mathbb{I} = w$ for all elements $w \in \Omega(\mathcal{A})$. The identity $\mathbb{I}\mathbb{I} = \mathbb{I}$ then implies

$$\mathrm{d}\mathbb{I} = 0\ . \tag{5}$$

We assume that d generates the spaces $\Omega^r(\mathcal{A})$ for $r > 0$ in the sense that $\Omega^r(\mathcal{A}) = \mathcal{A}\,\mathrm{d}\Omega^{r-1}(\mathcal{A})\,\mathcal{A}$. Using the Leibniz rule, every element of $\Omega^r(\mathcal{A})$ can be written as a linear combination of monomials $a_0\,\mathrm{d}a_1 \cdots \mathrm{d}a_r$. The action of d is then determined by

$$\mathrm{d}(a_0\,\mathrm{d}a_1 \cdots \mathrm{d}a_r) = \mathrm{d}a_0\,\mathrm{d}a_1 \cdots \mathrm{d}a_r\ . \tag{6}$$

---

[b] Though in many interesting cases one has $\Omega^r(\mathcal{A}) = \{0\}$ when r is larger than some $r_0 \geq 0$, one encounters examples where $\Omega(\mathcal{A})$ is actually an infinite sum. $\Omega(\mathcal{A})$ is then the space of all finite sums of arbitrary order.

[c] The elements of $\Omega^r(\mathcal{A})$, called *r-forms*, can be multiplied from left and right by elements of $\mathcal{A}$.

So far nothing has been said about commutation relations for elements of $\mathcal{A}$ and differentials. Indeed, in the largest differential calculus, the *universal differential envelope* $(\tilde{\Omega}(\mathcal{A}), \tilde{\mathrm{d}})$ of $\mathcal{A}$, there are no such relations. Smaller differential calculi are obtained by specifying corresponding commutation relations (which have to be consistent with the existing relations in the differential algebra, of course). The smallest differential calculus is $\Omega(\mathcal{A}) = \mathcal{A}$ where d maps all elements of $\mathcal{A}$ to zero.

3 Differential calculi on a finite set

Let $\mathcal{M}$ be a finite set and $\mathcal{A}$ the algebra of all $\mathbb{C}$-valued functions on it. $\mathcal{A}$ is generated by $\{e_i\}$ where $e_i(j) = \delta_{ij}$ for $i, j \in \mathcal{M}$. These functions satisfy the two identities

$$e_i\, e_j = \delta_{ij}\, e_j\,, \qquad \sum_i e_i = \mathbb{1}\,. \tag{7}$$

As a consequence of the identities (7) and the Leibniz rule, the differentials $\mathrm{d}e_i$ of a differential calculus on $\mathcal{A}$ are subject to the following relations,

$$\mathrm{d}e_i\, e_j = -e_i\, \mathrm{d}e_j + \delta_{ij}\, \mathrm{d}e_j\,, \qquad \sum_i \mathrm{d}e_i = 0\,. \tag{8}$$

Without additional constraints, we are dealing with the universal differential calculus $(\tilde{\Omega}(\mathcal{A}), \tilde{\mathrm{d}})$. Introducing the 1-forms

$$e_{ij} = e_i\, \tilde{\mathrm{d}}e_j \qquad (i \neq j) \tag{9}$$

one finds that they form a basis over $\mathbb{C}$ of the space $\tilde{\Omega}^1$ of universal 1-forms. Moreover, all first order differential calculi are obtained from the universal one by setting some of the e_{ij} to zero.

Let us associate with each nonvanishing e_{ij} of some differential calculus (Ω, d) an arrow from the point i to the point j :

$$e_{ij} \neq 0 \quad \Longleftrightarrow \quad i\bullet \longrightarrow \bullet j\,. \tag{10}$$

The universal (first order) differential calculus then corresponds to the complete digraph where all vertices are connected with each other by a pair of antiparallel arrows. Other first order differential calculi are obtained by deleting some of the arrows. The choice of a (first order) differential calculus on a finite set therefore means assigning a connection structure to it. The latter is mirrored in the following formula for the differential of a function $f \in \mathcal{A}$,[d]

$$\mathrm{d}f = \sum_{i,j} [f(j) - f(i)]\, e_{ij} \, . \tag{11}$$

Returning to the universal calculus, concatenation of the 1-forms e_{ij} leads to the *basic* $(r-1)$*-forms*

$$e_{i_1 \ldots i_r} := e_{i_1 i_2}\, e_{i_2 i_3} \cdots e_{i_{r-1} i_r} \qquad (r > 1) \, . \tag{12}$$

They constitute a basis of $\tilde{\Omega}^{r-1}$ over $\mathbb{C}$ and satisfy the simple relations

$$e_{i_1 \ldots i_r}\, e_{j_1 \ldots j_s} = \delta_{i_r j_1}\, e_{i_1 \ldots i_{r-1} j_1 \ldots j_s} \, . \tag{13}$$

Furthermore, we have

$$\tilde{\mathrm{d}} e_i = \sum_j (e_{ji} - e_{ij}) \tag{14}$$

$$\tilde{\mathrm{d}} e_{ij} = \sum_k (e_{kij} - e_{ikj} + e_{ijk}) \tag{15}$$

$$\tilde{\mathrm{d}} e_{ijk} = \sum_l (e_{lijk} - e_{iljk} + e_{ijlk} - e_{ijkl}) \tag{16}$$

$$\vdots$$

The first equation is a special case of (11). In a 'reduced' differential calculus (Ω, d) where not all of the e_{ij} are present, the possibilities to build (nonvanishing) higher forms $e_{i_1 \ldots i_r}$ are restricted and the above formulas for $\tilde{\mathrm{d}} e_{i_1 \ldots e_r}$ impose further constraints on them.

Example. The graph drawn in Fig. 1 determines a first order differential calculus with nonvanishing basic 1-forms $e_{12}, e_{23}, e_{14}, e_{43}$. Concatenation only leads to e_{123} and e_{143} as possible nonvanishing basic 2-forms. There are no nonvanishing r-forms with $r > 2$.

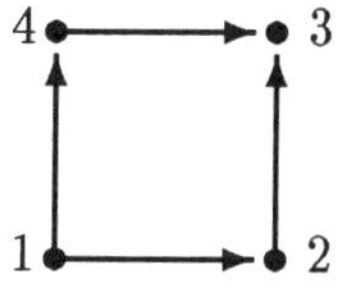

Fig. 1
The digraph associated with a special differential calculus on a set of four elements.

---

[d]More precisely, the summation runs over all i, j with $i \neq j$. Note that e_{ii} has not been defined. We may, however, set $e_{ii} := 0$.

The graph is obtained from the complete digraph by deletion of some arrows. In particular, an arrow from point 1 to point 3 is missing which corresponds to setting e_{13} to zero in the universal differential calculus. This leads to $0 = de_{13} = -e_{123} - e_{143}$. ■

A discrete set together with a differential calculus on it is called a *discrete differential manifold.* [8]

3.1 Representations of first order differential calculi on finite sets

As explained above, first order differential calculi on a set of N elements are in bijective correspondence with digraphs with N vertices and at most a pair of antiparallel arrows between any two vertices. On the other hand, in graph theory such a digraph is characterized by its *adjacency matrix* which is an $N \times N$-matrix $\mathcal{D}$ such that $\mathcal{D}_{ij} = 1$ if there is an arrow from i to j and $\mathcal{D}_{ij} = 0$ otherwise. One should then expect that the (first order) differential calulus determined by a digraph can be expressed in terms of $\mathcal{D}$. The simplest way to build a derivation $\mathrm{d} : \mathcal{A} \to \Omega^1(\mathcal{A})$ is as a commutator,

$$\mathrm{d}f := [\mathcal{D}, f] \tag{17}$$

which presumes, however, that the elements of $\mathcal{A}$ can be represented as $N \times N$-matrices. But this is naturally achieved via

$$f \quad \mapsto \quad \begin{pmatrix} f(1) & & 0 \\ & \ddots & \\ 0 & & f(N) \end{pmatrix} . \tag{18}$$

Comparison of (17) with our formula (11) shows that the basic 1-form e_{ij} is represented as the $N \times N$-matrix E_{ij} with a 1 in the ith row and jth column and zeros elsewhere. The adjacency matrix $\mathcal{D}$ represents $\sum_{i,j} e_{ij}$.

Proceeding beyond 1-forms, the above representation will not respect the $\mathbb{Z}_2$-grading of a differential algebra $\Omega(\mathcal{A})$. One may consider instead a 'doubled' representation [9]

$$e_i \mapsto \begin{pmatrix} E_{ii} & 0 \\ 0 & E_{ii} \end{pmatrix} , \qquad e_{ij} \mapsto \begin{pmatrix} 0 & E_{ij}^\dagger \\ E_{ij} & 0 \end{pmatrix} . \tag{19}$$

The grading can be expressed in terms of a grading operator which in our case is given by

$$\gamma := \begin{pmatrix} \mathbf{1} & 0 \\ 0 & -\mathbf{1} \end{pmatrix} . \tag{20}$$

It is selfadjoint and satisfies

$$\gamma^2 = \mathbf{1}\,, \qquad \gamma\hat{\mathcal{D}} = -\hat{\mathcal{D}}\,\gamma \qquad \gamma\hat{f} = \hat{f}\,\gamma \tag{21}$$

with

$$\hat{\mathcal{D}} := \begin{pmatrix} 0 & \mathcal{D}^\dagger \\ \mathcal{D} & 0 \end{pmatrix} \qquad \hat{f} := \begin{pmatrix} f & 0 \\ 0 & f \end{pmatrix} \tag{22}$$

where f has to be represented as in (18). In this way we do *not* in general obtain a representation of the first order differential calculus which we started with, however, but a representation of the corresponding 'symmetric' differential calculus where with $e_{ij} \neq 0$ there is also $e_{ji} \neq 0$ (so that the associated digraph is symmetric).

With the above representations of (first order) differential calculi we have established contact with Alain Connes' formalism[10] of noncommutative geometry. But in the present context $\mathcal{D}$ is *not*, in general, a *selfadjoint* operator (on the Hilbert space $\mathbb{C}^N$). The 'doubling' in (19) leads to a selfadjoint operator on the Hilbert space $\mathcal{H} = \mathbb{C}^{2N}$, however. $(\mathcal{A}, \mathcal{H}, \hat{\mathcal{D}})$ is an example of an *even spectral triple*, a central structure in Connes' approach to noncommutative geometry.[10,11] A *spectral triple* $(\mathcal{A}, \mathcal{H}, \hat{\mathcal{D}})$ consists of an involutive algebra $\mathcal{A}$ of operators on a Hilbert space $\mathcal{H}$ together with a selfadjoint operator $\hat{\mathcal{D}}$ satisfying some technical conditions. It is called *even* when there is a grading operator γ, as in our example.

4 Lattice differential calculus

Let $\mathcal{M} = \mathbb{Z}$. For $i, j \in \mathcal{M}$ we define a differential calculus by $e_{ij} \neq 0 \iff j = i+1$ following the rules described for finite sets in the previous section. This corresponds to the oriented linear lattice graph drawn in Fig. 2.

Fig. 2
An oriented linear lattice graph.

In this example we are dealing with an *infinite* set and thus infinite sums in some formulas which would actually require a bit more care. Introducing the lattice coordinate function

$$x := \ell \sum_j j\, e_j \tag{23}$$

with a real constant $\ell > 0$, one obtains

$$\mathrm{d}x = \ell \sum_i i\, \mathrm{d}e_i = \ell \sum_{i,j} i\,(e_{ji} - e_{ij}) = \ell \sum_i i\,(e_{i-1,i} - e_{i,i+1}) = \ell \sum_i e_{i,i+1} \tag{24}$$

and

$$[\mathrm{d}x, x] = \ell^2 \sum_{i,j} j\, [e_{i,i+1}, e_j] = \ell^2 \sum_i e_{i,i+1} = \ell\, \mathrm{d}x \tag{25}$$

using (13). Hence

$$[\mathrm{d}x, x] = \ell\, \mathrm{d}x\ . \tag{26}$$

In the limit $\ell \to 0$ the lattice coordinate function x naively approximates the corresponding coordinate function on the real line $\mathbb{R}$. From our last equation we then recover the familiar commutativity of ordinary differentials and functions. The above commutation relation makes also sense, however, on $\mathbb{R}$ when $\ell > 0$. We then have a *deformation* of the ordinary calculus of differential forms on $\mathbb{R}$ with deformation parameter ℓ. In the following we collect some properties of this deformed differential calculus. Written in the form

$$\mathrm{d}x\, x = (x + \ell)\, \mathrm{d}x\ , \tag{27}$$

the above commutation relation extends to the algebra $\mathcal{A}$ of all functions on $\mathbb{R}$ as

$$\mathrm{d}x\, f(x) = f(x + \ell)\, \mathrm{d}x\ . \tag{28}$$

Furthermore,

$$\begin{aligned} \mathrm{d}f &=: (\partial_{+x} f)\, \mathrm{d}x = \frac{1}{\ell}\, (\partial_{+x} f)\, [\mathrm{d}x, x] = \frac{1}{\ell}\, [(\partial_{+x} f)\, \mathrm{d}x, x] \\ &= \frac{1}{\ell}\, [\mathrm{d}f, x] = \frac{1}{\ell}\, (\mathrm{d}(f\, x - x\, f) - [f, \mathrm{d}x]) = \frac{1}{\ell}(\mathrm{d}x\, f - f\, \mathrm{d}x) \\ &= \frac{1}{\ell}[f(x+\ell) - f(x)]\, \mathrm{d}x \end{aligned} \tag{29}$$

so that the *left partial derivative* defined via the first equality turns out to be the right discrete derivative, i.e.,

$$\partial_{+x} f = \frac{1}{\ell}[f(x+\ell) - f(x)]\ . \tag{30}$$

Introducing a *right partial derivative* via $\mathrm{d}f = \mathrm{d}x\, \partial_{-x} f$, an application of (28) shows that it is the left discrete derivative, i.e.,

$$\partial_{-x} f = \frac{1}{\ell}[f(x) - f(x-\ell)]\ . \tag{31}$$

An *indefinite integral* should have the property

$$\int \mathrm{d}f = f + \text{‘constant’} \tag{32}$$

where 'constants' are functions annihilated by d. These are just the functions with period ℓ (so that $f(x+\ell) = f(x)$). It turns out that every function can be integrated.[12] Since the indefinite integral is only determined up to the addition of an arbitrary function with period ℓ, it defines a *definite integral* only if the region of integration is an interval the length of which is a multiple of ℓ (or a union of such intervals). Then one obtains

$$\int_{x_0-m\ell}^{x_0+n\ell} f(x)\,\mathrm{d}x = \ell \sum_{k=-m}^{n-1} f(x_0+k\ell) \tag{33}$$

and in particular

$$\int_{x_0-\infty\ell}^{x_0+\infty\ell} f(x)\,\mathrm{d}x = \ell \sum_{k=-\infty}^{\infty} f(x_0+k\ell)\ . \tag{34}$$

The integral simply picks out the values of f on a *lattice* with spacings ℓ and forms the Riemann integral for the corresponding piecewise constant function on $\mathbb{R}$. The point $x_0 \in \mathbb{R}$ determines how the lattice is embedded in $\mathbb{R}$.

Let now $\mathcal{M} = \mathbb{Z}^n$. For $a, b \in \mathcal{M}$ we define a differential calculus by

$$e_{ab} \neq 0 \quad \Longleftrightarrow \quad b = a + \hat{\mu} \qquad \text{where} \quad \hat{\mu} = (\delta^\nu_\mu)\ . \tag{35}$$

This corresponds to the oriented lattice graph drawn in Fig. 3.

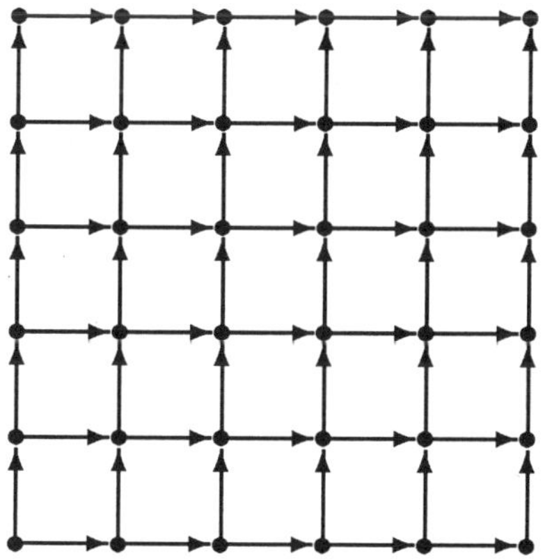

Fig. 3
A finite part of the oriented lattice graph.

Introducing the lattice coordinate functions

$$x^\mu := \ell^\mu \sum_a a^\mu\, e_a \tag{36}$$

with constants $\ell^\mu > 0$, one obtains

$$[dx^\mu, x^\nu] = \ell^\mu \, \delta^{\mu\nu} \, dx^\mu \tag{37}$$

using (13). $\{dx^\mu\}$ is a left $\mathcal{A}$-module basis of Ω^1. The above commutation relations define a deformation of the ordinary differential calculus on $\mathbb{R}^n$. This deformed differential calculus may be regarded as a basic structure underlying lattice field theories.[13]

5 A class of noncommutative differential calculi on a commutative algebra

Let $\mathcal{A}$ be the associative and commutative algebra over $\mathbb{R}$ freely generated by elements x^μ, $\mu = 1, \ldots, n$. For example, the x^μ could be the canonical coordinates on $\mathbb{R}^n$. The ordinary differential calculus on $\mathcal{A}$ has the property $[dx^\mu, x^\nu] = 0$, i.e., differentials and functions commute. Relaxing this property, there is a class of *noncommutative differential calculi* such that[e]

$$[dx^\mu, x^\nu] = C^{\mu\nu}{}_\kappa \, dx^\kappa \tag{38}$$

with *structure functions* $C^{\mu\nu}{}_\kappa(x^\lambda)$ which have to satisfy some consistency conditions. First, we have

$$\begin{aligned}[dx^\mu, x^\nu] &= (dx^\mu)\, x^\nu - x^\nu \, dx^\mu \\ &= d(x^\mu x^\nu - x^\nu x^\mu) - x^\mu \, dx^\nu + (dx^\nu)\, x^\mu = [dx^\nu, x^\mu] \, .\end{aligned} \tag{39}$$

Assuming the differentials dx^μ, $\mu = 1, \ldots, n$, to be linearly independent[f], this implies

$$C^{\mu\nu}{}_\kappa = C^{\nu\mu}{}_\kappa \, . \tag{40}$$

Furthermore,

$$\begin{aligned}0 &= ([dx^\mu, x^\nu] - C^{\mu\nu}{}_\kappa \, dx^\kappa)\, x^\lambda = [(dx^\mu)\, x^\lambda, x^\nu] - C^{\mu\nu}{}_\kappa \, (dx^\kappa)\, x^\lambda \\ &= [x^\lambda \, dx^\mu + C^{\mu\lambda}{}_\rho \, dx^\rho, x^\nu] - C^{\mu\nu}{}_\kappa \, (x^\lambda \, dx^\kappa + C^{\kappa\lambda}{}_\rho \, dx^\rho) \\ &= x^\lambda \, [dx^\mu, x^\nu] + C^{\mu\lambda}{}_\rho \, [dx^\rho, x^\nu] - x^\lambda \, C^{\mu\nu}{}_\kappa \, dx^\kappa - C^{\mu\nu}{}_\kappa \, C^{\kappa\lambda}{}_\rho \, dx^\rho \\ &= (C^{\mu\lambda}{}_\rho \, C^{\rho\nu}{}_\kappa - C^{\mu\nu}{}_\rho \, C^{\rho\lambda}{}_\kappa)\, dx^\kappa\end{aligned} \tag{41}$$

which leads to

$$C^{\lambda\mu}{}_\rho \, C^{\nu\rho}{}_\kappa = C^{\nu\mu}{}_\rho \, C^{\lambda\rho}{}_\kappa \tag{42}$$

---

[e] On the rhs of this equation we are using the summation convention.

[f] More precisely, we assume here that the dx^μ form a left $\mathcal{A}$-module basis of $\Omega^1(\mathcal{A})$.

or, in terms of the matrices C^μ with entries $(C^\mu)^\nu{}_\kappa := C^{\mu\nu}{}_\kappa$,

$$C^\mu \, C^\nu = C^\nu \, C^\mu \, . \tag{43}$$

For constant $C^{\mu\nu}{}_\kappa$ and $n \leq 3$, a classification of all solutions of the consistency conditions (40) and (42) has been obtained.[13,14] Besides the 'lattice differential calculus' discussed in the previous subsection, this includes other interesting deformations of the ordinary differential calculus on $\mathbb{R}^n$.[15,16] The relations (37) are obviously not invariant under (suitable) coordinate transformations. Invariance is achieved, however, with the form (38) of the commutation relations.

From the structure functions we can build

$$g^{\mu\nu} := \mathrm{Tr}(C^\mu \, C^\nu) \tag{44}$$

which for the lattice calculus (37) becomes $(\ell^\mu)^2 \, \delta^{\mu\nu}$, a kind of metric tensor. In the framework under consideration, the metric arises as a composed object. The set of structure functions $C^{\mu\nu}{}_\kappa$ is the more fundamental geometric structure.

6 An application in the context of integrable models

For two-dimensional σ-models there is a construction of an infinite sequence of conserved currents [5] which can be formulated very compactly in terms of ordinary differential forms. This then suggests to generalize the notion of a σ-model to noncommutative differential calculi such that the construction of conservation laws still works. In this way one obtains a simple though very much non-trivial application of the formalism of noncommutative geometry.[6,7]

6.1 Generalized integrable σ-models

Let $\mathcal{A}$ be an associative and commutative[g] algebra with unit $\mathbb{1}$ and (Ω, d) a differential calculus on it. Furthermore, let $\star : \Omega^1 \to \Omega^1$ be an invertible linear map such that

$$\star(w \, f) = f \star w \tag{45}$$

and

$$w \star w' = w' \star w \, . \tag{46}$$

In addition, we require that

$$\mathrm{d}w = 0 \quad \Rightarrow \quad w = \star\star \, \mathrm{d}\chi \tag{47}$$

[g] A generalization of the following to *non*commutative algebras seems to be possible if they admit an involution †. Then (45) has to be replaced by $\star(w \, f) = f^\dagger \star w$.

with $\chi \in \mathcal{A}$. Furthermore, let $a \in GL(n, \mathcal{A})$ and $A := a^{-1}\,\mathrm{d}a$. Then

$$F := \mathrm{d}A + AA \equiv 0 \tag{48}$$

since $\mathrm{d}a^{-1} = -a^{-1}\,(\mathrm{d}a)\,a^{-1}$. These definitions are made in such a way that the field equation of a *generalized σ-model*

$$\mathrm{d} \star A = 0 \tag{49}$$

and a construction of an infinite set of conservation laws in two dimensions[5] generalizes to a considerably more general framework. Let χ be an $n \times n$ matrix with entries in $\mathcal{A}$. Using the two relations (45), (46), and the field equation $\mathrm{d} \star A = 0$, we find

$$\mathrm{d} \star (A^i{}_j\, \chi^j{}_k) = \mathrm{d}(\chi^j{}_k \star A^i{}_j) = (\mathrm{d}\chi^j{}_k) \star A^i{}_j + \chi^j{}_k\, \mathrm{d} \star A^i{}_j = A^i{}_j \star \mathrm{d}\chi^j{}_k \tag{50}$$

and thus

$$\mathrm{d} \star D\chi = \mathrm{d} \star \mathrm{d}\chi + \mathrm{d}(\star A\chi) = \mathrm{d} \star \mathrm{d}\chi + A \star \mathrm{d}\chi = D \star \mathrm{d}\chi \tag{51}$$

where $D\chi := \mathrm{d}\chi + A\,\chi$. Let

$$\chi^{(0)} := \begin{pmatrix} \mathbb{1} & 0 & \cdots & 0 \\ 0 & \ddots & \ddots & \vdots \\ \vdots & \ddots & \ddots & 0 \\ 0 & \cdots & 0 & \mathbb{1} \end{pmatrix} \tag{52}$$

Then

$$J^{(1)} := D\chi^{(0)} = A \tag{53}$$

so that

$$\mathrm{d} \star J^{(1)} = 0 \tag{54}$$

as a consequence of the field equation. Hence, using (47),

$$J^{(1)} = \star\, \mathrm{d}\chi^{(1)} \tag{55}$$

with a matrix $\chi^{(1)}$. Now, let $J^{(m)}$ be a conserved current, i.e.,

$$J^{(m)} = \star\, \mathrm{d}\chi^{(m)} \,. \tag{56}$$

Then

$$J^{(m+1)} := D\chi^{(m)} \qquad (m \geq 0) \tag{57}$$

is also conserved since

$$\begin{aligned} d \star J^{(m+1)} &= d \star D\chi^{(m)} = D \star d\chi^{(m)} = DJ^{(m)} = D^2\chi^{(m-1)} \\ &= F\,\chi^{(m-1)} = 0 \qquad (m \geq 1) \end{aligned} \tag{58}$$

using (51) and the identity (48). Starting with $J^{(1)}$, we obtain an infinite set of (matrices of) conserved currents.[h] In case of the ordinary differential calculus on a two-dimensional Riemannian manifold, this construction reduces to the classical one.[5]

Let us (formally) define

$$\chi := \sum_{m \geq 0} \lambda^m \, \chi^{(m)} \tag{59}$$

with a constant $\lambda \neq 0$. Then (56) and (57) lead to

$$\star d\chi = \lambda \, D\chi \, . \tag{60}$$

As a consequence of this equation we get

$$0 = d \star D\chi^i{}_j = D \star d\chi^i{}_j + \chi^k{}_j \, d \star A^i{}_k \tag{61}$$

and

$$D \star d\chi = \lambda \, D^2 \chi = \lambda \, F \, \chi \, . \tag{62}$$

Using $A = a^{-1} \, da$, the integrability condition of the linear equation (60) is the field equation (49).

We have extended the definition of a class of σ-models to a rather general framework of noncommutative geometry, though still with the restriction to a commutative algebra $\mathcal{A}$ (which can always be realized as an algebra of functions on some space), but with *non*commutative differential calculi (where functions and differentials do not commute, in general). Already in this case a huge set of possibilities for integrable models appears.[6,7]

6.2 Example: recovering the Toda lattice

Let $\mathcal{A}$ be the (commutative) algebra of functions on $\mathcal{M} = \ell_0\mathbb{Z} \times \ell_1\mathbb{Z}$. Here $\ell_k\mathbb{Z}$ stands for the one-dimensional lattice with spacings $\ell_k > 0$. A special differential calulus $(\Omega(\mathcal{A}), d)$ on $\mathcal{A}$ is then determined by the following commutation

[h] There is no guarantee, however, that all these currents are really independent. For example, our formalism includes the free linear wave equation on two-dimensional Minkowski space. In that case, the higher conserved charges are just polynomials in the first one (which is the total momentum).

relations,

$$[\mathrm{d}t, t] = \ell_0 \, \mathrm{d}t \,, \quad [\mathrm{d}x, x] = \ell_1 \, \mathrm{d}x \,, \quad [\mathrm{d}t, x] = [\mathrm{d}x, t] = 0 \tag{63}$$

where t and x are the canonical coordinate functions on $\ell_0\mathbb{Z}$ and $\ell_1\mathbb{Z}$, respectively. This is our lattice differential calculus (37) for $n = 2$. As a consequence, we have

$$\mathrm{d}t \, f(\mathbf{x}) = f(\mathbf{x} + \ell_0) \, \mathrm{d}t \,, \quad \mathrm{d}x \, f(\mathbf{x}) = f(\mathbf{x} + \ell_1) \, \mathrm{d}x \tag{64}$$

where

$$\mathbf{x} := (t, x) \,, \quad \mathbf{x} + \ell_0 := (t + \ell_0, x) \,, \quad \mathbf{x} + \ell_1 := (t, x + \ell_1) \,. \tag{65}$$

Furthermore,

$$\mathrm{d}f = \frac{1}{\ell_0} \{ f(\mathbf{x} + \ell_0) - f(\mathbf{x}) \} \, \mathrm{d}t + \frac{1}{\ell_1} \{ f(\mathbf{x} + \ell_1) - f(\mathbf{x}) \} \, \mathrm{d}x \,. \tag{66}$$

Acting with d on (63), we obtain

$$\mathrm{d}t \, \mathrm{d}x = -\mathrm{d}x \, \mathrm{d}t \,, \quad \mathrm{d}t \, \mathrm{d}t = 0 = \mathrm{d}x \, \mathrm{d}x \,. \tag{67}$$

This familiar anticommutativity of differentials does not extend to general 1-forms, however. The differential calculus has the following property.

Lemma. Every closed 1-form is exact.

Proof: For $w = w_0(t,x) \, \mathrm{d}t + w_1(t,x) \, \mathrm{d}x$ the condition $\mathrm{d}w = 0$ means $\partial_{+t} w_1 = \partial_{+x} w_0$. For simplicity, we set $\ell_0 = \ell_1 = 1$ in the following. Let us define[i]

$$F(t,x) := \sum_{k=0}^{t-1} w_0(k,0) + \sum_{j=0}^{x-1} w_1(t,j) \,.$$

It satisfies

$$\partial_{+x} F = F(t, x+1) - F(t,x) = \sum_{j=0}^{x} w_1(t,j) - \sum_{j=0}^{x-1} w_1(t,j) = w_1(t,x)$$

[i]This function is obtained by integrating w along a path $\gamma : \mathbb{N} \to \mathbb{Z}^2$ first from $(0,0)$ to $(t,0)$ along the t-lattice direction, then from $(t,0)$ to (t,x) along the x-lattice direction. The result does not dependent on the chosen path. This follows from an application of Stokes' theorem.

and, using $\mathrm{d}w = 0$,

$$\partial_{+t} F = F(t+1, x) - F(t, x) = w_0(t, 0) + \sum_{j=0}^{x-1} \partial_{+t} w_1(t, j)$$

$$= w_0(t, 0) + \sum_{j=0}^{x-1} \partial_{+x} w_0(t, j) = w_0(t, 0) + w_0(t, x) - w_0(t, 0) = w_0(t, x) .$$

Hence $w = \mathrm{d}F$. ■

Let us now turn to the conditions for the $\star$-operator. First we introduce $g^{\mu\nu}$ via

$$\mathrm{d}x^\mu \star \mathrm{d}x^\nu = g^{\mu\nu} \, \mathrm{d}t \, \mathrm{d}x . \tag{68}$$

With $w = w_\mu \, \mathrm{d}x^\mu$, (46) becomes

$$[w_\mu(\mathbf{x}) \, w'_\nu(\mathbf{x} + \ell_\mu - \ell_\nu) - w'_\mu(\mathbf{x}) \, w_\nu(\mathbf{x} + \ell_\mu - \ell_\nu)] \, g^{\mu\nu} = 0 \qquad \forall \, w_\mu, w'_\mu . \tag{69}$$

It yields

$$g^{\mu\nu} = c^\mu \, \delta^{\mu\nu} \tag{70}$$

with arbitrary functions $c^\mu(\mathbf{x})$ which have to be different from zero in order for $\star$ to be invertible. This includes the metric (44). For the generalized Hodge operator we now obtain

$$\star \mathrm{d}t = c^0(\mathbf{x} - \ell_0) \, \mathrm{d}x , \qquad \star \mathrm{d}x = -c^1(\mathbf{x} - \ell_1) \, \mathrm{d}t \tag{71}$$

which extends to Ω^1 via (45).

In the following, we choose $g^{\mu\nu} = \eta^{\mu\nu}$ which in classical differential geometry are the components of the two-dimensional Minkowski metric with respect to an inertial coordinate system. We then have $\star\star w(\mathbf{x}) = w(\mathbf{x} - \ell_0 - \ell_1)$ which, together with the above Lemma, implies (47). Therefore, the construction of conservation laws does work in the case under consideration. Let us look at the simplest generalized σ-model where a is just a function (i.e., a 1×1-matrix). We write

$$a = e^{-q(t,x)} \tag{72}$$

with a function q and $q_k(n) := q(n\ell_0, k\ell_1)$. Then

$$A = \frac{1}{\ell_0}(e^{q_k(n) - q_k(n+1)} - 1) \, \mathrm{d}t + \frac{1}{\ell_1}(e^{q_k(n) - q_{k+1}(n)} - 1) \, \mathrm{d}x \tag{73}$$

$$\star A = -\frac{1}{\ell_0}(e^{q_k(n-1) - q_k(n)} - 1) \, \mathrm{d}x - \frac{1}{\ell_1}(e^{q_{k-1}(n) - q_k(n)} - 1) \, \mathrm{d}t \tag{74}$$

and the field equation $d \star A = 0$ takes the form

$$\frac{1}{\ell_0^2}\left[e^{q_k(n-1)-q_k(n)} - e^{q_k(n)-q_k(n+1)}\right] = \frac{1}{\ell_1^2}\left[e^{q_{k-1}(n)-q_k(n)} - e^{q_k(n)-q_{k+1}(n)}\right] \quad (75)$$

Replacing $\mathcal{A}$ with the algebra of functions on $\mathbb{R} \times \ell_1\mathbb{Z}$ which are smooth in the first argument, the limit $\ell_0 \to 0$ can be performed. This contraction leads to

$$\ddot{q}_k + \frac{1}{\ell_1^2}(e^{q_k - q_{k+1}} - e^{q_{k-1}-q_k}) = 0 \quad (76)$$

which is the *nonlinear Toda lattice* equation [17]. In particular, in this way a new and simple understanding of its complete integrability has been achieved. There is a 'noncommutative geometry' naturally associated with the Toda lattice equation. Generalizations of the Toda lattice are obtained by replacing the function a with a $GL(n, \mathcal{A})$-matrix.[6]

7 Metrics in noncommutative geometry

In the previous section we have introduced a generalized Hodge $\star$-operator. In classical (Riemannian) differential geometry, the Hodge operator contains information equivalent to a metric tensor which in turn has its origin in the problem of defining the length of a curve and the distance between points of a Riemannian space. On the basis of the formalism sketched in section 3.1, Connes proposed a generalization of the classical distance function to discrete and even noncommutative spaces.[4] Some examples are discussed in the following subsection.[19] The relation with a generalized Hodge operator or other generalized concepts of a metric still has to be understood, however.

7.1 *Connes' distance function associated with differential calculi on finite sets*

Let $(\mathcal{A}, \mathcal{H}, \hat{\mathcal{D}})$ be a spectral triple (cf section 3.1). A *state* on $\mathcal{A}$ is a linear map $\phi : \mathcal{A} \to \mathbb{C}$ which is positiv, i.e., $\phi(a^*a) \geq 0$ for all $a \in \mathcal{A}$, and normalized, i.e., $\phi(\mathbb{I}) = 1$. According to Connes [4], the distance between two states ϕ and ϕ' is given by

$$d(\phi, \phi') := \sup\{|\phi(a) - \phi'(a)| \; ; \; a \in \mathcal{A}, \; \|[\hat{\mathcal{D}}, a]\| \leq 1\} \, . \quad (77)$$

Given a set $\mathcal{M}$, each point $p \in \mathcal{M}$ defines a state ϕ_p via $\phi_p(f) := f(p)$ for all functions f on $\mathcal{M}$. The above formula then becomes

$$d(p, p') := \sup\{|f(p) - f(p')| \; ; \; f \in \mathcal{A}, \; \|[\hat{\mathcal{D}}, f]\| \leq 1\} \, . \quad (78)$$

Example 1. The universal first order differential calculus on a set of two elements p, q is described by a graph consisting of two points which are connected by a pair of antiparallel arrows. Its adjacency matrix is

$$\mathcal{D} = \begin{pmatrix} 0 & 1 \\ 1 & 0 \end{pmatrix} \tag{79}$$

so that

$$[\mathcal{D}, f] = \begin{pmatrix} 0 & f(p) - f(q) \\ f(q) - f(p) & 0 \end{pmatrix} . \tag{80}$$

Then

$$\begin{aligned} ||[\mathcal{D}, f]||^2 &= \sup_{||\psi||=1} ||[\mathcal{D}, f]\,\psi||^2 = \sup_{||\psi||=1} |f(p) - f(q)|^2 \,(|\psi_1|^2 + |\psi_2|^2) \\ &= |f(p) - f(q)|^2 \end{aligned} \tag{81}$$

for $\psi \in \mathbb{C}^2$. It follows that Connes' distance function defined with the adjacency matrix gives $d(p, q) = 1$. In this example, which appeared in models of elementary particle physics [18], there is no need for a 'doubling' of the representation as in (22). We may, however, replace $\mathcal{D}$ by $\hat{\mathcal{D}}$ also in this case. The result for the distance between the two points remains unchanged, however.

Example 2. [19] Let us consider the first order differential calculus on a set of N elements determined by the graph in Fig. 4.

Fig. 4
A finite oriented linear lattice graph.

The corresponding adjacency matrix is

$$\begin{pmatrix} 0 & 1 & 0 & \cdots 0 \\ \vdots & \ddots & \ddots & \vdots \\ \vdots & & \ddots & 1 \\ 0 & \cdots & \cdots & 0 \end{pmatrix} . \tag{82}$$

This matrix contains all the topological information about the lattice, i.e., the neighbourhood relationships. We can add information about the distances between neighbouring points to it in the following way. Let ℓ_k be the distance from point k to point $k+1$ (numbering the lattice sites by $1, \dots, N$). We define

$$\mathcal{D}_N := \begin{pmatrix} 0 & \ell_1^{-1} & 0 & \cdots 0 \\ \vdots & \ddots & \ddots & \vdots \\ \vdots & & \ddots & \ell_{N-1}^{-1} \\ 0 & \cdots & \cdots & 0 \end{pmatrix} . \tag{83}$$

With a complex function f we associate a real function F via

$$F_1 := 0 , \quad F_{i+1} := F_i + |f_{i+1} - f_i| \qquad i = 1, \dots, N-1 . \tag{84}$$

where $f_i := f(i)$. Then $|F_{i+1} - F_i| = |f_{i+1} - f_i|$ and

$$||[\hat{\mathcal{D}}_N, \hat{f}]\, \psi|| = ||[\hat{\mathcal{D}}_N, \hat{F}]\, \psi|| \tag{85}$$

for all $\psi \in \mathbb{C}^{2N}$. Hence, in calculating the supremum over all functions f in the definition of Connes' distance function, it is sufficient to consider *real* functions. Then $Q_N := [\hat{\mathcal{D}}_N, \hat{f}]$ is anti-hermitean and its norm is then given by the maximal absolute value of its eigenvalues. Instead of Q_N it is simpler to consider $Q_N\, Q_N^\dagger$ which is already diagonal with entries $0, \ell_1^{-2}(f_2 - f_1)^2, \dots, \ell_{N-1}^{-2}(f_N - f_{N-1})^2, \ell_1^{-2}(f_2 - f_1)^2, \dots, \ell_{N-1}^{-2}(f_N - f_{N-1})^2, 0$ on the diagonal. This implies

$$||[\hat{\mathcal{D}}_N, \hat{f}]|| = \max \{\ell_1^{-1}|f_2 - f_1|, \dots, \ell_{N-1}^{-1}|f_N - f_{N-1}|\} . \tag{86}$$

We have the obvious inequality

$$d(i, i+k) \leq \sup\{|f(i+1) - f(i)| + \ldots + |f(i+k) - f(i+k-1)| \,;\, ||[\hat{\mathcal{D}}_N, \hat{f}]|| \leq 1\} . \tag{87}$$

But a closer inspection shows that actually equality holds here. We conclude that $d(i, i+k) = \ell_i + \ell_{i+1} + \ldots + \ell_{i+k-1}$. [j] ■

Example 3. The graph in Fig. 1 has the adjacency matrix

$$\mathcal{D} = \begin{pmatrix} 0&1&0&1 \\ 0&0&1&0 \\ 0&0&0&0 \\ 0&0&1&0 \end{pmatrix} . \tag{88}$$

The norm of $[\hat{\mathcal{D}}, \hat{f}]$ is the positive square root of the largest eigenvalue of

$$[\hat{\mathcal{D}}, \hat{f}]\, [\hat{\mathcal{D}}, \hat{f}]^\dagger = \begin{pmatrix} [\mathcal{D}, f^*]^\dagger\, [\mathcal{D}, f^*] & 0 \\ 0 & [\mathcal{D}, f]\, [\mathcal{D}, f]^\dagger \end{pmatrix} . \tag{89}$$

[j] A different choice of $\hat{\mathcal{D}}$ has been made elsewhere[20] to define the distance on a lattice. See also Rieffel[21] for a reformulation of discrete metric spaces in Connes' framework.

It follows that

$$\|[\hat{\mathcal{D}}, \hat{f}]\| = \max\{\sqrt{|f_{21}|^2 + |f_{41}|^2}, \sqrt{|f_{32}|^2 + |f_{34}|^2}\} \tag{90}$$

where $f_{kl} := f_k - f_l$. Introducing

$$x := \frac{1}{2}(f_{21} - f_{41}), \quad y := \frac{1}{2}(f_{21} + f_{41}), \quad z := \frac{1}{2}(f_{32} + f_{34}), \tag{91}$$

we find the 'Euclidean' result

$$d(1,3) = \sup\{\frac{1}{2}|y + z| \, ; \, \max\{|x|^2 + |y|^2, |x|^2 + |z|^2\} \leq 2\} = \sqrt{2} \, . \tag{92}$$

■

Example 4. Let us consider the following digraph which is part of the lattice graph in Fig. 3.

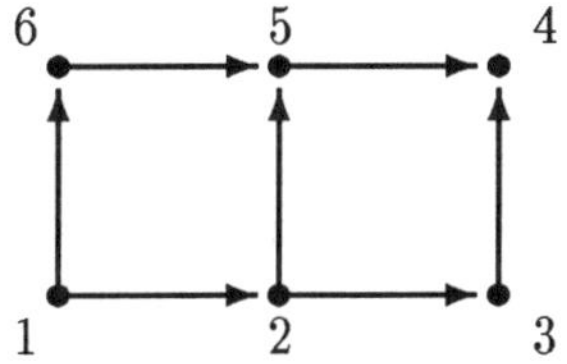

Fig. 5
The digraph associated with a special first order differential calculus on a set of six points.

A numerical evaluation of the distance function shows that $d(3,6) \lesssim 2 < d(1,4) < \sqrt{5}$ and thus deviates from the Euclidean value. ■

The last example shows that Connes' distance function defined in terms of the adjacency matrix of the n-dimensional oriented lattice graph with $n > 1$ (see Fig. 3) does *not* assign to it a Euclidean geometry, as might have been conjectured on the basis of our examples 2 and 3.

8 Final remarks

This work centered around examples which live on lattices. Such spaces do not at all exhaust the possibilities of noncommutative geometry of commutative algebras. In this case, and more generally in the case of discrete spaces, the generalized partial derivatives of a differential calculus are *discrete* derivatives, corresponding to an infinite sum of powers of ordinary partial derivatives. There are other differential calculi where the generalized partial derivatives are differential operators of finite order and some of them appear to be of relevance for an analysis of soliton equations, for example.[22] There is much more to mention in this context and we refer to a recent review [2] for further information and a guide to the relevant literature.

An important aspect of the formalism of noncommutative geometry is a technical one. On the level of generalized differential forms we have very compact expressions which are easy to handle thanks to the simple rules of differential calculus. Decomposed into components, however, we end up with rather complicated formulas, in general. This is precisely the experience which especially relativists make when they encounter the Cartan formalism in general relativity. Our generalization of the construction of conserved currents for (generalized) σ-models reviewed in section 6.1 is another nice example.

Acknowledgments

F M-H would like to thank the organizers of the 21st Johns Hopkins Workshop and in particular Professor Yishi Duan for the kind invitation and an enjoyable time in Lanzhou.

References

1. M. Dubois-Violette, Dérivations et calcul différentiel non commutatif *C. R. Acad. Sci. Paris* **307**, 403 (1988).
2. F. Müller-Hoissen, Introduction to noncommutative geometry of commutative algebras and applications in physics, in *Recent Developments in Gravitation and Mathematical Physics*, ed. A. Garcia *et al* (Science Network Publishing, Konstanz, 1997).
3. A. Dimakis and F. Müller-Hoissen, Discrete differential calculus, graphs, topologies and gauge theory *J. Math. Phys.* **35**, 6703 (1994).
4. A. Connes, *Noncommutative Geometry* (Academic Press, San Diego, 1994).
5. E. Brezin, C. Itzykson, J. Zinn-Justin and J.-B. Zuber, Remarks about the existence of non-local charges in two-dimensional models *Phys. Lett.* B **82**, 442 (1979).
6. A. Dimakis and F. Müller-Hoissen, Integrable discretizations of chiral models via deformation of the differential calculus *J. Phys.* A **29**, 5007 (1996).
7. A. Dimakis and F. Müller-Hoissen, Noncommutative geometry and integrable models *Lett. Math. Phys.* **39**, 69 (1997).
8. A. Dimakis, F. Müller-Hoissen and F. Vanderseypen, Discrete differential manifolds and dynamics on networks *J. Math. Phys.* **36**, 3771 (1995).
9. A. Dimakis and F. Müller-Hoissen, Differential calculus and gauge theory on finite sets *J. Phys.* A **27**, 3159 (1994).
10. A. Connes, Noncommutative geometry and reality *J. Math. Phys.* **36**, 6194 (1995).

11. A. Connes, Gravity coupled with matter and the foundations of non-commutative geometry *Commun. Math. Phys.* **182**, 155 (1996).
12. A. Dimakis and F. Müller-Hoissen, Quantum mechanics on a lattice and q-deformations *Phys. Lett.* B **295**, 242 (1992).
13. A. Dimakis, F. Müller-Hoissen and T. Striker, Noncommutative differential calculus and lattice gauge theory *J. Phys.* A **26**, 1927 (1993).
14. H.C. Baehr, A. Dimakis and F. Müller-Hoissen, Differential calculi on commutative algebras *J. Phys.* A **28**, 3197 (1995).
15. A. Dimakis and F. Müller-Hoissen, A noncommutative differential calculus and its relation to gauge theory and gravitation *Int. J. Mod. Phys. A (Proc. Suppl.)* **3A**, 474 (1993); Noncommutative differential calculus, gauge theory and gravitation, report GOE-TP 33/92; A. Dimakis and C. Tzanakis, Non-commutative geometry and kinetic theory of open systems *J. Phys.* A **29**, 577 (1996).
16. A. Dimakis and F. Müller-Hoissen, Stochastic differential calculus, the Moyal $*$-product, and noncommutative geometry *Lett. Math. Phys.* **28**, 123 (1993).
17. M. Toda, *Theory of Nonlinear Lattices* (Springer, Berlin, 1989).
18. A. Connes and J. Lott, Particle models and noncommutative geometry *Nucl. Phys. B (Proc. Suppl.)* **18**, 29 (1991).
19. A. Dimakis and F. Müller-Hoissen, Connes' distance function on one-dimensional lattices, preprint q-alg/9707016.
20. G. Bimonte, F. Lizzi and G. Sparano, Distances on a lattice from non-commutative geometry *Phys. Lett.* B **341**, 139 (1994).
21. M. Rieffel, Comments concerning non-commutative metrics, talk at an AMS special session at Texas A&M (1993).
22. A. Dimakis and F. Müller-Hoissen, Soliton equations and the zero curvature condition in noncommutative geometry *J. Phys.* A **29**, 7279 (1996).

CHERN–SIMONS FIELD THEORIES IN THE COULOMB GAUGE

FRANCO FERRARI* and IGNAZIO LAZZIZZERA **
*LPTHE, Université Paris VI and VII, 4 Place Jussieu,
F-75252 Paris CEDEX 05, FRANCE
E-mail: fferrari@lpthe.jussieu.fr

**Dipartimento di Fisica, Università degli Studi di Trento, Via Sommarive, 14
I-38050 Trento, Italy
E-mail: lazi@science.unitn.it

In this talk some recent results in the quantization of Chern-Simons field theories in the Coulomb gauge will be presented. In the first part, the consistency of the Chern–Simons field theories in this gauge is proven using the Dirac's canonical formalism for constrained systems. Despite the presence of non-trivial self-interactions in the gauge fixed functional, it will be shown that the commutation relations between the fields are trivial at any perturbative order in the absence of couplings with matter fields. If these couplings are present, instead, the commutation relations become rather involved, but it is still possible to study their main properties and to show that they vanish at the tree level. In the second part of the talk the perturbative aspects of Chern–Simons field theories in the Coulomb gauge will be analysed. In particular, it will be shown by explicit computations and in a regularization independent way that there are no radiative contributions to the n−point correlation functions. Finally the Feynman rules in the Coulomb gauge will be derived on a three dimensional manifold with a spatial section given by a closed and orientable Riemann surface.

1 Introduction

In the recent past, the Chern–Simons (C–S) field theories [1,2] have intensively been studied in connection with several physical and mathematical applications [3,4]. A convenient gauge fixing for these theories is provided by the Coulomb gauge. As a matter of fact, the presence of nontrivial interactions in the gauge fixed action allows perturbative computations. Perturbation theory is important in the calculations of the so-called link invariants[5,6,7,8] and whenever interactions are present, because in the latter case the C–S field theories are no longer exactly solvable. The advantage of the Coulomb gauge in this case is that the calculations are considerably simpler than in the covariant gauges and there are no radiative corrections. Moreover, with respect to the axial and light cone gauges, the Coulomb gauge can easily be imposed also on manifolds with non-flat spatial sections, like for instance Riemann surfaces. The absence of quantum contributions is a great advantage on non flat space-times, where the computation of Feynman integrals becomes technically difficult. Another important feature of the C–S field theories in the Coulomg gauge is that they can be considered as two dimensional models as it will be shown below.

Starting from the seminal works of refs. [2,9] and [10], the Coulomb gauge has been already applied in a certain number of physical problems involving C–S based models[4,11,12,13,14], but still remains less popular than the covariant and axial gauges. One of the main reasons is probably the fact that there are many perplexities concerning the use of this gauge fixing, in particular in the case of the four dimensional Yang–Mills theories [15,16,17,18]. Recently, also the consistency of the C–S field theories in the Coulomb gauge has been investigated using various techniques [19,12,20,21], and a detailed perturbative analysis of the non-abelian case has been done in [22].

In this talk some recent results in the quantization of C–S field theories in the Coulomb gauge will be presented following refs. [21,22,23] and avoiding technical details as much as possible. In the first part of the talk the C–S field theories are analysed by means of the Dirac's formalism for constrained systems. Besides some subtleties already noticed in [24], the derivation of the final Dirac brackets requires in the Coulomb gauge some care with distributions. Moreover, the final commutation relations (CR's) between the fields are derived both in the case of pure and interacting C–S field theories. With respect to the Yang–Mills field theories the CR's are rather involved. At a first sight, this is surprising in topological field theories with vanishing Hamiltonian and without degrees of freedom. However, at least in the pure C–S field theory, in which there are no interactions with matter fields, we show that this complexity is only apparent. As a matter of fact, taking into account the Gauss law and the Coulomb gauge fixing, the commutation relations between the gauge fields vanish identically at any perturbative order as expected. In this way the Chern–Simons field theories in the Coulomb gauge are not only perturbatively finite as has already been checked in the covariant gauges [25], but also free. This is not a priori evident, because in the Coulomb gauge the C–S functional contains non–trivial self–interaction terms. In the interacting case it is only possible to prove that the CR's are zero at the zeroth order approximation in perturbation theory. At higher orders however they are in general different from zero and have a very complicated expression. This is probably due to the fact that C–S field theories admit states with non-standard statistics.

In the second part of this talk the radiative corrections of the Green functions are computed at any loop order and it is shown in a regularization independent way that they vanish identically. No regularization is needed for the ultraviolet and infrared divergences since, remarkably, they do not appear in the amplitudes. The vanishing of the quantum corrections is in agreement with the triviality of the commutation relations found using the Dirac's canonical approach to constrained systems. It is important to notice that the absence of any quantum correction despite the presence of nontrivial self-interactions in the Lagrangian is a peculiarity of the Coulomb gauge that cannot be totally expected from the fact that the theories under consideration are topological, as finite renormalizations of the fields and of the coupling constants are always possible. For instance, in the analogous case of the covariant gauges, only the perturbative finiteness of the C–S amplitudes has been shown [26] in a regularization independent way exploiting BRST techniques [27]. Indeed, a finite shift of the C–S coupling constant has been observed in the Feynman gauges by various authors [28,29].

Finally, the Feynman rules of the C–S field theories will be derived also on a manifold whose spatial section is a Riemann surface of genus g

2 Canonical Quantization of the C–S field Theory in the Coulomb gauge

2.1 Notations

In this Section we will use the following notations. The Lagrangian of the pure $SU(N)$ C–S field theory is given by

$$L_{CS} = \frac{s}{8\pi}\epsilon^{\mu\nu\rho}\left(A^a_\mu \partial_\nu A^a_\rho - \frac{1}{3} f^{abc} A^a_\mu A^b_\nu A^c_\rho\right) \tag{1}$$

where s is a dimensionless coupling constant and A^a_μ is the gauge potential. Greek letters $\mu, \nu, \rho, \ldots = 0, 1, 2$ denote space–time indices, while the first latin letters $a, b, c, \ldots = 1, \cdots, N^2 - 1$ denote color indices. Moreover, the totally antisymmetric tensor $\epsilon^{\mu\nu\rho}$ is defined by the convention $\epsilon^{012} = 1$. The metric is given by $g_{\mu\nu} = \text{diag}(1, -1, -1)$. To derive the C–S Hamiltonian H_{CS} we have to compute the canonical momenta:

$$\pi^{\mu,a}(\mathbf{x}, t) - \frac{\delta S_{CS}}{\delta\left(\partial_0 A_\mu(\mathbf{x}, t)\right)} \tag{2}$$

where $S_{CS} = \int d^3x L_{CS}$, $t = x^0$ and $\mathbf{x} = \left(x^1, x^2\right)$. A straightforward calculation shows that:

$$H_{CS} = -\int d^2\mathbf{x} A^a_0 \left(D^{ab}_i \pi^{i,b} + \partial_i \pi^{i,a}\right) \tag{3}$$

In the above equation the following convention has been used for the spatial components of the covariant derivative: $D^{ab}_\mu \equiv \partial_\mu \delta^{ab} + f^{abc} A^c_\mu$. Finally, the nonvanishing equal time Poisson brackets (PB) among the canonical variables read as follows:

$$\left\{A_\mu^a\left(\mathbf{x},t\right),\pi_\nu^b\left(\mathbf{y},t\right)\right\}=\delta^{ab}g_{\mu\nu}\delta^{(2)}\left(\mathbf{x}-\mathbf{y}\right)$$

2.2 Constraints and intermediate Dirac brackets

From eqs. (1) and (2) we obtain the following primary constraints:

$$\varphi^{0,a}=\pi^{0,a} \tag{4}$$

$$\varphi^{i,a}=\pi^{i,a}-\frac{s}{8\pi}\epsilon^{ij}A_j^a \qquad i=1,2 \tag{5}$$

where ϵ^{ij}, $i,j=1,2$, is the two dimensional totally antisymmetric tensor defined by $\epsilon^{12}=1$. Following the Dirac procedure for constrained systems, the latter will be imposed in the weak sense: $\varphi^{\mu,a}\approx 0$. To this purpose, we construct the extended Hamiltonian:

$$\widetilde{H}_{CS}=H_{CS}+\int\lambda_\mu^a\varphi^{\mu,a}d^2\mathbf{x} \tag{6}$$

where the λ_μ^a's represent the Lagrange multipliers corresponding to the primary constraints $\varphi^{\mu,a}$.

From the consistency conditions $\dot{\varphi}^{\mu,a}=\left\{\varphi^{\mu,a},\widetilde{H}_{CS}\right\}\approx 0$, we obtain the secondary constraint:

$$\mathcal{G}^a=D_i^{ab}\pi^{i,b}+\partial_i\pi^{i,b}\approx 0 \qquad \text{(Gauss law)} \tag{7}$$

and two relations which determine the Lagrange multipliers λ_1 and λ_2:

$$\frac{s}{4\pi}\epsilon^{ij}\left(D_j^{ab}A_0^b-\lambda_j^a\right)\approx 0 \qquad i=1,2 \tag{8}$$

It is possible to see that the consistency condition $\dot{\mathcal{G}}^a\approx 0$ does not lead to any further independent equation. The operators $\mathcal{G}^a$ generate the $SU(N)$ group of gauge transformations only after eliminating the second class constraints $\varphi_i^a\approx 0$ of eq. (5). To this purpose, we introduce the intermediate Dirac brackets (DB's) $\{\ ,\ \}^*$ associated to these constraints. After some calculations one finds for the intermediate DB's among the canonical variables the following expressions:

$$\left\{A_i^a(t,\mathbf{x}),\pi^{j,b}(t,\mathbf{y})\right\}^*=\frac{1}{2}\delta^{ab}\delta_i^j\delta(\mathbf{x}-\mathbf{y}) \tag{9}$$

$$\left\{A_i^a(t,\mathbf{x}),A_j^b(t,\mathbf{y})\right\}^*=\frac{4\pi}{s}\delta^{ab}\epsilon_{ij}\delta(\mathbf{x}-\mathbf{y}) \tag{10}$$

$$\left\{\pi^{i,a}(t,\mathbf{x}),\pi^{j,b}(t,\mathbf{y})\right\}^*=\frac{s}{16\pi}\delta^{ab}\epsilon^{ij}\delta(\mathbf{x}-\mathbf{y}) \tag{11}$$

Exploiting the DB's (9)–(11), we obtain the relations:

$$\{\mathcal{G}^a(t,\mathbf{x}), A_i^b(t,\mathbf{y})\}^* = -D_i^{ab}(x)\delta(\mathbf{x}-\mathbf{y}) \tag{12}$$

$$\{\mathcal{G}[\psi], A_i^a(t,\mathbf{x})\}^* = D_i^{ab}(x)\psi^b(\mathbf{x}) \tag{13}$$

$$\left\{\mathcal{G}^a(t,\mathbf{x}), \mathcal{G}^b(t,\mathbf{y})\right\}^* = -f^{abc}\mathcal{G}^c(t,\mathbf{x})\delta(\mathbf{x}-\mathbf{y}) \tag{14}$$

where $\mathcal{G}[\psi] = \int d^2\mathbf{x}\mathcal{G}^a(t,\mathbf{x})\psi^a(\mathbf{x})$. This shows that the $\mathcal{G}^a(t,\mathbf{x})$ are the generators of the $SU(N)$ gauge transformations as desired.

2.3 Imposing the Coulomb gauge

At this point, we are left with the constraints given by eq. (4) and by the Gauss law (7). The former constraint, which is first class and involves the conjugate momentum of A_0^a, can be ignored. As a matter of fact, the field A_0^a just plays the role of the Lagrange multiplier associated to the Gauss law in the Hamiltonian (3) and has no dynamics. From eqs. (12)–(14) it turns out that the Gauss law (7) is a first class constraint. To make it second class, we introduce the Coulomb gauge fixing:

$$\partial_i A^{i,a} \approx 0 \tag{15}$$

and the new extended Hamiltonian:

$$\breve{H}_{CS} = \int d^2\mathbf{x}\left[-A_0^a\mathcal{G}^a + \frac{s}{8\pi}A_i^a\partial^i B^a + \lambda_0^a\pi^{0,a}\right] \tag{16}$$

From the condition $\{\partial_i A^{i,a}, \breve{H}_{CS}\}^* \approx 0$, we obtain an equation for A_0^a:

$$\partial^i D_i^{ab} A_0^b \approx 0 \tag{17}$$

Moreover, the requirement $\left\{\partial^i D_i^{ab} A_0^b(x), \breve{H}_{CS}\right\}^* \approx 0$ determines the Lagrange multiplier λ_0:

$$-\triangle\lambda_0^a - \left\{\partial_i(\mathbf{A}_i \times \mathbf{A}_0)^a, \breve{H}_{CS}\right\}^* \approx 0 \tag{18}$$

In the above equation the symbol $\triangle$ denotes the two dimensional Laplacian $\triangle = -\partial_i\partial^i$ and

$$(\mathbf{A}_i \times \mathbf{A}_0)^a \equiv f^{abc}A_i^b A_0^c$$

Another independent equation, which fixes the Lagrange multipliers B^a, is provided by the requirement $\dot{\mathcal{G}}^a \approx 0$:

$$\left\{\mathcal{G}^a, \breve{H}_{CS}\right\}^* \approx -\frac{s}{8\pi}D_i^{ab}\partial^i B^b \approx 0 \tag{19}$$

Let us notice that the above relations (7), (15) and (17)–(19) are compatible with the equations of motion of the gauge potentials:

$$\epsilon^{ij}(D_i^{ab}A_j^b - \partial_j A_i^a) = 0 \tag{20}$$

$$D_j^{ab}A_0^b - \partial_0 A_j^a = 0 \tag{21}$$

As a matter of fact (20) is equivalent to the condition $\mathcal{G}^a = 0$. Moreover, multiplying for instance eq. (21) with the differential operator $\epsilon_{ki}\partial^k$, we obtain the relation:

$$\partial_0\partial^k A_k^a - \partial^k D_k^{ab} A_0^b = 0$$

which is consistent with the Coulomb gauge and the condition (17) on A_0^a.

2.4 The final Dirac brackets and their properties

It is now possible to realize that the Gauss law (7) and the Coulomb gauge fixing (15) form a set of second class constraints, so that we can impose them in the strong sense computing the final Dirac brackets $\{\ ,\ \}_{DB}$. Putting

$$\chi_1^a = \mathcal{G}^a \qquad\qquad \chi_2^a = \partial_i A^{i,a}$$

with $\alpha, \beta = 1, 2$, and skipping all the technical details of the calculations that can be found in ref. [21], we have:

$$\{A_i^a(\mathbf{x}), A_j^b(\mathbf{y})\}_{DB} = -\frac{4\pi}{s}\delta^{ab}\epsilon_{ij}\delta(\mathbf{x}-\mathbf{y}) +$$

$$\frac{4\pi}{s}\epsilon_{ik}\partial_{\mathbf{x}}^k D_j^{bc}(\mathbf{y})\mathcal{D}^{ac}(\mathbf{x},\mathbf{y}) - \frac{4\pi}{s}\epsilon_{kj}D_i^{ac}(\mathbf{x})\partial_{\mathbf{y}}^k\mathcal{D}^{cb}(\mathbf{x},\mathbf{y}) \tag{22}$$

where

$$D_i^{ac}(\mathbf{x})\partial_{\mathbf{x}}^i\mathcal{D}^{cb}(\mathbf{x},\mathbf{y}) = \delta^{ab}\delta(\mathbf{x}-\mathbf{y}) \tag{23}$$

After imposing the constraints (7) and (15) in the strong sense, the Hamiltonian $\breve{H}_{CS}$ vanishes, but the commutation relations (CR's) between the fields remain complicated.

Let us study the main properties of the above DB's.

- **Antisymmetry**. The antisymmetry of the right hand side of eq. (22) is not explicit, but can be verified with the help of the relation:

$$\mathcal{D}^{ab}(\mathbf{x},\mathbf{y}) = \mathcal{D}^{ba}(\mathbf{y},\mathbf{x}) \tag{24}$$

The above identity is due to the fact that $\mathcal{D}^{ab}(\mathbf{x},\mathbf{y})$ is the Green function of the self-adjoint differential operator defined in eq. (23) [30]. Exploiting the above relation one finds that

$$\{A_i^a(\mathbf{x}), A_j^b(\mathbf{y})\}_{DB} = -\{A_j^b(\mathbf{y}), A_i^a(\mathbf{x})\}_{DB} \tag{25}$$

as expected

- **Consistency with the Coulomb gauge constraint.** The CR's (22) are consistent with the Coulomb gauge, i. e.:

$$\{A_i^a(\mathbf{x}), \partial^j A_j^b(\mathbf{y})\}_{DB} = \{\partial^i A_i^a(\mathbf{x}), A_j^b(\mathbf{y})\}_{DB} = 0$$

- **Covariance under the Poincaré group of transformations.** The proof that the C–S theory in the Coulomb gauge is invariant under the Poincaré group is not trivial due to the complicated CR's (22). A good strategy consists in evaluating the CR's among the generators of the Poincaré group using the intermediate DB's (9)–(11). In this way one finds that the Poincaré algebra is not closed due to "extra" terms which are proportional to the constraints. For instance, for the generators of the time and the space translations we obtain the following result:

$$\{P_0, P_k\}^* = \int d^2x A_0^a \partial_k \mathcal{G}^a$$

where $\mathcal{G}$ is given in (7). Clearly, all these unwanted terms disappear after imposing the final DB's (22) and the CR's between the generetors of the Poincaré group can be recovered.

- **Interactions.** For simplicity we have considered here pure Chern–Simons field theories. However, we stress that the form of the CR's (22) remains unchanged also adding to the lagrangian (1) interactions of the kind $L_I = \int d^3x A_\mu J^{\mu,a}$, where $J^{\mu,a}$ is a current associated to matter fields. The only differences occur in equations (7), (8) and (17)–(19), in which $J^{\mu,a}$ will appear as an external source. For instance the Gauss law (7) is modified as follows:

$$D_i^{ab}\pi^{i,b} + \partial_i\pi^{i,b} + J_0^a \approx 0 \tag{26}$$

2.5 The abelian case

The case of a Chern–Simons field theory with abelian gauge group $U(1)$ is particularly instructive in order to understand the meaning of the CR's (22).

Indeed, in this case the Green function $\mathcal{D}(\mathbf{x},\mathbf{y})$ has the following simple expression:

$$\mathcal{D}(\mathbf{x},\mathbf{y}) = -\frac{1}{2\pi}\log|\mathbf{x}-\mathbf{y}| \tag{27}$$

Let U_μ denote the abelian gauge fields. Substituting the right hand side of equation (27) in (22) and replacing the DB's with quantum commutators, we obtain:

$$[U_i(t,\mathbf{x}), U_j(t,\mathbf{y})] = 0$$

As a consequence the fields U_μ do not propagate. This result is in agreement with the fact that the theory is topological so that the fields have no dynamics. Indeed, exploiting the Gauss law, the Coulomb gauge fixing and eqs. (8), (17)–(19), it is easy to see that the the only possible solution of the equations of motion is $U_i = U_0 = \lambda_\mu = B = 0$. On the other side, the triviality of the CR's holds also in the presence of interactions, i. e. when the theory becomes no longer topological and the solutions of the equations of motion U_μ, λ_μ, B are in general different from zero.

2.6 The non-abelian case

In non-abelian C–S field theories the equations of motion of the constraints are nonlinear and can be solved only using a perturbative approach. At the zeroth order, the Green function $\mathcal{D}(\mathbf{x},\mathbf{y})$ is given again by eq. (27). Thus the CR's (22) are zero at this order. At higher orders, however, the right hand side of eq. (22) is in general different from zero, apart from the case in which there are no interactions. The vanishing of the CR's for the pure C–S theories, proven at any perturbative order in the coupling constant $\frac{1}{s}$ in [21], is in agreement with the fact that these theories are topological. Finally, we notice that the CR's (22) are particularly complicated with respect to the usual Yang–Mills field theories. This is probably related to the fact that field theories coupled to a C–S term exhibit a non-abelian statistics.

2.7 Final Remarks

- The C–S theories become in the Coulomb gauge two dimensional models. Only the fields A_i^a, for $i = 1, 2$, have in fact a dynamics, which is governed by the commutation relations (22). Moreover, the latter do not contain time derivatives, so that the time can be considered as an external parameter.

- If no interactions with matter fields are present, the CR's (22) vanish at any perturbative order. Thus the C–S field theories in the Coulomb gauge are not only finite, but also free. A natural question that arises at this point is if analogous conclusions can be drawn for the covariant gauges. For this reason it would be interesting to repeat the procedure of canonical quantization developed here also in this case.

- If the interactions with other fields are switched on, the CR's (22) still remain trivial in the abelian case. Thus, if we quantise the C–S theory replacing the DB's with quantum commutators, we obtain that

$$\left[A_i^a(\mathbf{x}), A_j^b(\mathbf{y})\right] = 0 \tag{28}$$

In the non-abelian case the above relation is valid only at the zeroth order in the coupling constant $\frac{1}{s}$, while at higher orders the CR's do not vanish and are rather complicated. Let us notice that there is no contradiction between eq. (28) and the fact that, starting from the Lagrangian (1), it is possible to derive a non-zero propagator for the C–S fields. In fact, going from the Hamiltonian normal-ordered formalism to the Lagrangian time-ordered formalism it is know that contact terms may arise, which contain distributions in the time variable. Indeed, the components of the propagator computed in the next section will have exactly the form of contact terms of this kind [31].

- Finally, we have shown that the CR's (22) are perfectly well defined and do not lead to ambiguities in the quantization of the C–S models in the Coulomb gauge. They are consistent with the constraints and the Poincaré covariance of the theory. Moreover, in the pure C–S field theory the CR's vanish at any perturbative order in $\frac{1}{s}$.

3 Perturbative Analysis of the C–S Field Theory in the Coulomb Gauge

3.1 Derivation of the Feynman rules

In this section we consider the following gauge fixed C–S action:

$$S_{CS} = S_0 + S_{GF} + S_{FP} \tag{29}$$

whith

$$S_0 = \frac{s}{4\pi}\int d^3x\epsilon^{\mu\nu\rho}\left(\frac{1}{2}A_\mu^a\partial_\nu A_\rho^a - \frac{1}{6}f^{abc}A_\mu^a A_\nu^b A_\rho^c\right) \tag{30}$$

$$S_{GF} = \frac{is}{8\pi\lambda}\int d^3x\left(\partial_i A^{a\,i}\right)^2 \tag{31}$$

and

$$S_{FP} = i\int d^3x\, \overline{c}^a \partial_i \left(D^i [A] c\right)^a \tag{32}$$

In (31) λ is a real gauge fixing parameter. With respect to the previous section the metric is now Euclidean and is of the form $g_{\mu\nu} = \mathrm{diag}(1,1,1)$. Moreover, the covariant derivatives are now defined as follows

$$D_\mu^{ab}[A] = \partial_\mu \delta^{ab} - f^{abc} A_\mu^c$$

In eq. (29) the Coulomb gauge constraint is weakly imposed and the proper Coulomb gauge fixing

$$\partial_i A^{a\,i} = 0 \qquad\qquad i = 1,2 \tag{33}$$

of the previous section is recovered setting $\lambda = 0$ in eq. (31).

From (29) the components of the gauge field propagator $G_{\mu\nu}^{ab}(p)$ in the Fourier space are given by:

$$G_{jl}^{ab}(p) = -\delta^{ab}\frac{4\pi\lambda}{s}\frac{p_i p_l}{\mathbf{p}^4} \tag{34}$$

$$G_{j0}^{ab}(p) = \delta^{ab}\left(\frac{4\pi}{s}\epsilon_{0jk}\frac{p^k}{\mathbf{p}^2} - \frac{4\pi\lambda}{s}\frac{p_j p_0}{\mathbf{p}^4}\right) \tag{35}$$

$$G_{0j}^{ab}(p) = -\delta^{ab}\left(\frac{4\pi}{s}\epsilon_{0jk}\frac{p^k}{\mathbf{p}^2} + \frac{4\pi\lambda}{s}\frac{p_0 p_j}{\mathbf{p}^4}\right) \tag{36}$$

$$G_{00}^{ab}(p) = -\delta^{ab}\frac{4\pi\lambda}{s}\frac{p_0^2}{\mathbf{p}^4} \tag{37}$$

where $\mathbf{p}^2 = p_1^2 + p_2^2$. Let us notice that the variable p_0 appears only in the longitudinal contributions to the propagator and disappears after choosing the proper Coulomb gauge. Also the ghost propagator $G_{gh}^{ab}(p)$ is independent on p_0:

$$G_{gh}^{ab}(p) = \frac{\delta^{ab}}{\mathbf{p}^2} \tag{38}$$

Finally, the three-gluon vertex and the ghost-gluon vertex are respectively given by:

$$V_{\mu_1\mu_2\mu_3}^{a_1a_2a_3}(p,q,r) = -\frac{is}{3!4\pi}(2\pi)^3 f^{a_1a_2a_3}\epsilon^{\mu_1\mu_2\mu_3}\delta^{(3)}(p+q+r) \tag{39}$$

and

$$V_{\text{gh}\,i_1}^{a_1 a_2 a_3}(p,q,r) = -i(2\pi)^3\,(q)_{i_1}\,f^{a_1 a_2 a_3}\delta^{(3)}(p+q+r) \tag{40}$$

In the above equation we have only given the spatial components of the ghost-gluon vertex. From eq. (32), it is in fact easy to realize that in the Coulomb gauge its temporal component is zero. As we see, the presence of p_0 remains confined in the vertices (39)–(40) and it is trivial because it is concentrated in the Dirac δ–functions expressing the momentum conservations. As a consequence, the CS field theory can be considered as a two dimensional model in the proper Coulomb gauge.

3.2 Potential divergences

At this point, we study the divergences that may arise in the computation of the Feynman diagrams. The potential divergences are of three kinds: ultraviolet, infrared and spurious.

1. Ultraviolet divergences (UV). The naive power counting gives the following degree of divergence $\omega(G)$ for a given Feynman diagram G:

   $$\omega(G) = 3 - \delta - E_B - \frac{E_G}{2} \tag{41}$$

   with [b]

   (a) δ = number of momenta which are not integrated inside the loops

   (b) E_B = number of external gluonic legs

   (c) E_G = number of external ghost legs

   Eq. (41) shows that UV divergences are possible in the two and three point functions, both with gluonic or ghost legs. Moreover, there is also a possible logarithmic divergence in the case of the four point interaction among two gluons and two ghosts. In principle, we had to introduce a regularization for these divergences but in practical calculations this is not necessary. As a matter of fact, we will see below that there are no UV divergences in the quantum corrections of the Green functions.

2. Infrared (IR) divergences. The pure C–S field theories are known to be free of infrared divergences [2] so that there is no need to discuss them.

[b] We use here the same notations of ref. [32]

3. Spurious divergences. These singularities appear because the propagators (34)–(38) are undamped in the time direction and are typical of the Coulomb gauge. To regularize spurious divergences of this kind, it is sufficient to introduce a cutoff $\Lambda_0 > 0$ in the domain of integration over the variable p_0:

$$\int_{-\infty}^{\infty} dp_0 \to \int_{-\Lambda_0}^{\Lambda_0} dp_0 \tag{42}$$

The physical situation is recovered in the limit $\Lambda_0 \to \infty$. As we will see, this regulatization does not cause ambiguities in the evaluation of the radiative corrections at any loop order. In fact, the integrations over the temporal components of the momenta inside the loops turn out to be trivial and do not interfere with the integrations over the spatial components.

3.3 Perturbative analysis at one loop order

In this Section we compute the n–point correlation functions of C–S field theories at one loop order. From now on, we choose for simplicity the proper Coulomb gauge, setting $\lambda = 0$ in eq. (31). In this gauge the gluon-gluon propagator has only two nonvanishing components:

$$G_{j0}(p) = -G_{0j}(p) = \delta^{ab}\frac{4\pi}{s}\epsilon_{0jk}\frac{p^k}{\mathbf{p}^2} \tag{43}$$

The following observation greatly reduces the number of diagrams to be evaluated:

Observation: Let $G^{(1)}$ be a one particle irreducible (1PI) Feynman diagram containing only one closed loop. Then all internal lines of $G^{(1)}$ are either ghost or gluonic lines.

A proof of the above observation can be found in ref. [22]. An important consequence is that, at one loop, the only non–vanishing diagrams occur when all the external legs are gluonic. Hence we have to evaluate only the diagrams describing the scattering among n gluons.

This can be done as follows. First of all, we consider the diagrams with internal gluonic lines. After suitable redefinitions of the indices and of the momenta, it is possible to see that their total contribution is given by:

$$V_{i_1 \dots i_n}^{a_1 \dots a_n}(1; p_1, \dots, p_n) = C\left[-i\left(2\pi\right)^3\right]^n \frac{n!(n-1)!}{2}\delta^{(2)}(\mathbf{p}_1 + \dots + \mathbf{p}_n) \tag{44}$$

$$f^{a_1 b'_1 c'_1} f^{a_2 b'_2 b'_1} \dots f^{a_n c'_1 b'_{n-1}} \int d^2\mathbf{q}_1 \frac{\left[q_1^{i_1} \dots q_n^{i_n} + q_1^{i_2} \dots q_j^{i_{j+1}} \dots q_{n-1}^{i_n} q_n^{i_1}\right]}{\mathbf{q}_1^2 \dots \mathbf{q}_n^2}$$

where $C = (2\Lambda_0)^{2n}$ is a finite constant coming from the integration over the zeroth components of the momenta and

$$\begin{array}{ll} q_2 = q_1 + p_1 + p_n + p_{n-1} + \dots & + p_3 \\ \vdots & \vdots \\ q_j = q_1 + p_1 + p_n + p_{n-1} + \dots & + p_{j+1} \\ \vdots & \vdots \\ q_n = & q_1 + p_1 \end{array} \tag{45}$$

for $j = 2, \dots, n-1$.

The case of the Feynman diagrams containing ghost internal lines is more complicated. After some work, it is possible to distinguish two different contributions to the Green functions with n gluonic legs that cannot be reduced into one by renaming indices and momentum variables:

$$V_{i_1 \dots i_n}^{a_1 \dots a_n}(2a; p_1, \dots, p_n) = -C\left[-i(2\pi)^3\right]^n \frac{n!(n-1)!}{2}$$
$$\delta^{(2)}(\mathbf{p}_1 + \dots \mathbf{p}_n) f^{a_1 b'_1 c'_1} f^{a_2 b'_2 b'_1} \dots f^{a_n c'_1 b'_{n-1}} \int d^2\mathbf{q}_1 \frac{q_1^{i_1} \dots q_n^{i_n}}{\mathbf{q}_1^2 \dots \mathbf{q}_n^2} \tag{46}$$

and

$$V_{i_1 \dots i_n}^{a_1 \dots a_n}(2b; p_1, \dots, p_n) = -C[-i(2\pi)^3]^n \frac{n!(n-1)!}{2} f^{a_1 b'_1 c'_1} f^{a_2 b'_2 b'_1} \dots f^{a_n c'_1 b'_{n-1}}$$
$$\delta^{(2)}(\mathbf{p}_1 + \dots + \mathbf{p}_n) \int d^2\mathbf{q}_1 \frac{q_n^{i_1} q_1^{i_2} \dots q_j^{i_{j+1}} \dots q_{n-1}^{i_n}}{(\mathbf{q}'_1)^2 \dots (\mathbf{q}'_n)^2} \tag{47}$$

In the above equations the variables $q_2, \dots, q_n$ and the constant C are the same as in eq. (44).

As it is possible to see from eqs. (44).(46) and (47), the only nonvanishing components of the n points functions are those for which all tensor indices $\mu_1, \dots, \mu_n$ are spatial. We notice here that eq. (47) has been obtained after a shift of the integration variable q_1. However, it is not difficult to verify that the right hand sides of eqs. (44)–(47) are neither IR nor UV divergencent for $n \geq 3$, so that a shift of q_1 is not dangerous. At this point we can sum equations (44).(46) and (47) together. It is easy to realize that the total result is zero, i. e.:

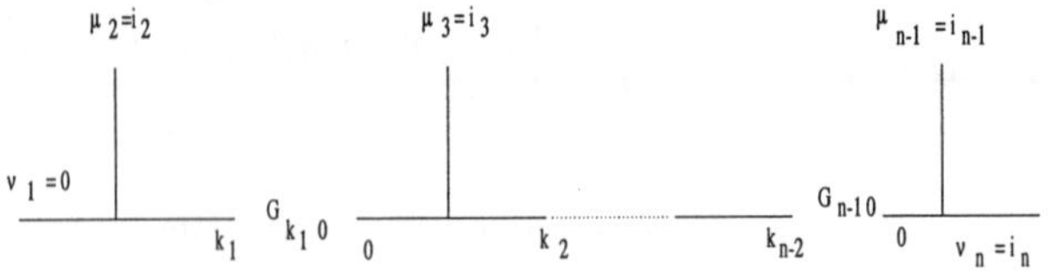

Figure 1: This figure shows that in an arbitrary tree diagram $T_{\nu_1\nu_2\ldots\nu_{n-1}\nu_n}$ constructed in terms of the gauge fields propagator (43) and the three gluon vertex (39), only one component in the space-time indices ν_i, $i = 1, \ldots, n$, can be temporal.

$$V_{i_1\ldots i_n}^{a_1\ldots a_n}(1;p_1,...,p_n)+V_{i_1\ldots i_n}^{a_1\ldots a_n}(2a;p_1,...,p_n)+V_{i_1\ldots i_n}^{a_1\ldots a_n}(2b;p_1,...,p_n)=0 \quad (48)$$

For $n \geq 3$ this result is regularization independent since the Feynman integrals are IR and UV convergent.

Only the case $n = 2$ needs some more care and will be treated separately. After a few calculations one obtains that for $n = 2$ the total contribution to the gluonic propagator to one loop is given by:

$$V_{ij}^{ab}(1;p_1,p_2) + V_{ij}^{ab}(2a;p_1,p_2) + V_{ij}^{ab}(2b;p_1,p_2) =$$

$$(2\pi)^6 (2\Lambda_0)^2 N\delta^{ab}\delta^{(2)}(\mathbf{p}_1 + \mathbf{p}_2) \int d^2\mathbf{q} \frac{[q_i(p_1)_j - q_j(p_1)_i]}{\mathbf{q}^2 (\mathbf{q}+\mathbf{p}_1)^2} \quad (49)$$

The integrand appearing in the rhs of (49) is both IR and UV finite. Moreover, a simple computation shows that the integral over $\mathbf{q}$ in (49) is zero. As a consequence, there are no contributions to the Green functions at one loop.

3.4 The higher order radiative corrections

Now we are ready to consider the higher order corrections of the n points Green functions. At two loop, a general Feynman diagram $G^{(2)}$ can be obtained contracting two legs of a tree diagram $G^{(0)}$ with two legs of a one loop diagram $G^{(1)}$. As previously seen, the latter have only gluonic legs and their tensorial indices are all spatial. Consequently, in order to perform the contractions by means of the propagator (43), there should exist one component of $G^{(0)}$ with at least two temporal indices, but this is impossible. To convince oneself of this fact, it is sufficient to look at fig. (1) and related comments. The situation does not improve if we build $G^{(0)}$ exploiting also the ghost-gluon vertex (40), because it has no temporal component. As a consequence, all the Feynman graphs vanish identically at two loop order. Let us notice that it is possible to verify their vanishing directly, since the number of two loop diagrams is relatively small in the Coulomb gauge and one has just to contract the space-time indices without performing the integrations over the internal momenta. However, this procedure is rather long and will not be reported here.

The proof that also higher order diagrams vanish can be done by induction. First of all, a diagram with $N+1$ loops $G^{(N+1)}$ has at least one subdiagram $G^{(N)}$ containing $N-$loops. Supposing that $G^{(N)}$ is identically equal to zero because it cannot be constructed with the Feynman rules (38)–(40) and (43), also $G^{(N+1)}$ must be zero. As we have seen above, there are no Feynman diagrams for $N = 2$. This is enough to prove by induction that C–S field theories have no radiative corrections in the Coulomb gauge for any value of N.

3.5 Final remarks

- In the Coulomb gauge the C–S field theories do not have quantum corrections at any loop order, has as been shown by explicit computations.

- IR and UV divergences are absent in the calculations of the Feynman diagrams. Only spurious singularities are present, related to the fact that the propagators are undamped in the time direction. They are similar to the singularities observed in the four dimensional Yang–Mills field theories [15], but in the C–S case appear in a milder form. In fact, after introducing the regularizarion (42) and integrating over the time component of the momenta in a given amplitude, the total contribution at any loop order reduces to an overall constant factor. The remaining calculations consist of finite two dimensional Feynman integrals over the space variables. As a consequence, the results obtained here are regularization independent.

- The vanishing of the quantum contributions described in Section 3 is a peculiarity of the Coulomb gauge that does not strictly depend the fact that the C–S field theories are topological. In fact, finite remormalizations of the fields and of the coupling constant s are always possible as it happens in the case of the covariant gauges. An analogous situation in which there are no radiative corrections occurs in the light cone gauge in the presence of a boundary. In that case, radiative corrections arise in principle due to the interactions of the fields with the boundary [33] but all the related Feynman diagrams vanish identically.

- The C–S field theories in the proper Coulomb gauge can be considered as two dimensional models. This has been shown in the previous section and has been confirmed here by the fact that the dependence on the the time component of the momenta in the propagators and vertices is trivial.

- Contrary to what happens using the covariant gauges or the axial gauges, the Coulomb gauge can easily be applied also when space-times with non-trivial spatial section are considered, like for instance a Riemann surface. The absence of radiative corrections particularly useful in this case, where the momentum representation does not exist and this the evaluation of Feynman diagrams becomes forbiddenly difficult.

- The C–S field theories can be considered as a good laboratory in order to study the possible remedies of pathologies that appear in similar ways in the more complicated four dimensional gauge field theories. For example, it would be interesting to apply to the Yang–Mills case the regularization (42) introduced here for the spurious singularities. Let us notice that a different regularization has been recently proposed in [18].

4 Chern–Simons field theories in the Coulomb Gauge on Curved Space–Times

In this section we consider a manifold M_3 with a Robertson-Walker metric and Euclidean signature of the kind

$$g_{00} = 1 \qquad g_{z\bar{z}}(z,\bar{z},t) = g_{z\bar{z}}(z,\bar{z},t) = a(t)h(z,\bar{z}) \qquad g^{z\bar{z}}g_{z\bar{z}} = 1 \tag{50}$$

$g^{z\bar{z}}$ is the metric on a Riemann surface Σ_g of genus g and z and $\bar{z}$ are local coordinates on Σ_g: We suppose that $a(t) > 0$ for each values of the time t. Thus M_3 correspond to an expanding universe having the Riemann surface Σ_g as spatial section.

The gauge fixed Chern-Simons action (29) becomes in complex coordinates $S_{\rm CS} = S_{\rm free} + S_{\rm int}$, where:

$$S_{\rm free} = \int_{M_3} d^2z dt \left[2i\left(A_0^a \partial_{\bar{z}} A_z^a + A_z^a \partial_0 A_{\bar{z}}^a + A_{\bar{z}}^a \partial_z A_0^a - \text{c.c.}\right)\right.$$

$$\left. + (a(t)\lambda)^{-1} g^{\bar{z}z} (\partial_z A_{\bar{z}}^a + \partial_{\bar{z}} A_z^a)^2 + 2\bar{c}^a \partial_z \partial_{\bar{z}} c^a\right] \tag{51}$$

$$S_{\rm int} = \int_{M_3} d^2z dt \left[\epsilon^{\mu\nu\rho} f^{abc} A_\mu^a A_\nu^b A_\rho^c - f^{abc} \bar{c}^a \left(A_z^b \partial_{\bar{z}} + A_{\bar{z}}^b \partial_z\right) c^c\right] \tag{52}$$

and $d^2z = \frac{1}{2i} dz \wedge d\bar{z}$. The factor $2i$ in eq. (51) comes from the form of the $\epsilon^{\mu\nu\rho}$ tensor in complex coordinates. In fact, the Levi-Civita tensor $[\epsilon]^{\mu\nu\rho} = g^{-\frac{1}{2}}\epsilon^{\mu\nu\rho}$ becomes in these coordinates:

$$[\epsilon]^{0z\bar{z}} = -2ig^{z\bar{z}}a^{-1}(t) \tag{53}$$

All the other components can be obtained from eq. (53) permuting the indices 0, z and $\bar{z}$ and changing the sign according to the order of the permutation. In this Section it will be useful to denote a sum over the complex indices with the first letters of the Greek alphabet α, β, γ and so on. For example, the Coulomb gauge condition becomes now $\partial^\alpha A^a_\alpha = 0$. Using the metric (50) to rise and lower the indices, this equation reads:

$$\partial_z A^a_{\bar{z}} + \partial_{\bar{z}} A^a_z = 0 \tag{54}$$

Eq. (54) does not contain the metric explicitely. This means that the Coulomb gauge condition is compatible with the transition functions at the intersections of the open sets covering the Riemann surface Σ_g. Therefore eq. (54) is globally valid on M_3.The gauge fields $(A^a_z, A^a_{\bar{z}}, A^a_0)$ are connections on the trivial principal bundle

$$P(M_3, \mathrm{SU(N)}) = M_3 \otimes \mathrm{SU(N)}$$

This bundle is trivial due to the fact that SU(N) is a simply connected Lie group. One can show as in the flat case that the Coulomb gauge (54) is a good gauge fixing without Gribov ambiguities [35] at least in the perturbative approach (see ref. [23] for details. We are now ready to compute the propagators of the gauge fields

$$G^{ab}_{\mu\nu}(z, w; t, t') =< A^a_\mu(z, \bar{z}, t) A^b_\nu(w, \bar{w}, t) >$$

where now $\mu, \nu = 0, z, \bar{z}$. The equations satisfied by the above propagator are:

$$-4i\partial_z G^{ab}_{\bar{z}0}(z, w; t, t') + 4_i\partial_{\bar{z}} G^{ab}_{z0}(z, w; t, t') = \frac{8\pi}{s}\delta^{ab}\delta^{(2)}_{z\bar{z}}(z, w)\delta(t - t') \tag{55}$$

$$-4i\partial_{\bar{z}} G^{ab}_{0w}(z, w; t, t') + 4i\partial_0 G^{ab}_{\bar{z}w}(z, w; t, t') - 2\frac{a^{-1}(t)}{\lambda}\partial_{\bar{z}}\left[g^{z\bar{z}}\partial_{\bar{z}} G^{ab}_{zw}(z, w; t, t') + \right.$$

$$\left. + g^{z\bar{z}}\partial_z G^{ab}_{\bar{z}w}(z, w; t, t')\right] = \frac{8\pi}{s}\delta^{ab}\delta^{(2)}_{\bar{z}w}(z, w)\delta(t - t') \tag{56}$$

Another equation can be obtained from (56) permuting the indices z and $\bar{z}$ and substituting the index w with $\bar{w}$. There are still other relations relating the various components of the propagators together:

$$\partial_z G^{ab}_{\bar{z}\alpha}(z, w; t, t') - \partial_{\bar{z}} G^{ab}_{z\alpha}(z, w; t, t') = 0 \tag{57}$$

$$-4i\partial_\alpha G^{ab}_{00}(z, w; t, t') + 4i\partial_0 G^{ab}_{\alpha 0}(z, w; t, t') -$$

$$-\frac{a^{-1}(t)}{\lambda}\partial_\alpha\left[g^{z\bar{z}}\partial_{\bar{z}} G^{ab}_{z0}(z, w; t, t') + g^{z\bar{z}}\partial_z G^{ab}_{\bar{z}0}(z, w; t, t')\right] = 0 \tag{58}$$

where $\alpha = w, \bar{w}$. Eq. (57) implies that the propagators $G^{ab}_{z\bar{z}}(z,w;t,t')$ and $G^{ab}_{\bar{z}z}(z,w;t,t')$ do not have transverse components.. Finally we have:

$$-4i\partial_{\bar{z}} G^{ab}_{0\bar{w}}(z,w;t,t') + 4i\partial_0 G^{ab}_{\bar{z}\bar{w}}(z,w;t,t') -$$

$$-\frac{a^{-1}(t)}{\lambda}\partial_{\bar{z}}\left[g^{z\bar{z}}\partial_{\bar{z}} G^{ab}_{z\bar{w}}(z,w;t,t') + g^{z\bar{z}}\partial_z G^{ab}_{\bar{z}\bar{w}}(z,w;t,t')\right] = 0 \qquad (59)$$

Again it is possible to get another independent relation from eq. (59) interchanging the two indices z and $\bar{z}$ and substituting $\bar{w}$ with w. Eqs. (55–59) are the equivalent of the equations defining the propagator in the flat case. However, they are still uncomplete, because in deriving them we have neglected the zero mode contributions. In fact, we should remember that due to a theorem stating that the total charge on a Riemann surface (like in any other two dimensional compact manifold) is always zero, an isolated δ function $\delta^{(2)}(z,w)$ is not allowed. Therefore, in the right hand sides of eqs. (55–56) there must be also terms containing zero modes, whose expressions will be uniquely determined below. Since it is very difficult to solve equations (55–59) for any value of λ, we choose here the proper Coulomb gauge taking the limit $\lambda \to 0$. In this case drastic simplifications occur, so that the above equations reduce to the following two relations:

$$\partial_z G^{ab}_{\bar{z}0}(z,w;t,t') - \partial_{\bar{z}} G^{ab}_{z0}(z,w;t,t') = \frac{4\pi i}{s}\delta^{ab}\delta^{(2)}(z-w)\delta(t-t') + \text{zero modes} \qquad (60)$$

$$\partial_{\bar{z}} G^{ab}_{z0}(z,w;t,t') + \partial_z G^{ab}_{\bar{z}0}(z,w;t,t') = 0 \qquad (61)$$

These equations describe exactly the main requirement of the Coulomb gauge, i.e. the fact that only the transverse fields in the two dimensional spatial section Σ_g of M_3 propagate. The transverse fields in complex coordinates satisfy in fact the following condition: $A^a_z = \overline{(A^a_z)} = -A^a_{\bar{z}}$. The solution of eqs. (60) and (61) is provided by the following Green functions:

$$< A^a_z(z,t)A^b_0(w.t') >= \frac{2\pi i}{s}\delta^{ab}\partial_z K(z,w)\delta(t-t') \qquad (62)$$

and

$$< A^a_{\bar{z}}(z,t)A^b_0(w,t') >= -\frac{2\pi i}{s}\delta^{ab}\partial_z K(z,w)\delta(t-t') \qquad (63)$$

where $K(z,w)$ is the usual propagator of the scalar fields on a Riemann surface satisfying the equations (see ref. [36] for more details):

$$K(z,w) = \delta^{(2)}_{z\bar{z}}(z,w) + \frac{g_{z\bar{z}}}{\int_{\Sigma_g} d^2u g_{u\bar{u}}} \qquad (64)$$

$$\partial_z \partial_{\bar{w}} K(w,z) = -\delta^{(2)}_{z\bar{w}}(z,w) + \bar{\omega}_i(\bar{z})\left[\text{Im }\Omega\right]^{-1}_{ij} \omega_j(w) \tag{65}$$

$$\int_{\Sigma_g} d^2 z g_{z\bar{z}} K(z,w) = 0 \tag{66}$$

In eq. (65) the $\omega_i(z)dz$, $i = 1, \ldots, g$, denote the usual holomorphic differentials and Ω_{ij} represents the period matrix. It is important to stress here that $K(z,w)$ is a singlevalued function on Σ_g. Using the propagators (62) and (63) it is easy to see that eq. (61) is trivially satisfied. Therefore, the Coulomb gauge requirement (54) is fulfilled and the above defined propagators describe exactly the transverse components of the gauge fields. Still there is an ambiguity in the solutions (62) and (63) due to the zero mode sector of the fields A^a_z and $A^a_{\bar{z}}$. In order to remove this ambiguity, we have to require that the above propagators are singlevalued along the nontrivial homology cycles of the Riemann surface. Otherwise, the propagators are not well defined on M_3, but in one of its coverings. Therefore, the propagators should obey the following relations:

$$\oint_\gamma dz < A^a_z(z,t) A^b_0(w,t') >= \oint_{\bar{\gamma}} d\bar{z} < A^a_{\bar{z}}(z,t) A^b_0(w,t') >= 0 \tag{67}$$

along any nontrivial homology cycles γ. Due to the properties of singlevaluedness of the Green function $K(z,w)$, eq. (67) is trivially satisfied by the propagators given in eqs. (62) and (63). In this way these two propagators are well defined and also the freedom in the zero mode sector is removed. Now we insert their expressions in eq. (60) in order to get the exact form of the zero mode terms appearing in the right hand side of this equation:

$$\partial_z G^{ab}_{\bar{z}0}(z,w;t,t') - \partial_{\bar{z}} G^{ab}_{z0}(z,w;t,t') =$$

$$\frac{4\pi i}{s}\delta^{ab}\delta^{(2)}(z,w)\delta(t-t') + 4\pi i s \frac{g_{z\bar{z}}}{\int_{\Sigma_g} d^2 u g_{u\bar{u}}}\delta(t-t') \tag{68}$$

The fact that the propagators in the Coulomb gauge must obey eq. (67) can be understood also decomposing the fields by means of the Hodge decomposition of the gauge fields in a coexact, exact and harmonic part:

$$A^a_z = i\partial_z \varphi^a + \partial_z \rho^a + A^{\text{har}}_z \tag{69}$$

$$A^a_{\bar{z}} = i\partial_{\bar{z}} \varphi^a + \partial_{\bar{z}} \rho^a + A^{\text{har}}_{\bar{z}} \tag{70}$$

ϕ^a and ρ^a represent two real scalar fields. The above decomposition is allowed since the gauge invariance has been completely fixed by the choice of the Coulomb gauge, at least in the perturbative approach, and the G−bundle $P(M_3, \mathrm{SU(N)})$ is trivial as we previously remarked. In the Coulomb gauge, the only components of the gauge fields which are allowed to propagate are the coexact differentials, i.e. the 1−forms obtained differentiating the scalar fields φ^a in eqs. (69) and (70). Therefore, the requirement (67) is a pure consequence of the fact that the coexact forms have vanishing holonomies around the nontrivial homology cycles.

Let us notice that the zero mode term appearing in the right hand side of eq. (68) is totally irrelevant. To eliminate it it is sufficient to introduce new gauge fields, let say $\tilde{A}_z$, $\tilde{A}_{\bar{z}}$, differing from the old ones by the fact that they are normalized to zero at a point $(0,0)$ of the Riemann surface[c]:

$$\tilde{A}^a_z(z,\bar{z},t) = A^a_z(z,\bar{z},t) - A^a_z(0,0,t) \tag{71}$$

$$\tilde{A}^a_{\bar{z}}(z,\bar{z},t) = A^a_{\bar{z}}(z,\bar{z},t) - A^a_{\bar{z}}(0,0,t) \tag{72}$$

Using the above new fields it is easy to check that the second term in the right hand side of eq. (68), which is a zero mode contribution, cancels out.

We finish this Section providing the explicit form of the other correlation functions of Chern-Simons field theory. The propagator of the ghost fields becomes:

$$G^{ab}_{\mathrm{gh}}(z,w;t,t') = \delta^{ab} K(z,w)\delta(t-t')$$

The vertex coming from the cubic interaction between the gauge fields reads instead:

$$V^{abc}_{z_1 00}(z_1,z_2,z_3;t,t',t'') = \frac{2\pi^2 is}{3}\int_{\Sigma_g} d^2z f^{abc}\partial_{z_1}K(z_1,z)\left[\partial_z K(z_2,z)\partial_{\bar{z}}K(z_3,z)-\right.$$
$$\left.\partial_{\bar{z}}K(z_2,z)\partial_z K(z_3,z)\right]\delta(t-t'')\delta(t'-t'')$$

The simple integration in the variable t has been already carried out in the above expression of the vertex. The component $V^{abc}_{\bar{z}_1 00}(z_1,z_2,z_3;t,t',t'')$ of the vertex can be simply obtained replacing the derivative ∂_{z_1} in the above equation with its complex conjugate. Finally, the vertex describing the interaction between ghost and gauge fields has only one component which is given by:

$$V^{abc}_{0\ \mathrm{gh}}(z_1,z_2,z_3;t,t',t'') = \frac{2\pi i}{sa(t)}\int_{M_3} d^2z f^{abc}K(z_1,z)\left[\partial_z K(z_2,z)\partial_{\bar{z}}K(z_3,z)\right.$$
$$\left.-\partial_{\bar{z}}K(z_2,z)\partial_z K(z_3,z)\right]\delta(t-t')\delta(t'-t'')$$

It is easy to check that the above expressions of the vertices are real as it should be.

---

[c]On M_3 this implies that the new fields are normalized to zero along the whole line of the time. This is possible to do since the three dimensional manifold is flat in the time direction.

5 Conclusions

In summary, our study indicates that the Coulomb gauge is a convenient and reliable gauge fixing, especially in the perturbative applications of C-S field theory. Let us remember that, despite of the fact that the theory does not contain degrees of freedom, the perturbative calculations play a relevant role, for instance in the computation of knot invariants [5,6,7,8,29]. Contrary to what happens in the covariant gauges, where it becomes more and more difficult to evaluate the radiative corrections as the loop number increases [29,6,34], in the Coulomb gauge only the tree level contributions to the Green functions survive. This feature is particularly useful in the case of non-flat manifolds, where the momentum representation does not exist. As an application, the Feynman rules of C–S field theories on Riemann surfaces have been derived in section 4. Moreover, the analysis performed using the Dirac's formalism has shown that the the CR's (22) are perfectly well defined and do not lead to ambiguities in the quantization of the C–S models in the Coulomb gauge. In particular, it has been verified the consistency of the CR's (22) with the constraints and the covariance of the theory under the Poincaré group.

Despite of these positive results, there are still many open questions concerning the use of the Coulomb gauge. For instance we have seen that, in this gauge, the C–S theories become two dimensional models, so that it is lecit to ask how it is possible to compute three dimensional link invariants. At the lowest order, where the link invariant is the simple Gauss invariant, one can check that the results obtained in the Coulomb gauge are consistent with those obtained in the covariant gauge (see appendix).

Another problem already mentioned is the derivation of the gauge field propagator starting from the commutation relations (22).

Finally one would also apply the prescription (42) also to the more complicated case of the four dimensional Yang–Mills field theories.

Acknowledgments

The work of F. Ferrari has been supported in part by the European Union, TMR programme, under grant ERB4001GT951315.

Appendix: Wilson loop in the abelian case

Let us consider for instance in the abelian case the vacuum expectation value

$$\langle W(C)\rangle = \langle e^{i\oint_C dx^\mu A_\mu}\rangle$$

for a single closed loop C. At the lowest order we have:

$$\langle W(C)\rangle \sim -i\frac{2\pi}{s}\varphi(C) \tag{73}$$

where

$$\varphi(C) = \frac{1}{4\pi}\oint_C dx^0 \oint_C dy^i \epsilon_{ij}\partial^j \log|\mathbf{x}-\mathbf{y}|\delta(x^0-y^0)$$

If the loop C lies on a plane, it is easy to see that the Wilson loop (73) is trivial. If the loop is not planar, using Stokes theorem, we obtain:

$$\varphi(C) = \frac{1}{4\pi}\oint_C dx^0 \int_{\Sigma_0} d^2y\delta^{(3)}(x-y) \tag{74}$$

where Σ_0 is the projection on the plane x_1, x_2 of a surface Σ spanned by the loop C. This result is to be compared with what we would obtain in the covariant gauge:

$$\varphi(C)_{cov} = \frac{1}{4\pi}\oint_C dx^\mu \int_\Sigma d^2S_\mu\delta^{(3)}(x-y)$$

where now d^2S_μ is the infinitesimal area element on the surface Σ. Introducing a framing[5] with framing contour C_f in equation (74) we have:

$$\varphi_f(C) = \frac{1}{4\pi}\oint_{C_f} dx^0 \int_{\Sigma_0} d^2y\delta^{(3)}(x-y)$$

After a few calculations one finds that the above integral is exactly a Cauchy integral counting how many times the loop C_f is intersecting the loop C, which is exactly the Gauss link invariant as expected.

References

1. R. Jackiw and S. Templeton, *Phys. Rev.* **D23** (1981), 2291; S. Deser, R. Jackiw and S. Templeton, *Phys. Rev. Lett.* **48** (1983), 975; J. Schonfeld, *Nucl. Phys.* **B185** (1981), 157.
2. C. R. Hagen, *Ann. Phys.* (NY) **157** (1984), 342.
3. G. Moore and N. Seiberg, *Phys. Lett.* **B220** (1989), 422; J. Fröhlich and C. King, *Comm. Math. Phys.* **126** (1989), 167; J. M. F. Labastida and A. V. Ramallo, *Phys. Lett.* **238B** (1989), 214; M. Bos and V. P. Nair, *Int. Jour. Mod. Phys.* **A5** (1990), 989; W. Chen, G. W. Semenoff and Y. S. Wu, *Mod. Phys. Lett.* **A5** (1990), 1833 *Phys. Rev.* **D46** (1992), 5521; E. Guadagnini, M. Martellini and M. Mintchev, *Nucl. Phys.* **B336**

(1990), 581; G. W. Semenoff, *Phys. Rev. Lett* **61** (1988), 517; E. Fradkin *Phys. Rev. Lett.* **63** (1989), 322; M. Lüscher, *Nucl. Phys.* **B326** (1989), 557; E. Witten, *Nucl. Phys.* **B311** (1988), 46; Y. H. Chen, F. Wilczek, E. Witten and B. I. Halperin, *Int. Jour. Mod. Phys.* **B3** (1989), 1001;

4. R. Iengo and K. Lechner, *Phys. Rep.* **213** (1992), 179.
5. E. Witten, *Comm. Math. Phys.* **121** (1989), 351.
6. E. Guadagnini, M. Martellini and M. Mintchev, *Phys. Lett.* **B227** (1989), 111.
7. P. Cotta Ramusino, E. Guadagnini, M. Martellini and M. Mintchev, *Nucl. Phys.* **B330** (1990), 557.
8. S. Axelrod and I. M. Singer, in *Proceedings of the XXth International Conference on Differential Geometrical Methods in Theoretical Physics*, New York 1991, edited by S. Catto and A. Rocha (World Scientific, Singapore, 1992).
9. C. R. Hagen, *Phys. Rev.* **D31** (1985), 2135.
10. S. Deser, R. Jackiw and S. Templeton, *Ann. Phys.* (N. Y.) **140** (1984), 372.
11. A. Desnieres de Veigy and S. Ouvry, *Phys. Lett.* **387B** (1993), 91; D. Bak and O. Bergman, *Phys. Rev.* **D51** (1995), 1994; O. Bergman and G. Lozano, *Ann. Phys.* (NY) **229** (1994), 229.
12. F. P. Devecchi, M. Fleck, H. O. Girotti, M. Gomes and A. J. da Silva, *Ann. Phys.* **242** (1995), 275.
13. A. Foerster and H. O. Girotti, *Statistical transmutations in* $2+1$ *dimensions*, in J. J. Giambiagi Festschrift, eds. H. Falomir, R. E. Gamboa Saravi, P. Leal Ferreira and F. A. Schaposnik (World Scientific, Singapore, 1990), p. 161; *Phys. Lett.* **B230** (1989), 83; *Nucl. Phys.* **B342** (1990), 680.
14. A. Bellini, M. Ciafaloni and P. Valtancoli, *Nucl. Phys.* **B454** (1995), 449; **B462** (1996), 453.
15. P. J. Doust and J. C. Taylor, *Phys. Lett.* **197B** (1987), 232; P. J. Doust, *Ann. Phys.* (N. Y.) **177** (1987), 169.
16. H. Cheng and E. C. Tsai, *Phys. Rev. Lett.* **57** (1986), 511.
17. G. Leibbrandt, *Noncovariant Gauges*, World Scientific, Singapore, 1994.
18. G. Leibbrandt and J. Williams, *Nucl. Phys.* **B475** (1996), 469 and references therein.
19. F. Ferrari, *Phys. Rev.* **D50** (1994), 7578.
20. K. Haller and E. L. Lombridas, *Ann. Phys.* **246** (1996), 1; M.-I. Park and Y.-J. Park, *Phys. Rev.* **D50** (1994), 7584; S.-S. Feng, H.-S. Zong, Z.-X. Wang and X.-J. Qiu, *Induced Electronic Interactions in Chern–Simons Systems*, *Int. Jour. Theor. Phys.*, **36** (8), (1997), 1743.

21. F. Ferrari and I. Lazzizzera, *Phys. Lett.* **B395**, (1997), 250.
22. F. Ferrari and I. Lazzizzera, *Perturbative Analysis of the Chern–Simons Field Theory in the Coulomb Gauge*, Preprint UTF 387/96, PAR-LPTHE 96-44. *Int. Jour. Mod. Phys.* **A**, in press.
23. F. Ferrari, *On the Quantization of the Chern–Simons Field Theory on Curved Space–Times: the Coulomb Gauge Approach*, Preprint LMU-TPW 96-5, hep-th/9303117.
24. Q.-G. Lin and G.-J. Ni, *Class. Quantum Grav.* **7** (1990), 1261.
25. L. Alvarez–Gaumé, J. M. F. Labastida and A. V. Ramallo, *Nucl. Phys.* **B334** (1990), 103; W. Chen, G. W. Semenoff and Y. S. Wu, *Mod. Phys. Lett.* **A5** (1990), 1833 *Phys. Rev.* **D46** (1992), 5521; D. Birmingham, M. Rakowsky and G. Thompson, *Phys. Lett.* **B251** (1990), 121; E. Guadagnini, M. Martellini and M. Mintchev, *Phys. Lett.* **B227** (1989), 111; A. Blasi and R. Collina, *Nucl. Phys.* **B345** (1990), 472; M. Asorey and F. Falceto, *Phys. Lett.* **B241** (1990), 31; F. Delduc, C. Lucchesi, O. Piguet and S. P. Sorella, *Nucl. Phys.* **B346** (1990), 313.
26. A. Blasi and R. Collina, *Nucl. Phys.* **B354** (1990), 472; F. Delduc, C. Lucchesi, O. Piguet and S. P. Sorella, *Nucl. Phys.* **B346** (1990), 313.
27. O. Piguet and S. P. Sorella, *Algebraic Renormalization*, Springer Verlag 1995.
28. G. Giavarini, C. P. Martin and F. Ruiz Ruiz, *Phys. Lett.* **B332** (1994), 345; *Phys. Lett.* **B314** (1993), 328; *Nucl. Phys.* **B381** (1992), 222; C. P. Martin, *Phys. Lett.* **B241** (1990), 513; M. Asorey, F. Falceto, J. L. Lopez and G. Luzón, *Phys. Rev.* **D49** (1994), 5377; M. Asorey and F. Falceto, *Phys. Lett.* **B241** (1990), 31.
29. L. Alvarez–Gaumé, J. M. F. Labastida and A. V. Ramallo, *Nucl. Phys.* **B334** (1990), 103;
30. J. Schwinger, *Phys. Rev.* **125** (1962), 1043.
31. F. Ferrari and I. Lazzizzera, work in progress.
32. C. Itzykson and J.-B. Zuber, *Quantum Field Theory*, McGraw–Hill, Singapore 1980.
33. S. Emery and O. Piguet, *Helv. Phys. Acta* **64** (1991), 1256.
34. M. Chaichian and W. F. Shen, *Two loop Finiteness of Chern–Simons Field Theory in Background Field Method*, hep-th/9607208.
35. V. N. Gribov, *Nucl. Phys.* **B139** (1978),1; P. van Baal, *Nucl. Phys.* **369** (1992), 259.
36. E. Verlinde, H. Verlinde, *Nucl. Phys.* **B288** (1987), 357.